£3-49
(10) 46

There are also 3 Sorts of *Axes*, which I shall take Notice of, *viz. Felling*-Axes, *House*-Axes, and *Lopping*-Axes.

There are 4 Sorts of *Felling*-Axes, which are sold from 1*s* 5*d* to 2*s* a *piece*; *viz.* N° 1. at 1*s* 5*d*; N° 2. at 1*s* 6*d*; N° 3. at 1*s* 10*d*; and N° 4. at 2s a *piece*.

E. Hoppus, **Practical Measuring... *London,* 1765**

Whereas many ironmongers are in the habit of offering to supply tools at "NURSE & Co.'s prices" but supplying articles that never came from us, we would particularly request that all orders be sent to us *direct,* when they will receive our most careful and prompt attention, and customers can rely on receiving goods of the highest possible quality only.

C. Nurse & Co., Walworth Road, London, c.1905

Tools bearing our **Name** or **Trade Marks** are **Warranted** and any proved faulty in Material or Workmanship will be exchanged... We do not exchange worn-out Tools. Cast Iron Goods cannot be warranted against Fracture. **Please do not mutilate this Catalogue. Spare copies of any page can be supplied by return.**

Turner, Naylor & Co. Ltd. Sheffield, 1928

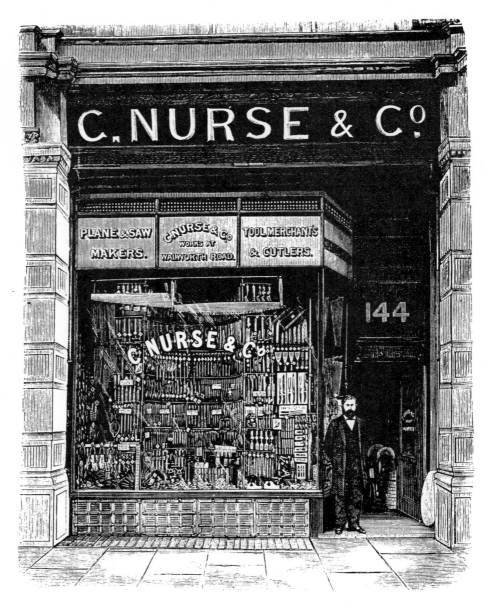

The above is an Engraving from a Photograph of OUR CITY BRANCH at

144, Bishopsgate St. Without, London, E.C.

*In 1905 when this picture was made, **Charles Nurse & Co.** had three retail shops in the City of London and the shop and planemaking factory in the Walworth Road. The firm had an unusual history, starting as planemakers in Maidstone, Kent before moving to Walworth Road in 1887. It soon became one of the capital's principal planemakers and tool sellers, continuing in business until 1937. The shop window display is typical of many tool shops with a "pack 'em in like sardines" approach.*

TOOLS

A Guide for Collectors

Jane & Mark Rees

ROY ARNOLD

Published by
Roy Arnold
77 High Street, Needham Market
Suffolk IP6 8AN

ISBN 0 904638 10 3

A CIP catalogue record for this book is available from the British Library

Text and layout prepared by Jane & Mark Rees
Output by Avonset, Midsomer Norton, Bath
Printed and bound in Great Britain by
The Ipswich Book Company Ltd., Suffolk

— ▸◆◂ —

Acknowledgements

The text has been edited by Julanne Arnold. This has been undertaken with her normal professional skill and we are very grateful to her.

Some drawings are by the authors and some are by Harry Arnold. We much appreciate his help in producing these.

We would like to acknowledge the assistance of David Stanley, who has permitted us to use information from his auction sale catalogues and results.

CONTENTS

Recently we attended an auction where a single English moulding plane sold for £2750. Out in the car park, other tools were changing hands for as little as £10 – and the purchasers outside seemed to be just as happy and interested in their items as the big spenders. This must say something about the fascination of old tools – you can empty your pocket as much or as little as you like and still buy something of interest – but perhaps out in the car park there was a moulding plane that will eventually sell for… so read on, knowledge is all.

The illustrations

Most of the illustrations in this book have been taken from a variety of manufacturers' and distributors' catalogues, dating from around 1900, a time when the quality and range of British hand tools reached their zenith.

Glass boring outfit [Buck & Ryan, 1930]. This engraving clearly demonstrates both the fineness and detail that can be achieved with engravings.

From the moment we first saw an old trade catalogue, illustrated with engraved blocks, we have been attracted by the skill of the trade engravers. The representation of the items in the engravings made by them gives greater detail than any other method – even photography.

Trade catalogues – their history, use, methods of production and their place in the marketing of tools – are discussed in two sections *Marketing*, p. 42 and *Old Books and Catalogues*, p. 216. A listing of the catalogues from which the illustrations have been taken is given on p. 237.

Round scooper graver (engraving tool) and handle [Ward & Payne, 1911]

Price Guides

• The information on prices has been taken from auctioneers' catalogues and the prices asked by leading dealers. For common items – and this in general means goods in the lower price range – we have condensed the information into general price guidance.

• The auction prices cited exclude those we consider to have been excessively high or low and therefore unlikely to be repeated.

• Descriptions of condition follow the classification established by *The Fine Tool Journal*, published in the United States. This is set out on p.15. Always remember that condition can radically alter the value, and in spite of the standard assessment of condition, quality lies in the eye of the buyer.

• Where no condition is cited, it is G+.

• The prices shown include the auctioneers' buyers' premium where this is levied.

• [DP] indicates a dealers' price. [AP] an auction price, other than a David Stanley auction.

• Since October 1983 David Stanley has conducted twenty-six auctions of selected tools for each of which a detailed illustrated catalogue was prepared. We have drawn extensively on this pool of information. Prices taken from these auctions are given as follows: [DS 24/329] indicates sale number (24)/lot number (329).

• In the price guides, makers are generally indicated only by surname and the following abbreviations used: m.p. = moulding plane; H&R = hollow and round planes; o/a = overall; d/t = dovetailed; C. = century.

• You may think that the prices quoted are too high or too low, but in the case of auction prices, *that* was **the** price fetched on *that* day.

Where we were unable to find catalogue illustrations, the items have been illustrated by drawings. The drawings on pp. 99, 188, 206, 212, 213, 214, 235 and 236 are by Harry Arnold. The remainder are by the authors.

Addresses

Roy Arnold, Books & Tools, 77 High St., Needham Market, Ipswich, IP6 8AN. Tel. 01449 720110.

Mark Rees Tools, Barrow Mead Cottage, Rush Hill, Bath, BA2 2QP. Tel. 01225 837031.

David Stanley Auctions, Stordon Grange, Osgathorpe, Leics. LE12 9SR. Tel. 01530 222320.

Twenty-five years ago the aspiring tool buyer had very little information to help him. Whether anyone realised that some of the tools in kits of old joiners' tools could be of 18th century origin is doubtful. The few serious collectors mostly bought continental planes which were decorated with carving; they had to be old – the date was often included. The plain English tools were largely ignored.

The last two decades have seen a flow of information so that today's collector can find detailed information about almost all aspects of tools and trades. This, combined with a maturing market, has helped to define what is rare and desirable. Published information breeds confidence amongst buyers even though their motivations may be varied.

Information is also the weapon to counter the unscrupulous. I well remember Ultimatum (brass framed) braces being sold as "presentation pieces" with the inference that they were exceptional and extremely rare. Some varieties of Ultimatums may be very rare but as a species they are not.

From acorns ...

Thirtysomething years ago we moved into our first house, a Victorian job, ignored in subsequent reigns but now home. I exchanged several garden tools – wedding presents from a misinformed aunt who thought there was a garden – and purchased my first, very necessary, woodwork tools. I still have the Stanley No. 4 smooth plane and three of the black plastic-handled chisels.

The next house was worse, neglected since the 1930s. Woodworm and dry rot had consumed the ground floor and were starting on the first floor. Had the roof not been bombed and then repaired in 1947 under the war damage scheme, it is doubtful if the house would have survived at all.

The need for more tools to tackle the huge amount of work was pressing and some second-hand tools were acquired. Amongst these were a few that just seemed too good to use. My interest was kindled and trips to the Portobello Road in search of antiques yielded instead more tools although these were still intended for use.

Conversion came in that road when I bought some things I knew I would never use. It caused me unease for several weeks and I did not feel entirely happy until, in 1975, R.A. Salaman's *Dictionary of Tools used in the Woodworking and Allied Trades* was published and I realised that there was a world of tools awaiting exploration.

In 1983 the Tool & Trades History Society was formed by a group of about 100 enthusiasts and what had previously been a rather solitary pursuit became more social and, equally important, the production of literature and the exchange of ideas between collectors burgeoned.

Gent's No. 4 oak tool chest, typical of the chests of tools sold by many manufacturers for the amateur/domestic market. The tools were generally of second quality [Wingfield, Rowbotham. 1904].

INTRODUCTION

Second-hand traditions

In the past, when a man died, it was common practice to auction his tools in the workshop and to send the money to his widow. The philosophy was that his mates were doing the widow a good turn but, like many workshop practices, there was some benefit to the organisers. It's not unknown today when attending country auctions to hear harsh things being said about collectors who take tools away from "real users" – the feeling that tools somehow belong to the craftsman as of right and he should get them on the cheap lives on.

"Tool money" was another workshop practice; the young apprentice was forced to pay a fine when he used a tool he had not used before, and the proceeds were spent by his mentors on beer.

Today's market in second-hand tools is the continuation of a trade with a long history – the 18th century trade cards of **F. Styles**, Old Street, and **John Willshire**, Cow Cross, London, specifically mention that they buy, sell and take old tools in part exchange. There is, perhaps, one new influence on today's market. Whereas in the past the quality and range of hand tools was improving, today the opposite is the case.

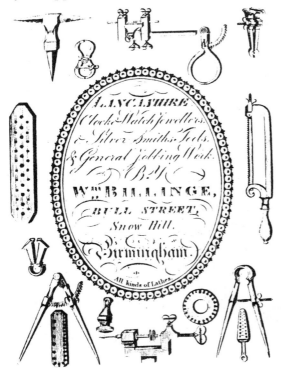

Trade card, dating from the 1820s, of a Birmingham manufacturer of small steel tools who is mainly illustrating what are usually known as "Lancashire" tools – small clockmakers' and engineers' tools.

Commercially made

Collectors of commercially made tools are comforted by the notion that these are more definable as to date and quality than the craftsman-made product and that prices can be charted for similar items. But there will be a considerable number of similar, if not identical, items. The purchase of craftsman-made pieces, each of which is unique, does, however, require more judgement on the part of the buyer.

Moulding planes would appear to hover half-way between these two extremes – they were commercially made but the variations in style and the simplicity of the earlier planes have their own artistic appeal. And they have one advantage – quantity – but how can this be advantageous?

Collectors like to buy and will not, in general, collect in areas where they only find something once in a blue moon. There are thousands, even millions, of moulding planes in existence so they can always be found. The one found may be a bit wider/older/better than the others but it will complement the collection somehow. The same can be said of tools generally and this may be part of the reason why tool collecting has become so popular today.

Collecting tip

> *Many tools will be found engraved with the name of the original owner. This does not indicate a "presentation piece". It was normal practice for owners to put their name on their tools. A handsomely engraved name will usually increase the value of the tool.*

Take a bow Stanley

There are already comprehensive price guides to Stanley tools (notably John Walter's *Antique and Collectible Stanley Tools*) so we have only touched very briefly on their products.

This is no slight to Stanley; their planes have dominated the world for the last eighty years. Only when other manufacturers finally gave up the struggle to produce their own designs and made what were essentially copies of the Stanley planes did these other companies start to prosper. But for the tool buff, the years of struggle and development were productive of some rare and now collectable items.

A combined drill/brace [Wingfield, Rowbotham, 1904], worth around £35. If this were a simple brace, the value would be about £5. In auction, the result would be unpredictable – although this is rare it might not appeal to a wide audience, being relatively modern.

Deciding value

The value of anything is what it can be sold for on a particular day and there is no clearer or harsher arbiter of this than the auction sale. I well remember attending an auction, about fifteen years ago, when there were three identical lots of antique silver. The auctioneer had divided a set of thirty-six plates because he didn't think that anyone would want them all. How right he was! The first lot fetched £2200, the next £1600 and the last £1400, each going to a different buyer. So what was the correct price? What had happened was that, in this instance, each buyer wanted only one set and, as he had been satisfied, he had been removed as a contender. There was presumably a fourth buyer who was only prepared to pay £1350. But the opposite also sometimes happens – a group of buyers who don't really know what to pay. So the first lot goes fairly cheaply and succeeding lots increase in price as buyers realise that their last chance has come.

The parable of the glass eyes

There is a story in the antiques trade of the avid collector of glass eyes. For many years he pursued them in markets, begged them from the bereaved, bought them wherever he could. He stored his collection in numerous boxes. There were thousands of eyes and he could tell the difference between the products of different makers – his was *the collection* and *he* was *the expert*. When he died suddenly, the whole collection was sold in three lots and fetched about £500, a tiny proportion of what they had cost. The fact is that there aren't many collectors of glass eyes – they may be difficult to find and the dealers who sell the odd one will say "they can't remember when they had another one so they must be rare" – but no dealer wants several hundred of anything other than at a very cheap price.

The tool market is a small section of the general market in antiques; if you only collect planes you confine yourself to a sub-section and if you only collect planes made in Birmingham in the 18th century you are probably the only specialist collector in your chosen field. Who is going to buy your hundred planes when you decide to sell? Too much specialisation can damage your financial health particularly if you have let it be known that you will always buy in your chosen field and at auction always outbid rivals.

Collecting tips

➤ *It's worth repeating: The value of your tool is what someone is prepared to pay **today** – by tomorrow he may have bought one from another source, by next week be impoverished or by next month be collecting in another field.*

Price Guide

There must have been more than one person who wanted these!

◆ An early French cormier wood smoother, 10½" x 2¼", dated 1688 with foliate carving to the mouth escapement. The front tote nicely carved in the form of a serpent's head, crack to wedge-retaining lug. G **£2200** [DS 23/121]

◆ A d/t steel mitre plane by **Norris** London. The snecked iron, which has been re-tipped, is probably original. This plane is almost certainly 19th C. and is the only known Norris mitre plane with a wedge. Some discolouration to steel. G **£3300** [DS 19/1396]

◆ A rare brass button pad beech brace with heavy brass plates by **Pilkington Pedigor & Storr** with brass faced ebony head. Also marked "Her Majesty's Royal Letters Patent". F **£990** [DS 20/1441]

The advertisement read "Auction Sale: Tools, workshop equipment and materials. The property of a recently deceased restorer". For many years he had taught restoration at a leading college and had also run courses for amateurs. The sale contained plenty to attract – a sprinkling of pieces of furniture that he had never got round to, a quantity of wood, both new and second-hand, and boxes of cabinet ironmongery.

Almost everything fetched good money. His tools were mostly good but, apart from one or two bench planes, there wasn't much for the collector. But the prices were astonishingly high with many lots fetching two or three times their retail value. It seemed that buyers wanted this man's tools not only because they were a momento but because they thought that some of his skill would pass to them – what other reason could there be for chisels making three times their retail value? But this is pure sorcery. I tell the tale merely to illustrate the complexity of the motives that may drive the buyer.

Bought for use

Many people do buy old, and even antique, tools for use. This is perfectly logical for in a modern world, where machine production of almost everything is normal, the range of new hand tools available has become very limited. For many specialist users, there is no alternative but to buy second-hand and this demand undoubtedly drives a significant part of the market.

Incidentally many amateurs, and professionals for that matter, will claim that they are buying a tool for use when, often, what they are saying is that they hope to use it one day – not at all the same thing – but why should we be denied our fantasies!

Artistic merit

Some manufactured tools have evolved over the years with both user's and maker's input whilst others are the result of the industrial designer's skill. Whether tools from either of these stables can be considered to have real artistic merit is a subject for debate. Certainly many made by craftsmen do. Frequently this may be a naive feeling; sometimes the simplicity coming from the reuse of materials or bits garnered from the work environment – a joiner's square made from a piece of brass coffin plate or from an engraved copper printing plate – also recovered from the deceased, in this instance, the billhead of a defunct woollen mill.

Decoration

This section will no doubt cause some squirming amongst the purists. Tools do make fine decoration to a home and – prepare to squirm some more – theme pubs and restaurants. What other use could be found for 12 ft. long frame saws or plumb levels? Were it not for the decorator's market many over-large objects would have gone to the rubbish dump long since.

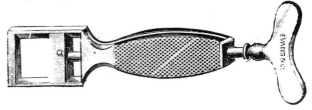

A finely made gent's adjustable wrench? See foot of column.

Curiosity

The antiquarian's cabinet of curiosities was the origin of today's museum so interest in the curious has a good pedigree. An object may appeal because it has a known but intriguing use or because you really don't know what it was used for – these are the tools of intellectual satisfaction or curiosity.

A lot of people seem to enjoy this type of stimulation as curiosities find a ready sale and the tool collecting fraternity has even created a word to describe them – "The what's it".

Association

Not a few collectors are motivated to buy the now obsolete tools of their trade. Doctors and dentists are drawn to the finely-made instruments of an age when anaesthetics and antiseptics were unknown and bleeding was still a normal treatment – the

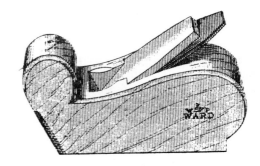

Boxwood chariot plane [Ward & Payne, 1911].

Allingham's screw pile clamp [Evans and Wormull, 1889].

plastic and chromium plate of today may be practical but lacks the collecting appeal of polished steel and horn chequered handles. So, too, with other occupations, the tools of the woodworking trades being the most widely sought, for even today many occupations touch on these trades.

Twelve-blade scarificator – an instrument for bleeding humans. The blades are flicked over by a heavy spring. [Evans & Wormull, 1889]. These were still included in the Allen & Hanbury catalogue of 1930.

Tradition

Mass production results in millions of similar items each of which is devoid of human input; or, to put it more accurately, the human skill went into building the machine and into designing a suitable product. Thereafter the machine produces widgets as long as it's fed. Yes, I know that machines produce things that are wholly impossible to make by hand and to tolerances and repeatability that our grandfathers could only dream of but, so far, that plastic injection moulded widget hasn't become collectible.

Many buyers, even if they can't explain it clearly, are attracted by the quality of the handmade object. The attraction can be simplicity; why else will collectors pay substantial prices for a country item little more than a stick cut from a hedge. There may be patination, or it may have rustic simplicity; the real reason is that it is a thatcher's tool and thus it connects the collector with the traditions of that trade.

Fantasy

With these tools, you will become the ultimate craftsman and will build that Chippendale desk that you always wanted – no, on second thoughts, a dining suite of a triple pedestal table and a dozen chairs should be the first project!

The lifeboat-men

Their mission in life is the salvation of numerous tools and the *complete* contents of disused workshops because otherwise "they will be lost". Indeed, they spend a great deal of time rescuing innocent spanners, sash cramps and the occasional half brace from the perdition of the car boot sale. The saved are transported home and somehow stowed away in the attic/ shed/ garage/ workshop/ barn/ lock-up/ conservatory/ dining room/ bedroom to await cleaning and conservation which somehow never comes as there are so many other souls out there awaiting rescue.

Amongst these salvationists there are just a few with real foresight and the nose for something worthwhile – these are those rare folks who save *everything* from the last workshop making rules by hand – a task executed in one day before the demolition men arrived and, by the way, they also saved three hundredweight of papers giving a complete picture of the firm's operations over the last hundred years.

Oh well, dream on – but the saving of some part of Britain's trade heritage may well fall to one of the lifeboat-men.

Investment

Every book about collecting seems to contain strictures about not buying for investment purposes – you must surely be disappointed if this is your object. If you are a true collector, buying only what pleases you, one day, when you decide to sell, surprise! surprise! Your collection may turn out to be a wonderful investment.

If investment is your aim, however, it is essential that you follow a few very important rules. First, whatever you buy must be in perfect or near perfect condition. Chips, breaks, scratches, repairs may not seem important to you but will certainly reduce value when you come to sell. Second, however minor the item, make sure that it is the very best of its kind. Third, if you are the only person that wants something, it has no value – avoid glass eyes in quantity.

Twenty years ago an acquaintance of mine moved into a house he had just bought. Under the stairs, was a box of old tools. He was intrigued; many of them he could not identify – this was the start of a voyage of discovery. He now has a large collection, displayed in several rooms of his house. If he and his wife had not decided to buy that house would he ever have become a tool collector? Who knows – I suspect that most of us started in some equally chance way.

Over the years I have seen many collections; some, the product of the lifeboat-men, are very unstructured but most collectors, after a brief period as vacuum cleaners, start to apply some sort of theme. This can be as broad or as narrow as you wish to make it.

By date: This often means early tools – those dating from the 18th century being the earliest that are available in any quantity.

By place of manufacture: Wooden planes were made in many towns throughout Britain so they are fertile ground for localised collections.

By manufacturer: Many collectors limit themselves to one particular maker; **Preston**, **Norris** and **Mathieson** spring to mind. These all published extensive catalogues that have been reprinted in recent years. Incidentally I have never met anyone who collected Marples as a theme: perhaps their tools are just too prolific to provide a challenge.

By trade: Coopering, thatching, and leather working seem to be favourite trades as they all have specialist and identifiable tools. The woodworking trades are more difficult for collectors with this approach as many tools are common to a range of trades.

By product: Some products, for example the Strowger (an old-fashioned mechanical telephone exchange), required specialist tools to install and maintain. So did the sash window and the piano.

By material: Here I confess to a liking of crème brûlée and boxwood – both should be golden in colour. Other collectors are strongly attracted to the gleam of polished steel. One collection I know contains only objects of which ebony is a major part.

By artistic merit: A few discerning collectors eschew all commercially made tools. Their interest is in tools which express the artistic taste or, perhaps more accurately, the innate artistry of the craftsman-maker. There are not too many tools that meet the high ideals of these collectors and the best will achieve surprising prices in the specialist auctions. In recent years, plumb bobs and trammels are both areas where prices for craftsman-made examples have outstripped the commercially made products.

By tool type: This is a fairly obvious theme. The most numerous of the collectors-by-type are the plane collectors who often limit themselves further by sticking to a particular location, period, maker or purpose. Braces, drilling tools and rules are other areas of collection by type.

By technical interest: The technology represented in a tool is of interest to many. Tool patents and registered designs are numerous but for a considerable number there are no surviving examples, proof, if any be needed, that the development of useful new tools is far harder than it looks. The failed, and therefore rare, provide interest for collectors.

Price Guide

• **Date:** A fine and rare pair of 18th C. French 21" hand forged dividers dated 1747 with "fleur de lys" decoration to hinge, the curved points protrude from stylised animal mouths at the base of the legs, some minor pitting. **£1200** [DS 24/951]

• **Place:** A pair of fully boxed snipe bills by **Barnes, Worcester.** **£30** [DP]

• **Manufacturer:** A **Preston** violin makers' gun metal plane, 1¾" x 1", round sole with both plain and toothed iron. G++ **£165** [DP]

• **Trade:** A good quality chequered ebony trepanning tool handle with three adjustable brass bits. G++ **£125** [DS 22/153]

• **Material:** A brass faced ebony mitre square by **Marples** with 9" arm. **£35** [DP]

• **Artistic merit:** A unique 19" fruitwood trying plane, probably Austrian, the front carved in the form of a mythical dolphin, the tote as the tail, the body beautifully carved with a full length portrait of a gentleman and the wedge monogrammed G.H. in a floral border. Cupid's bows carving to the escapement and a cupid's bow shaped whalebone insert to the mouth. This plane was used by a Glaswegian boat builder and later displayed in the Fisheries Museum in Anstruther. G+ **£2450** [DS 20/1003]

• **Tool type:** A spill plane of moulding plane form by **Watkins,** Bradford. F **£35** [DP]

• **Technical interest:** A combination ratchet brace and bevel gear drill by **H.S.B. & Co.** with rosewood head and handle. G++ **£60** [DS 22/1103]

General collecting tips

➤ *Condition is always important. Original finish/unused condition can double or triple value.*

➤ *Buy what appeals to you. If you like it others will probably also do so.*

➤ *Resist the temptation to buy a damaged item – you won't be able to sell it when a really good one turns up.*

➤ *Don't over-specialise. This can be financially dangerous and indicates greater obsession than normal – even for tool collectors!*

➤ *Don't undertake repairs unless you are certain of a satisfactory outcome.*

➤ *Do look after your tools. It is amazing how many collectors watch them deteriorate.*

➤ *Read the literature; the more you know, the more your collection will mean to you.*

➤ *Tools are rarely a suitable medium for financial investment.*

➤ *Buying dirty tools is always risky. Cleaning often reveals cracks that can make your purchase almost worthless.*

➤ *Beware rust! After cleaning, pits often look much worse than when the tool was dirty.*

➤ *Never break up sets – think carefully – sometimes apparently diverse items may be a set.*

➤ *If you find a complete kit of any age keep it together – one day you may be pleasantly surprised.*

➤ *Never throw away any original packing however tatty.*

The shop of John Hughes, wholesale and retail ironmonger, at 26 Worcester Street, Birmingham, from his trade card of c. 1810. In the windows can be seen a wide variety of items including a panel saw, a compass saw, a Lancashire pattern hacksaw, a billhook, a bed winch and an assortment of kitchen items including mincing knives.

The auction

A huge folklore surrounds the auction. The image of buyers in evening dress bidding in millions for old masters is all part of this mythology; the other end of the market is the uncatalogued auction sale held in a muddy field whilst drizzle soaks both goods and bidders. The stories of thousands bid, bargains bought for a few pounds, trade rings re-auctioning goods and competitor's devious tricks to distract a bidder at the critical moment are all good stories – the truth is that most auctions are conducted in a business-like manner and it isn't that easy to buy a thousand pound lot by scratching one's nose!

Selling

For the seller, the auction is an easy way of disposing of tools, particularly if there is a really large quantity – few dealers are financially able to take on such a burden or have the immediate market for a wide variety of tools of differing quality. Dealers want to buy what sells quickly at a satisfactory profit – some items may be extremely rare and will sell for a high price to the right specialist collector but there may be only a handful of people who really want to acquire that rarity and it could be years before one of them arrives in the dealer's shop. On the other hand, the dealer pays for the goods immediately – quite possibly in cash – whereas the specialist sale may be months away and there is always the possibility that your lots may fail if you have set reserves or may sell for well below your expectations if you have not.

Tips for Sellers

➤ *Rarities and high value tools will usually fetch the best price at auction – even when the auctioneers' charges have been taken into account.*

➤ *Established dealers will often give as much as, if not more than, will be obtained at auction for middle range tools of standard type and quality.*

➤ *Mixed or multiple lots at auction are mostly bought by dealers so if that's what you are selling, it's better to offer them direct.*

➤ *Don't auction items of small value – the auctioneers' minimum charge for each lot can consume most of the proceeds.*

In recent years there has been a considerable increase in auctioneers' fees with the introduction of buyers' premiums, and costs are now often in the 18% to 25% range when all the charges are added together. It matters not how these are divided between the buyer and seller; they will be reflected in the sum the vendor will receive.

Valuations

Household insurance policies normally include a cap both on the value of a collection taken as a whole and on individual items with a value exceeding a specified sum – often as low as £500 – so attention to insurance arrangements is of importance if you wish to cover your collection properly.

Have a complete list and valuation prepared before disaster strikes. This is where the established dealer or specialist auctioneer is needed but you will have to pay a fee for this service. It could be money well spent – try convincing your insurance company that three moulding planes were worth £1000 **after** they have disappeared.

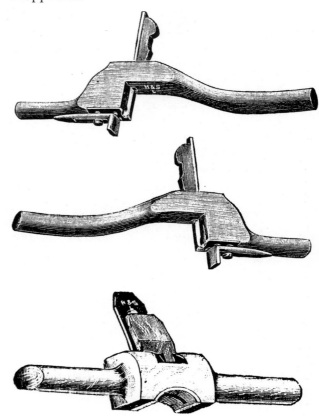

A pair of coachmakers' pistol routers and a coachmakers' jarvis [Harding, 1903].

Buying

In Britain several auctioneers now conduct specialist tool auctions. These sales normally include many hundreds of lots and, for the quality sales, highly detailed catalogues with photographs of virtually every lot are prepared. These are distributed throughout the English speaking world.

The Anglo/American influence in both the design and manufacture of hand tools is dominant in many parts of the world and today's overseas collectors are attracted by this cultural heritage resulting in numerous absentee bids.

For the buyer, the specialist auction sale provides a good opportunity to see and handle a large number of tools and a unique opportunity to compare quality and design. Even if you are unable to attend, the informative catalogues of the major sales give classifications of condition and estimates of what each lot is likely to fetch.

If attending an auction you do, however, need to do your homework. English auctions are conducted at a fast pace – particularly tool auctions when 1200 lots may be sold in one day. So you need to have decided which lots interest you before going to the sale room – you can't *thoroughly* view a thousand or more lots. Leaving bids, even if you are attending the auction, is a good discipline; it avoids that common auction syndrome of "onemoreandiamsuretogetititis". If you are unable

to attend, the now widely used system of "condition classification for antique hand tools" gives clear guidance. This system, started some years ago in America by *The Fine Tool Journal*, has now become the accepted standard in both Britain and America.

"Egg beater" type steel drill brace [Harding, 1903].

Below: The Fine Tool Journal *Condition Classification for Antique Hand Tools. An additional category, G++, is used by some dealers and auctioneers.*

	Category	Usable	Finish	Wear	Repair	Rust	Misc.
N	New	totally	100%	none	none	none	+ orig. pkg.
F	Fine	totally	90-100%	minimal	none	trace	
G+	Good+	yes	75-90%	normal	minor or none	light	some nicks or scratches OK
G	Good	yes	50-75%	normal - moderate	minor	light	minor chips
G–	Good–	probably	30-50%	moderate - heavy	correct	moderate	chips OK
Fr	Fair	no	0-305	excessive	major	moderate - heavy	
P	Poor	no	n/a	excessive	damaged	heavy	

Finding the goods

Stories of how and where interesting and, occasionally, valuable tools have been found are legion and when recounted between collectors often raise many unworthy emotions. The reality is that tools, particularly old and dull-looking early tools, are in general little regarded by most people, so they turn up almost anywhere – in attics, at car boot fairs, in country auctions. The one common feature is that tools, when no longer of productive use, get left and forgotten in disused workshops and tool boxes until some change of circumstances brings about disposal.

But lady luck lives just around the corner – maybe even next door. For twenty years we lived next door to a family whose grandfather had been a builder. Over the years we exchanged stories about the building industry. It must have been ten years later, after he had died, that the family decided to move and, in clearing out thirty years' accumulation, found a box containing all the tools that the old man had put away because he couldn't find much use for them. His working tools, that had done a lifetime's work, were almost valueless. But there were some good pieces including several

ebony thumb planes and a rare compassed sash plane, left unused for years. I was delighted with the contents of the box and, to the tool collector, having provenance is an added bonus.

Whilst good tools may come your way as the result of luck, for the less experienced and for those for whom time is scarce, there is much to be said for purchasing from an established and experienced dealer who has a reputation to guard and who will give you the benefit of his advice.

Tool cabinet and bench (fitted with Tools of Best Quality) [Ward & Payne, 1911]. This was available in three sizes. The price of the largest size, 5' 6" long, complete with all the tools, was £28. It seems that not many amateur craftsmen bought these as compendium benches are rare today.

The Wheelwright from A Book of Trades *published in London in 1804 by Tabart & Co.*

These are the principal associations for collectors and others interested in the history of hand tools. All publish informative literature.

In America there are also a number of collectors' clubs organised on a regional or interest basis.

THE TOOL & TRADES HISTORY SOCIETY

The Administrator, 60 Swanley Lane, Swanley, Kent BR8 7JG, U.K.

Publications: Newsletter (quarterly) and Tools & Trades (occasional journal).

Aims: to advance the education of the general public in the history and development of hand tools and their use and of the people and trades that used them. Membership is open to all who share in these aims.

EARLY AMERICAN INDUSTRIES ASSOCIATION, INC.

John S. Watson, Treasurer, E.A.I.A. P.O. Box 143, Delmar, New York 12054, U.S.A.

Publications: Shavings (quarterly newsletter) and Chronicle (quarterly journal).

Aims: to encourage the study and better understanding of early American industries in the home, in the shop, on the farm and on the sea; also to discover, identify, classify, preserve and exhibit obsolete tools, implements and mechanical devices that were used in Early America.

MID-WEST TOOL COLLECTORS ASSOCIATION, INC.

Secretary: Mel Ring, 35 Orchard Lane, Huntingdon, IN 46750, U.S.A.

Publication: The Gristmill

THE TOOL GROUP OF CANADA

Peter Wood, 7 Tottenham Road, Don Mills, Ontario, Canada L9G3L1

Publication: Yesterday's Tools

HAND TOOL PRESERVATION ASSOCIATION OF AUSTRALIA

Secretary: Frank J. Ham, 21 Adeney Avenue, Kew, Victoria, Australia.

Publication: The Tool Chest

AMBACHT & GEREEDSCHAP

H.K. Rude, Seminarieweg 23, 4854PA Bavel, Netherlands.

Publication: Newsletter

Books

Eaton, Reg. *The Ultimate Brace.* **Reg Eaton, Kings Lynn, Norfolk 1989.** A most detailed book on the brass-framed brace, describing its construction and history. It does not cover other types of brace.

Gaynor, J.G. & Hagedorn, N.L. *Tools: Working Wood in Eighteenth-Century America.* **The Colonial Williamsburg Foundation, Williamsburg, Virginia, U.S.A. 1993.** Published to accompany the important exhibition of this name held in Colonial Williamsburg, the book contains the most comprehensive view yet of both the American and the English toolmaking and selling trade in the 18th century.

Goodman, W.L. *The History of Woodworking Tools.* **Bell & Hyman Ltd., London, 1964.** Traces the history of many types of woodworking tools from earliest times in Britain and Europe. It has been reprinted many times but is at present out of print.

Goodman, W.L. *British Planemakers from 1700,* **3rd ed. revised by Jane & Mark Rees. Roy Arnold, Needham Market, Suffolk, 1993.** The third and completely revised edition of the book that no plane collector should be without. One thousand, six hundred and seventy British planemakers are listed together with their marks and biographical details. The text describes the development of the English wooden plane.

Landis, Scott. *The Workbench Book.* **The Taunton Press Inc. Newtown, Conn. U.S.A. 1987.** Whilst principally about modern work benches used by woodworkers today, the initial chapters contain valuable descriptions and discussions of period work benches from the 16th century onwards.

Nicolle, George. *The Woodworking Trades – A Select Bibliography.* **Twybill Press, Plymouth, 1993.** This book will save you much time if you wish to investigate the world of woodworking trades in the recent or more distant past. A selected bibliography by an author with considerable knowledge of tools and trades.

Pollak, Emil & Martyl. *A Guide to American Wooden Planes and their Makers,* **3rd Ed. The Astragal Press, Morrristown, New Jersey, U.S.A. 1994.** Lists all identified American planemakers with reproductions of their marks and some biographical details where these are known.

Rees, Jane and Mark, ed. *The Tool Chest of Benjamin Seaton 1797.* **The Tool & Trades History Society, Swanley, Kent. 1994.** Written by a group of TATHS members, this monograph illustrates, describes and discusses the chest and tools of this 18th century cabinet maker from Chatham, Kent.

Roberts, Kenneth D. *Some 19th Century English Woodworking Tools.* **Ken Roberts Publishing Co., Fitzwilliam, New Hampshire, U.S.A. 1980.** This large format book, printed from typescript with signs of rapid production, includes trade advertisements and extracts from catalogues. Its principal interest is in the listings of Sheffield edge and joiners' toolmakers, extracted from period trade directories, and its citation of brace patents.

Salaman, R.A. *Dictionary of Tools used in the Woodworking and Allied Trades c.1700 to 1970.* **2nd ed. revised by Philip Walker, Unwin Hyman Ltd., London, 1989.** The most comprehensive and well illustrated listing of British woodworkers' tools available. The majority of the tools are listed individually but for a few special trades they are described in groups.

Salaman, R.A. *Dictionary of Leather-working Tools c. 1700 – 1950.* **George Allen & Unwin, London, 1986.** A companion volume to the above, covering leather-working tools. It is always worth consulting as many difficult-to-identify tools are from leather-working trades.

Sellens, Alvin. *Dictionary of American Hand Tools.* **Alvin Sellens, Augusta, Kansas, U.S.A. 1990.** Illustrates and describes American tools for the widest range of trades. Includes many original and patent tools not illustrated elsewhere.

Walker, Philip, *Shire Album No. 50: Woodworking Tools.* **Shire Publications Ltd., Princes Risborough, 1980.** An excellent and factual, if short, introduction to the world of old tools.

Walter, John. *Antique and Collectible Stanley Tools – A Guide to Identity and Value.* **The Tool Merchant, Akron, Ohio, U.S.A. 1990.** A paperback listing of most U.S.A.-made Stanley tools complete with outline descriptions, period of manufacture and a 1990 price guide.

Whelan, John M. *The Wooden Plane: Its History Form and Function*. **The Astragal Press, Morrristown, N.J. U.S.A. 1993.** A detailed description of the types and uses of both general and specialised wooden planes worldwide. Includes an extensive classification system of moulding planes by profile.

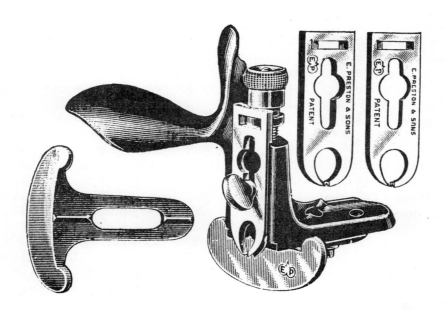

Patent adjustable iron quirk router, suitable for both straight and circular work [Preston 1909].

Reprints of trade catalogues

Howarth, James & Sons, Sheffield, 1884. Reprinted 1988 by Roy Arnold, Needham Market, Suffolk. Although principally steel makers, their JH mark is to be found on a wide range of tools, notably chisels. Like many firms they also acted as factors (merchant wholesalers) for other Sheffield toolmakers and their catalogue is therefore a general tool catalogue.

Millers Falls Company, Massachusetts, Catalogue No. 35. 1915. Reprinted 1989 by Roger K. Smith, Athol, Mass. U.S.A. A comprehensive illustrated listing of this company's often ingenious products from the heyday of the American hand tool manufacturers. Other small catalogues from this firm have also been reprinted.

Preston, Edward & Sons, Birmingham, 1909 (with 1912 supplement). Reprinted 1991 by The Astragal Press, Morristown, N.J. U.S.A. Although now principally remembered for their metal spoke-shaves and planes, the firm were also wooden planemakers and rule and level makers. All these products are listed and illustrated in this comprehensive catalogue. The reprint includes an illustrated history of the firm by Mark Rees.

Rabone, John & Sons, Birmingham 1892. Reprinted 1982 by Ken Roberts Publishing Co., Fitzwilliam, N.H. U.S.A. This publication has a nice period feel given by the reproduction of the original simple colour illustrations. The listing of rules and measuring equipment of all types is extensive. There is also an alphabetical list of Birmingham rule makers extracted from trades directories.

Stanley Rule & Level Co. Connecticut, Catalogue No. 34, 1915. Reprinted 1985 by The Stanley Publishing Co. Westborough, Mass. U.S.A. Facsimile of the 1915 catalogue and supplementary list.

Stanley Rule & Level Co. Connecticut, Catalogue No. 139, 1939. Reprinted 1988 by Roger K. Smith, Lancaster, Mass. U.S.A. Facsimile reproduction of the 1939 catalogue. A retailer's catalogue which includes point-of-sale display stands for Stanley products. *Note:* A number of other Stanley catalogues have also been reprinted.

Timmins, Richard, Birmingham *Pattern Book of Tools, 1845.* Reprinted 1994 in reduced size by Studio Editions Ltd., London. A fine example of a Birmingham "toy maker's" catalogue (see p.34) printed from copper plates which include the original prices. Introduction and commentary by Philip Walker.

The Handsaw Catalogue Collection (1910 – 1919) **Makers: Spear & Jackson (Sheffield) 1915, Disston 1918, Atkins (Indianapolis) 1919, Simmonds (Massachusetts) 1910.** Printed 1994 by The Astragal Press, Morrristown, N.J. U.S.A. A compendium selection of the handsaw sections of the principal sawmakers of this period.

Scottish and English Metal Planes by Spiers and Norris 2nd ed. printed 1991 by Ken Roberts Publishing Co., Fitzwilliam, N.H. U.S.A. The principle contents are reproductions of catalogues of **Stewart Spiers**, 1909 and c. 1930 and **T. Norris and Son,** 1914 and 1928. It includes a short history of both firms.

John Wyke, Liverpool 1758-70. *Catalogue of Tools for Watch and Clock Makers.* Reprinted 1978 by University Press of Virginia, U.S.A. for Henry Francis du Pont Winterthur Museum. Although now out of print, this finely illustrated catalogue is a notable first, being the earliest illustrated tool catalogue known.

Original trade catalogues

We have not listed original trade catalogues in general as these are now collected in their own right (see *Old Books & Catalogues*. pp. 216–218) and are not, in general, easy to find. However, catalogues by the following firms can be found quite easily and purchased for small sums.

Buck & Hickman, London. Their first catalogue was a 32 page paperback published in 1867. The large (approx. 1200 pages) red hardback catalogue was first published in the 1890s and continued to be published intermittently until 1964. Editions are quite easy to find and are excellent value considering the huge range of items illustrated.

Moore & Wright, Sheffield. Paperback catalogue of "British Precision Engineers Tools". The firm started trading in 1909. Useful reference for British made engineering measuring tools.

The L.S. Starrett Company, Athol, Massachusetts, U.S.A. Paperback catalogue of "Fine Mechanical Tools" from one of America's leading manufacturers. The catalogue was issued intermittently; No. 13 dates from 1895, No. 24 from 1927 and No. 25 from 1930. The 1895 catalogue has been reprinted. These catalogues illustrate how advanced American manufacture of precision tools was in the early decades of this century.

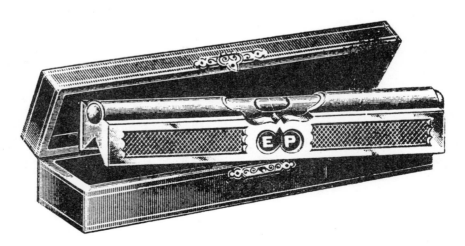

Engineers' nickel plated spirit level, available either 4" or 6" long, with patent cat's eye tube [Preston 1909].

Approach

Collectors often ask "How should I clean tools?" In many cases what they are really asking is "To what extent should I clean these tools?" The choice lies between the two extremes. At the one end, trying to return the tool to the condition when it left the factory or at the other, almost complete inaction. There is some justification for the latter course. After all, if the last working owner was happy to use it in that condition there is probity in leaving it thus – as long as appropriate steps have been taken to stop further deterioration.

It is probable, however, that many tools will have languished in sheds for decades before coming into your hands and are not, therefore, in the same condition as when last used so some cleaning is appropriate. Most collectors will buy and keep an object because they believe it to have some intrinsic artistic merit; it is therefore logical to clean to the extent that the beauty of the object is shown to best advantage.

Good workmen were concerned about their tools and their bosses knew this. Even in my lifetime in the building industry I have seen trade foremen asking to see a man's carry-box before hiring him.

Saw grinders at the Sheffield Grinding Mills from "A Day at the Sheffield Cutlery Works", The Penny Magazine, 27 April 1844. Grinding operations always leave marks and are a clue to the methods of manufacture.

Cleaning tip

➤ *We think that tools should be cleaned to the state in which they would have been when used regularly by a good tradesman – and no further.*

Original finish

All methods of finishing leave marks which both contribute to the overall character and, more importantly, give some indication of how they were made. Cleaning methods should, as much as possible, conserve these marks.

Wooden planes were finished with planes (which didn't leave much clue) and files (which leave tell-tale marks). Brass work was often finished with files or on sanding machines, leaving a fine grained surface of parallel marks, and was very rarely polished to a glossy finish. Metal planes were finished either on lathes – large radius tool marks are the clue here – or on grinding machines, often with quite coarse wheels. Chisel blades were ground and glazed (a form of fine grinding with abrasive applied to wooden wheels) but the neck and bolster were transversely ground giving an attractive contrast at the shoulder.

Pocket knives and early engineers' precision tools were ground and polished to high standards which, if perfectly preserved, can to the unknowledgeable seem like a plated finish.

All these different types and qualities of finish are distinct and appropriate to the type and quality of tool. Heavy cleaning, amounting to refinishing may, on occasions, improve the appearance of a tool but something of the original is always lost, so think carefully before you get going with emery paper, paint stripper or other harsh methods. As we shall detail, there are more satisfactory methods.

Rust removal

The electrolytic removal of rust is an effective method that does not damage iron and steel. For this you need a large non-metallic container (such as a plastic washing-up bowl), a 6/12 volt battery charger of minimum 4 amp capacity, a piece of stainless steel, some caustic soda or, preferably, sodium carbonate and a wire brush.

The method is as follows: Make an electrolyte by mixing the caustic soda with the water in the container (approximately two teaspoons to a quart of water). Connect the piece of stainless steel to the **positive** clip of the charger and suspend it in the water. Connect the tool to be cleaned to the negative clip of the charger and suspend it in the water. The clip must not be immersed in the water, so it may be necessary to clean the two ends of a tool as separate actions.

Select the voltage on the charger to give a reading of approximately 2 amps (this may also involve adjusting the distance between the stainless steel and the tool). At this stage there should be evidence of bubbling around the electrodes – if not, check the electrical connections.

The time that the tool should remain in the water depends on its size and amount of rusting and can be anything between 10 minutes and 3 hours. The exact time is not important as extra time will do little harm.

Wipe the tool to remove the black residue and, if necessary, carefully loosen the rust flakes with a blunt chisel or knife. The tool should finally brushed with a wire brush. If done under water, this obviates any dust. After thorough drying the tool should either be lightly oiled or waxed with restoring wax.

An alternative method can be followed that does not require the item to be immersed in the electrolyte, although it is more time consuming. A cloth, soaked in the chosen electrolyte is laid on the

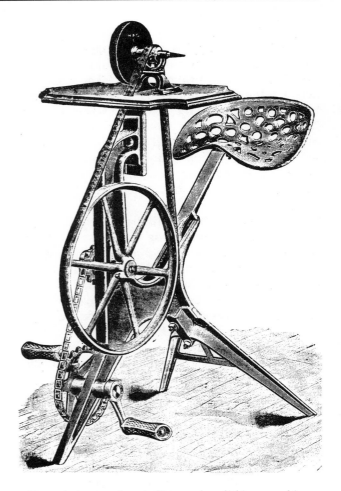

Barnes' improved grinding and polishing machine [Harding 1903]. Buffing machines are nothing new!

part of the tool to be cleaned and a strip of sacrificial metal laid on top of the cloth. The negative clip should be connected to the work and the positive clip to the stainless steel. It is important that the current passes only through the electrolyte-soaked cloth.

The buffing machine

Dismissively known by this name amongst tool buffs (no connection – or is there?) this is more correctly described as a mop head metal polishing machine. Of recent years, these have been much used by dealers to clean both metal and wood. No machine has a greater potential to cause irreversible damage to tools than these ferocious beasts and they will also savage the hand that feeds them unless proper precautions are taken. The mops rotate at very high speed and are fed with buffing soap – actually an abrasive compound in a wax carrier. Brass, even when completely black, can be speedily brought to a high polish but

---

Warning

Chemicals, water and electricity are a potent combination so take great care when using this method of cleaning. Rubber gloves and eye goggles should be worn at all times. Fumes will be given off; therefore the process should always be carried out in a well-ventilated location. Any accidental splashes should be washed in plenty of water.

corners and edges are worn down and original finish and even lettering disappears in a trice. Applied to wood, buffing machines will rip away the patination and then drive black waxy dirt into the grain of the wood. I have seen literally hundreds of boxwood carpenters' folding rules that have been buffed – the brass fittings gleam unnaturally bright whilst a black tidemark is left on the boxwood and any traces of the original shellac polish have been removed so the brass will quickly become black.

If there is a proper place for buffing machines in the world of tool collecting, it is on those items that were originally sold in a highly burnished state or on bright steel items such as chisel blades that were "glazed" to a high finish. They can be of use to clean items that are almost impossible by hand. But don't use them where inappropriate, and this means on anything that is small, soft (brass or nickel silver), relies on sharp well-finished arises for appeal or has an original finish that will be removed by buffing.

Cleaning tip

➤ *Cover any areas that you don't want to buff with masking tape.*
➤ *Apply wax polish to wooden areas adjacent to metal before you buff. This will reduce penetration of the dirt into the wood.*

Shock! Horror!

A word of warning – I number amongst my tool buff friends several who have been savaged by buffing machines and rather than type the usual boring safety injunction I thought that you should know about the friend who told me how he had two fingers left literally dangling by shreds – it was quite difficult to get the fingers out of the glove. It didn't look too bad when I visited him in hospital – at least a bandage covered the injury – but it suddenly seemed essential to sit down. **If you use a buffing machine – do read and follow the safety instructions.**

Woodworm

Although various theories have been advanced from time to time about destroying woodworm by freezing or by cooking, the only effective method that we would recommend is a proprietary woodworm fluid. This should be used following the manufacturer's instructions and always in a well-ventilated space.

Once the woodworm has been treated the holes should be filled. There are a number of reasons for this, not least because if more holes appear, you will know to repeat the treatment. It also prevents loose powder falling out of the holes which is unsightly and also can lead to uncertainty about the success of the treatment. It also prevents breaking up of the wood between holes if the tool has been badly infected. Wood that has been so

Thirty-six-tool presentation case of carving tools [Ward & Payne, 1911]. These tools are amateur pattern.

eaten that it becomes "spongy" can be treated either with linseed oil or, in very severe cases, with thinned PVA, both of which will harden and protect damaged wood.

Recording your collection

What records you keep about when and where you acquired your tools and what they cost is a matter of personal preference. My only plea is that, if you collect tools from known sources, do record this in some way. A collection of tools from a known boat builder or cabinet maker may, in the future, be of huge significance to trade or local historians. Most tools will be devoid of provenance by the time you get them, so this makes those that do have a background even more significant. **Write it down:** from where, from whom and what date – otherwise you will forget and when you are dead your relatives won't know!

Storing your tools

Wooden tools are best kept in a dry shed or garage of heavyweight construction where the temperature will vary only slowly. Highly unsuitable are light-construction wooden sheds where the temperature goes up and down quickly causing condensation on metal surfaces. Centrally heated homes can be too drying for wooden tools and will lead to shrinkage. Provided a shed is dry, metal tools will not rust. Allowing metal parts to rust will not only present you with a difficult cleaning job but can also lead to permanent damage – rust staining, irons jammed into planes, wooden parts blown apart by rust expansion. The traditional tool chest is a good option – they tend to even out environmental changes. However, bright polished metal items are best in the house as they do need to be kept warm and dry.

Patent Tool Sharpener.

This reverses the usual order—as will be readily seen by the Illustration. The Tool is rigidly held in the hand while motion of Stone sharpens the Tool.

No. 328.

Price, **12/6** each.
Washita Oilstones extra from **2/-** each.

We claim for this method of sharpening Tools that it will preserve the bevel of Plane Irons, etc., and enable one to sharpen the tools more effectively for many workmen, and quicker than the regular method.

Arrange the length of strap by means of the slip buckle, so that the foot placed in the loop nearly touches the floor at the end of each stroke.

Hold the tool to be sharpened against this upper surface of the rest with the left hand, and with the right press it **gently** down on the stone.

For ordinary work the Machine needs no fixing to the bench, but if it is desired to do so, there are two holes in the framework through which screws can be inserted into the bench.

Patent tool sharpener [Melhuish, 1899]. Patent but in some way impractical devices are a feature of the history of tools. Such contrivances, made in very small quantities, often pass unidentified.

Wire wool

The best general purpose abrasive for cleaning wood as it doesn't leave scratch marks. "Double O" (OO), fine, and "Triple O" (OOO), very fine, are the most useful grades; some manufacturers make "Quadruple O" (OOOO) which is even finer. However, the thickness of grades varies from one manufacturer to another.

Some wire wool sold for domestic purposes is treated with mineral oil so buy trade material from a polish house – their material isn't.

Warning

The finer grades of wire wool are combustible, particularly when impregnated with solvents – some authorities say spontaneous combustion is possible – so dispose of waste safely.

Emery cloth

Emery cloth needs to be used with considerable care or bright scratches will be made in the metal surface. Not, therefore, an advised or necessary part of the tool restorers' armoury. See *Sandpaper*.

Sandpaper

Used on wood, it will remove any patina there is. Occasionally you may come across the odd rough patch where sandpaper may be of some use but on oiled or polished surfaces it clogs quickly – see *Wet and dry paper*. For the restorer of tools, sandpaper, either medium or medium fine grade, is more useful in cleaning iron and steel. Unless applied with the utmost vigour, it will not scratch these surfaces but will remove dirt and rust.

Wet and dry paper

Made with a fine emery abrasive bonded to a paper backing, it is available in a range of grits from 120 which is similar to a fine grade of sandpaper to 600 which is so fine that the surface has a floury feel. Used over-enthusiastically, wet and dry paper has the same capability as emery paper to leave scratch marks on the surface of metals although these will be smaller. However, used with discretion as to grade and application it can be an effective remover of that mixture of rust, oil, paint, resins etc. that often seems to coat metal surfaces.

Because of the fine abrasive size, wet and dry paper will clog up very quickly if what is being removed is at all "gummy" – so a lubricant is essential. As implied by the name, these papers can be used with water but on tools this is really not a good idea as any residue left on metal will promote rust and on wood, will swell the surface. When cleaning tools you will often be working on metal and wood in close proximity, so a hydrocarbon (oil or spirit) lubricant is preferable. Paraffin (in US, kerosene) is probably the best. It is cheap, has a high flash point – important for safety – and dries quickly but not too quickly. Most importantly, it is not so solvent of dry linseed oil, shellac or wax polishes that it will strip away that all-important surface patina that we seek to retain but it is volatile enough to dry quickly from a surface or, at the worst, overnight from a wooden object that has become saturated.

Rottenstone and pumice powders

Rottenstone is a very fine grey powder and pumice a somewhat coarser white powder. Both are useful for reducing the shine on newly applied French polish which may look far too bright. Wetted with a drop or two of paraffin or water they make a mild abrasive paste. Jewellers' rouge, available in stick or powder form, is also a fine, abrasive polishing agent that can be of use when cleaning fine items such as drawing instruments.

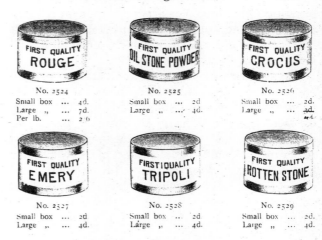

Polishing compounds from Isaac [Isaac, c. 1900].

Linseed oil

Available raw or boiled. Raw linseed oil is more penetrative than boiled but it takes much longer to dry. We have considerable reservations about the advisability of using linseed oil on wooden planes that already have a surface finish of old oil (see *Cleaning planes*, pp. 28–29). One of the drawbacks of linseed oil is that, when applied to tools kept in

The Wellington Works of John Oakley & Sons Ltd. in Westminster Road, London from a billhead dated 1876. Oakleys were leading manufacturers of every kind of abrasive papers, powders and compounds.

anything but the best of environments, it provides a nutritious feed for mould growths. A light application of raw linseed oil will however work wonders on dry, neglected-looking wood by bringing out colour. For a lighter application, thin the oil with turpentine or paraffin. This will make a more penetrative mixture which, once dry, will leave little evidence on the surface.

It may surprise many of our readers to learn that we consider linseed oil to be more useful as a finish to iron and steel more than wood. Applied to iron or steel objects it forms a varnish, protecting against rust, and any remaining particles of red rust are darkened to an almost black colour, so becoming much less obtrusive. Importantly, even dry linseed oil remains soluble or at least softenable in methylated spirits, so it is reasonably easy to remove any coating.

Petroleum grease and oils

Many new tools were greased up before leaving the manufacturer; the military were particularly keen on this, though in recent years wax-based protective coatings have become more usual in both the civil and military fields. Grease and oil are fine for preserving engineering tools which don't have wooden parts but for any tools that do, petroleum-based oils or greases should be avoided like the plague as they permeate the wood, turning it black and uninteresting and making it impossible to develop a decent polish. There must be tools of some trades where the appropriate look is black and oily but for the woodworking trades this is to be avoided.

"Unembalming fluid"

We don't use this ourselves but it has received so much commendation from good tool fiends – who are also friends – that we must include it here. Usually passing under the name of restorers' mixture, its purpose is to clean, feed and provide a finish all together. Mix in equal proportions linseed oil, white spirit or turpentine and vinegar in a bottle and shake well. This mixture can be applied by rag or with wire wool which speeds the removal of heavy grime. After a while apply a final burnish with a rag. If you use real turpentine in the mixture your workshop will be filled with one of the finest and most evocative of aromas but white spirit is more economical.

Wax polishes

All wax polishes intended for wood have two basic constituents: wax, which can be of various sorts, and a solvent to soften the wax and make it easier to spread. The solvent evaporates, the wax hardens and you can burnish it with a cloth giving a beautiful and, hopefully, much admired shine. By repeated applications a fine polish is built up. The most traditional and expensive polishes are formulated with beeswax.

The restorer of antique tools is usually faced with a different problem. The surface is likely to be coated with a layer of grime held together with the residues of oils and greases. If the beauty of the wood is to be revealed, this needs to be removed.

Restorers' wax polish, applied with fine wire wool, is the best method to achieve this in a controlled manner. These polishes are available in a wide range of colours but the real significance to the tool restorer is that they are compounded with an active solvent and the tool grime is softened by these polishes. Apply with fine wire wool and continue to rub until the required amount of the dirt has been removed by the action of the wire wool on the softened grime. The value in this method is that it is completely controllable, the action is not too fast and as soon as the required colour is achieved, the item can be aside to allow the wax to harden – half an hour is adequate but overnight is better – and then polished off with a soft rag.

In general, wax polishes inhibit mould growth and, in our experience, waxed tools stored in unheated conditions will not grow mildew. Although restoring polishes are made in a wide range of colours, it is not necessary to match colours closely and, on previously finished surfaces, it will have little effect on the overall colour. Restoring polishes do have a significant effect on raw woods, and here the colour used does need to considered with care.

To the tool restorer, wax polishes are also of considerable value as an application to iron or steel surfaces, particularly where these are not bright ground. Once an iron object has been cleaned, an application of restorers' wax will tone down any remaining rust, help to fill small pits and, if care is taken not to burnish too much of the wax away, provide a useful protective coating.

Fillers

Worm, and previous owners, have a nasty habit of leaving holes in the tools we collect. Worm holes are always of British standard wormhole size and the easiest way of filling them is with wax filling which is sold in small sticks. Most manufacturers have a range of about fifteen colours designed to match wood colours. If you are going to attend to more than the odd item, it is best to buy a box containing a range. This will allow you to experiment to find the best possible colour or to mix sticks if necessary. From experience, we find that the least conspicuous result is obtained by choosing a colour that is a little darker than the surface in general. The colour will then be a close match with the darker parts of the grain. If there are a large number of holes, a good tip is to use several slightly different colours. You will be surprised how much this deceives the eye.

Wax filler melts at a low temperature. Heat a metal item (a medium-weight chisel is ideal) until the wax runs when touched by it and let the wax filler drop onto the holes. With the chisel reheated as necessary, run the wax into the holes but do not worry about trying to level it off. Allow to cool and then pare or scrape away the excess. Finish by rubbing with wire wool.

Hard filling, which goes under a variety of names (beaumontage, or cabinet stopping) is a shellac-based mixture which melts at a higher temperature than wax filler and is sold as pre-coloured sticks in a variety of colours. It is more suitable than wax filler for larger size holes and has the advantage that shellac-based polishes can be applied over it but it is more difficult to coax into place and to level down.

Occasionally you may want fill a hole, or make good, in a position where a stronger material is needed. This can be done using an iso-cyanurate type of glue (two part resin such as Araldite) to which are added powder colours or wood powder from the sander. Colours can only be arrived at by experiment and remember that any significant quantity of filler will start to reduce the strength but this technique has its uses.

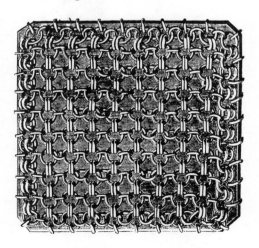

Chain burnisher, still a useful tool cleaning aid [Melhuish, 1904].

Cleaning a moulding plane

First remove the iron – easy to say, often not at all easy to do! The correct method is to take the plane in the left hand holding both the stock and wedge. With a wooden mallet strike the rear of the plane a good sharp blow. As the wedge loosens, a slightly hollow sound will be heard. If, after four or five blows, the wedge is still tight, place the plane in the vice at an angle so that the wedge is horizontal just above the jaws. Take a piece of wood around nine inches long and, using this as a punch against the underside of the wedge, encourage it to come out. This is a slightly risky technique as thin wedges may break but try to arrange the stick punch to be as axial as possible to the wedge. If this has failed to shift the wedge you are in real trouble. If you can afford the time, place the plane somewhere really dry and warm for a couple of weeks and try again. Another technique that sometimes works is to strike the top of the iron and drive it forward thus breaking the bond that can be caused by rusting of the iron.

The stock

A plane in good order will require little attention other than cleaning with wire wool and restoring wax. If the colour of the plane is dark, be particularly careful how much of the dark patina you remove – most planes have quite a few dents and scratches which you will not be able to clean out and the result of too much cleaning will be to emphasise these and to give a spotty look.

Boxing

Take out any displaced or loose boxing, clean the housing and the box pieces until they fit easily and then glue back into position. The glue originally used was scotch (animal) glue and this is to be preferred although there are very practical modern glues.

The "boxing" in moulding planes is, of course, boxwood. Less obvious is that, in order to give maximum wear resistance, it was cut at an angle so that the grain is at 45° to the sole of the plane. If you need to refinish it, start from the back of the plane so that you are working with the grain. Boxwood is pretty intractable stuff to work – files are a more certain route to success than planes which have a habit of tearing out bits.

Should you need to make good any missing pieces of boxing, this is best done with wood recovered from planes that really have come to the end of their days – a box donation is their final service to their race. The boxing on fillisters, both moving and sash, can be extremely complex, with up to five dovetails holding it in place in the beechwood stock. It has been said that these joints are some of the most accurate woodwork to be found anywhere and it would be a confident restorer who attempted serious repair to one of these, so our advice is – do not buy planes with complex but defective boxing.

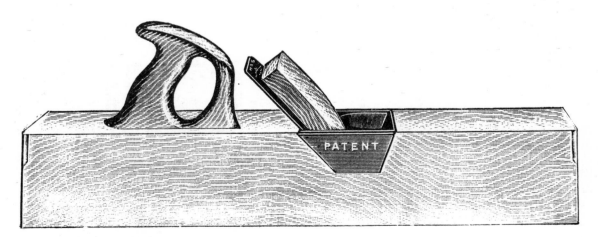

A 22" try plane with the David Kimberley patent iron mouth [Harding, 1903]. This design is one of the very few attempts by English planemakers to make an improved form of wooden plane. Similar castings are also to be found fitted to jack planes and smoothing planes.

The iron

The cleaning of a moulding plane iron is straight-forward. With a sharp chisel – yes! a chisel, as a tool collector you will have plenty of these – pare or scrape away any surface rust. This method is quick and minimises scratches to the metal. Follow up with wire wool and restorers' wax polish. The result will be an iron that is clean but not overclean, with a coating of wax which will help to prevent future rusting or sticking of the wedge.

From time to time, the reader may come across a plane that has no iron. The only solution is to adjust an iron from a defunct or less valuable plane. Try to choose one that is as near as possible to the original – remember, the older the plane the thinner the original iron was likely to have been. If the shape of the iron is fairly simple, for example, a round, it is comparatively easy to regrind it to shape on the electric bench grinder. If it is more complex, say an iron for a triple reed, it is not possible to grind the shape without specialist grinding equipment so it is necessary to revert to the original method of manufacture.

The irons were filed to shape by the planemaker and carefully fitted to the profile. They were then hardened by heating to cherry red and quenched. To reverse this process, heat the iron to bright heat in a propane torch and cool slowly – this will soften it so it can be filed to shape; getting the right shape with files is quite easy. Then re-harden.

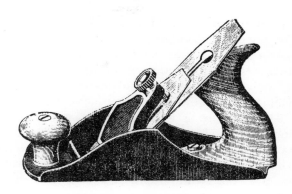

The "Suffolk" smooth plane, bright sides, 8½" long with 2" double iron [Wingfield Rowbotham, 1904]. Made in Sheffield, these planes had no knob or lever adjustment.

Cleaning a cast iron metal plane

When cast iron rusts, the iron oxide formed adheres tightly to the surface so that the rate of rusting slows down. For this reason, unless they have been kept in very bad conditions, most cast iron planes will not prove to be too badly pitted when cleaned. The same cannot be said for steel which, in similar conditions, will rust into deep pits which will be cleaned out when the restorer gets to work and will thereafter remain as gaping blemishes.

The quick way to remove any rust that has erupted above the surface is to pare it away with a sharp chisel. As we have previously said, this method, properly done, is the least likely to make scratches in the surface of the metal and will help to leave rust that is below the surface where it is – it may be rust but at least the surface will look flatter than if all pits are excavated.

Cast iron is best cleaned with fine or medium grade sandpaper. This, unlike emery paper, will not leave unsightly scratches in the surface and, unless used with extreme zeal, will leave the well-adhered oxides on the surface where they not only continue to protect but also look good, giving that glowing black sheen that is the mark of old, but well cared for, metal. If you must have a brighter appearance, use wet and dry paper, 320 grade or finer, lubricated with paraffin. To finish apply wax polish.

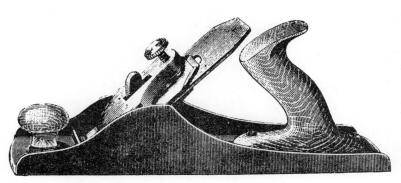

An iron jack plane, 14" long with 2⅛" double iron [Wingfield Rowbotham, 1904]. English made, this plane has a distinctive hook for securing the screw cap, but not having knob or lever adjustment, it was not a good seller and is today rare.

Boxwood rules

Remember that, on large print rules, the numbers are only printed onto the surface and that even on rules made by the traditional process the markings are only lightly scribed into the surface and you will understand the need for care when cleaning.

The first step is to remove any paint spots or lumps of glue by very careful scraping – the safest tool is a thumb nail but, occasionally, a chisel wielded with caution will be needed. Then clean the boxwood with restorers' wax applied either with a rag or if necessary with **very fine** wire wool – OOO or OOOO. Only rub just enough to remove the surface dirt and, hopefully, not the shellac polish which was put onto the surface of both the boxwood and the brass fittings when the rule was made. If the rule is light in colour, use a clear wax but on darker rules a mid-brown colour will sometimes give better results and will freshen up markings that have become worn.

On well-used rules, most of the original shellac polish will have worn away and the brass will be tarnished or even completely black. We would advise you not to polish the brass on rules in this condition but to clean away dirt and thick oxides with wire wool and to stop when the colour of the metal gleams through. Highly polished brass needs regular cleaning to keep it bright; this is not only a chore but will soon give the rule a well-worn look – it is no secret that rule collectors are attracted by original or little-restored condition.

Any iron parts may have erupted into raised patches of rust. The best way to remove these is to pare them away with a small sharp chisel applied at a low angle. This method minimises scratches and leaves a flat surface even if a little rust remains.

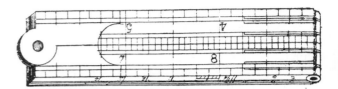

A 1ft. 4-fold ivory rule with nickel silver arch joint [Preston 1909]. Ivory rules of this length are relatively common as they were sold to the gent's market. Often carried in pockets for long periods, many found today are worn.

Ivory rules

Clean ivory rules in the same way as described above but use only clean rag to apply, sparingly, clear wax. Be extremely careful when cleaning any metal parts as the oxides can become impregnated into the ivory. Use masking tape to protect the ivory if necessary.

Collecting tips

➤ *It is not possible to glue broken ivory or boxwood rules without the repair being apparent to the careful observer.*

➤ *Boxwood and ivory rules will often be found with iron stains from the fittings. These are deep in the wood or ivory and seem almost completely resistant to bleaching or cleaning. So, if you cannot live with the stain – don't buy the rule.*

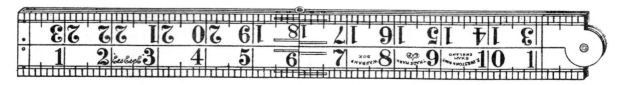

The engagingly named "EesEseE" large print rule introduced by Edward Preston in 1912 [Preston, 1909].

Toolmaking

TOOLMAKING CENTRES

London

The earliest identified commercial planemakers, Thomas Granford and Robert Wooding, were operating in London at the end of the 17th and beginning of the 18th centuries. Thereafter London remained an important centre of wooden planemaking until the trade died in the 20th century. *British Planemakers from 1700* traces the development of the trade starting in the City and then, in the middle/late 18th century, moving to peripheral areas such as Westminster and around Old Street.

The history of other toolmaking in the capital is at present less well documented. London was the largest manufacturing centre in Britain until the Industrial Revolution and there can be no doubt that there must have been considerable production of all sorts of tools.

Evidence suggests that sawmaking was a separate and well-established trade in the 18th century. The sawmakers seem to have favoured two areas. **William Squire** was working in Soho as a spring sawmaker by 1754 until at least 1760. His

business was continued by **John Peters** until the end of the century.

In the other area around Old Street, **Alice White** was working as a sawmaker by 1753 and **Lewis Powell** by 1768. **William Moorman** worked in Golden Lane from 1772 later taking over Powell's business and becoming the most prominent London sawmaker of the period.

There were also edge-tool makers working in London during the 18th century. **William Gilpin** is one of the most interesting, working in Southwark in the 1780s. At some time later, he relocated his business to Cannock in Staffordshire.

With the advent of the Industrial Revolution and the move to methods of production that required ever more power, the problems of the London manufacturers became apparent. The horse mills and steam engines (in use by 1807) could not compete with the water power and cheap coal available in Sheffield and the other great manufacturing centres. The edge-tool trade drifted away. The **Addis** family, making carving tools (see *Carving Tools*, p.74), were one of the few to remain.

Sheffield

During the 17th century iron making was taking place in many areas. Steel making was more demanding. Moxon, writing in 1680, records steel being made in Gloucestershire, Sussex, the Weald of Kent and Yorkshire but states that "the best is made about the Forest of Dean..." He also names Flemish, Swedish, Spanish and Damascus steel, the implication being that some was being imported.

By the middle of the 18th century the advantages of Sheffield – ample water power, local gritstone for grinding wheels, local coal and a long-term expertise in forging – were beginning to outweigh the undoubted transport difficulties, and the goods produced there, and this now included saws and joiners' edge-tools, were taking over countrywide markets. But the domination of the market eventually achieved by the Sheffield manufacturers, in virtually every branch of the tool trade, did not happen overnight.

The reluctance of the 18th and 19th century tradesman to be inducted into the factory system together with the reluctance of the masters to pay regular wages should not be underestimated. The result was that, until as late as the 1960s, piecework was still a normal method of payment for many workers in manufacturing. Sheffield industry, which often turned out thousands of the same item, was ideally suited to this type of system. The worker, whether working in the factory or taking work away from the employer's premises, was, to a greater or lesser extent, a sub-contractor.

The particular Sheffield system of "little mesters" (masters) developed over a long period. These tradesmen might act as sub-contractors, carrying out a specific part, sometimes a very specific part, in the manufacture of a product. Equally, they might be making and selling finished products either direct to customers or to a larger manufacturer who did not find it economic to make small numbers or who was already fully committed. Or the little mesters might be doing both.

What developed was a very efficient system of production by specialists and manufacturers interlocked in the system. If there was a bad side to this, it was that, when trade was slack, the smaller men saw their orders speedily dwindle whilst the larger kept the work for themselves.

During the 19th century Sheffield manufacturers grew to dominate the British hand tool trade. Their wares were exported to every part of the world reached by British mercantile influence.

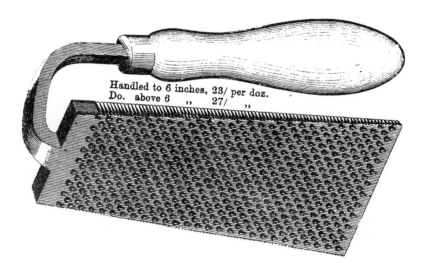

Handled to 6 inches, 23/ per doz.
Do. above 6   „   27/   „

Bread rasp: used to file off the overcooked or ash-covered bottoms of loaves cooked in traditional brick bread ovens, fired with faggots. This illustration is from the Sheffield Standard [Illustrated] List of 1862. The Sheffield Lists are an indication of the co-operation that existed between the many small manufacturers and merchants in Sheffield. These lists, published from the early 19th century until the early years of the 20th century, do not bear the name of any one manufacturer but were intended to show the range of goods that could be had from Sheffield. From the 1880s onwards, the rise of larger manufacturers who published their own catalogues undermined the list system.

York

From 1750, when **Richard Nelson** established himself as a planemaker at Helperby near the city, until 1904 when the Varvill business closed, York was an important centre of wooden planemaking.

The reason why a planemaking trade, which grew to sell throughout Britain, should have developed in the city is not at all obvious. Geography would seem to be against it. Possibly the enterprise of one man, **John Green** (working 1768-99), followed by his son, also named John, (working 1787-1808) was the foundation. The Green business certainly made a huge number of planes; today's collector will have little difficulty in finding examples of its work.

In the 19th century the Varvill firm, founded in 1793 by **Michael Varvill**, became the largest in the city prospering into the 20th century, by which time it was one of the few British wooden planemakers still operating. Varvill-made planes are also plentiful.

Lancashire

You may see references to Lancashire tools. These were tools originally used in the making of clocks and watches, but soon found useful in light metal-working generally.

By the middle of the 18th century, the area around Prescot had become a centre for the manufacture of clock and watch parts which were then distributed throughout the country to regional clockmakers who finished and assembled them into complete movements. Tools were also made in the area. These included not only specialist clockmakers' tools such as turns and spring winders but also more general tools such as small files, pliers in great assortment, punches and hammers.

The trade was founded on a home-working system with individuals specialising in specific groups of tools, organised by merchants. One of these, **John Wyke** of Liverpool, who worked from 1740-87, scored a notable first in publishing, between 1758 and 1770, the earliest illustrated tool catalogue. This catalogue, with finely engraved plates showing the tools in considerable detail, has been reprinted. (See *Further Reading*, p.20.)

Another notable figure in the Lancashire hand-tool industry, whose name will be more familiar today, was **Peter Stubs** who, although already a successful brewer, decided in 1800 to enter the tool business. He did not live long but under the direction of his sons, the business prospered. It was also an early publisher of a tool catalogue (1805). The principal product was files, for which it had a reputation for the best. By 1826, the firm had established a steelworks at Rotherham near Sheffield which was not closed until the 1950s.

Billhead from Peter Stubs, 1876. Many tools made by Stubs are marked only with PS.

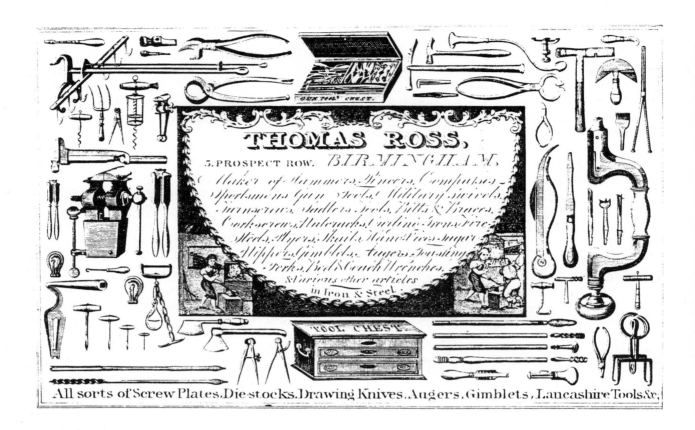

All sorts of Screw Plates, Die-stocks, Drawing Knives, Augers, Gimblets, Lancashire Tools &c.

Birmingham

Edge-tool making was already established in Birmingham by 1750 when **Robert Moore** is first recorded. Quite a number of plane irons and several octagonal socketed chisels marked with his name have been noted.

By the middle of the 18th century, Birmingham was rapidly growing; what had been a village was becoming the heart of a manufacturing area which, over the next hundred years, became the centre for the making of ironmongery, locks, guns, jewellery, silver and plated wares and Birmingham "toys" – small items forged in steel and iron, generally bright finished.

The 1845 catalogue of a leading maker of Birmingham toys – **R. Timmins & Sons** (later to become **Wynn & Timmins**) illustrated some small tools such as turnscrews, pliers, etc. that came into this category and also wooden braces and bits, a speciality of Birmingham makers in the early decades of the 19th century, though later this trade

Trade card of Thomas Ross, Birmingham "toy maker", c.1820. Whilst a wide range of small tools and domestic items are shown, the importance of the Birmingham gun making trade is apparent by the number of specialist gun tools illustrated.

seems to have been taken over by Sheffield manufacturers.

During the 19th century, Birmingham and Wolverhampton came to supply virtually all the market in rules – both boxwood and ivory.

George Darby, working by 1750, is the earliest planemaker known to have worked in Birmingham. The trade grew steadily, and by the 1850s, twenty-five makers were working there, equalling the number in London. Of particular prominence was the firm of **Edward Preston** .

The Black Country was also the site of several of the larger firms in the heavy edge tool trade (spades, forks, billhooks, etc.) such as **William Gilpin** at Cannock in Staffordshire, **Isaac Nash** at Stourbridge and **Swindell & Co**. at Dudley.

Wm. Marples & Sons

There is no better known or more prolific producer of Sheffield-made tools than **Wm. Marples & Sons**. In an industry that once boasted many famous but generally small firms, Marples has proved to be a survivor and still trades today as a division of the Record Ridgeway Group.

The origins can be traced back to 1828 when 21-year-old William Marples (junior – his father was also a Sheffield toolmaker) started as a joiners' toolmaker. In Sheffield trade terms, this means that he made such items as marking and mortice gauges, squares, bevels and spokeshaves – a whole range of tools which include a significant element of wood.

Wm. Marples trade marks in 1909.

By 1875 the firm had become one of the strongest toolmakers in Sheffield and was making, or at least selling (there was always a great deal of factoring – wholesaling – sub-contracting of production in the Sheffield trade) not only every tool that the joiner, carpenter or cooper might need but also tools for many other trades.

There followed a period of further expansion as **Wm. Marples & Sons** acquired its weaker competitors: **Turner Naylor & Co.** in 1876; **Thomas**

Price Guide

♦ A **Marples** No. 6810 corner cramp with trade label in original box for two. N **£15** [DS 23/78]

♦ A **Marples** 10" rosewood parallel side level, brass plated top and bottom. **£26** [DP]

♦ An Ultimatum brass framed beech brace with ivory ring in ebony head by **Marples**. This is the transitional type c.1860, before "the Sons" joined the firm, engraved A. Randle. **£319** [DS 23/1392]

♦ A set of three brass and one steel mitre templates by **Marples**. G **£28** [DP]

♦ A "fine quality" handled beechwood plough plane with boxwood stems and fence by **Marples**. G++ **£120** [DP]

♦ A set of six side beads, ⅛" to ⅝" by **Marples** (fully boxed in smaller sizes). **£42** [DP]

♦ A beech open front router plane (old woman's tooth) by **Marples**. G **£18** [DP]

♦ A gent's drawknife, 10" long, with boxwood handles by **Marples**. **£28** [DP]

Ibbotson & Co. in 1905; **John Moseley & Son**, circa 1883 and **John Sorby & Son** in 1932.

A word of warning is perhaps appropriate here. The acquisition of companies did not necessarily mean that their name and trade mark disappeared; indeed, quite the contrary was usually the case and the published dates of acquisition may be equally misleading. On the Sheffield scene, firms might have been effectively controlled by a larger firm either because they had become financially dependent on them or were virtually exclusive sub-contractors to them.

With such a long period of operation it is important that the collector considering a purchase looks carefully at any Marples tools to determine their age. The trade mark, the clover leaf, derives from the name of the factory, Hibernia Works, situated in Westfield Terrace. The company first occupied these premises in 1859 and did not finally vacate them until 1972.

Wm. Marples & Sons Hibernia Works, Westfield Terrace, Sheffield [Marples, 1909].

Holtzapffel

1794: John Holtzapffel. 1804: Holtzapffel & Deyerlain. 1828: Holtzapffel & Co. 1923: Holtzapffel & Co. Ltd. Today usually thought of as the makers of high class lathes used by amateur turners, the Holtzapffel business, particularly in the earlier days, was at the forefront of precision engineering technology and was the supplier of tools and equipment to a number of trades particularly the gun trade. "Gent's" tools (smaller than normal versions of tools or tools intended for household or amateur use) and tools made by other manufacturers were also part of their trade.

The business operated in London from 1794 until the late 1930s by which time they were selling model railways, and successor firms continued in this business until the 1970s.

The Holtzapffels were good publicists providing advice and encouragement to their customers in a series of five volumes titled *Turning & Mechanical Manipulation* published at intervals between 1843 and 1884

(perhaps they just got tired of answering all those questions). These contain much worthwhile material for both the turner and the historian of technical practices. Reprints are now available.

Contrary to popular perception, only about one-third of the lathes sold were intended for ornamental turning. Today these lathes are cherished, and occasionally used, by their amateur owners. If a variety of work is to undertake, ornamental lathes need a considerable number of

Tools and Instruments for

ARCHITECTS.	COPPERSMITHS.	MASONS.	SEAL ENGRAVERS.
BOOKBINDERS.	ENGINEERS.	MILLWRIGHTS.	SILVERSMITHS.
BRUSHMAKERS.	ENGRAVERS.	MODELLERS.	SMITHS.
BUILDERS.	GARDENERS.	OPTICIANS.	SURVEYORS.
CABINETMAKERS.	GUNMAKERS.	PAINTERS.	TINSMITHS.
CARPENTERS.	HARNESSMAKERS.	PLASTERERS.	TURNERS.
CARVERS.	HATTERS.	PLUMBERS.	WATCHMAKERS.
CLOCKMAKERS.	JEWELLERS.	PRINTERS.	WHEELWRIGHTS.
COACHMAKERS.	MACHINISTS.	SADDLERS.	WIREDRAWERS.

Cutlery of every Description.

The list of trades for which Holtzapffel provided tools [Holtzapffel 1847].

accessories so the lathes were normally supplied with an extensive kit. Items separated from their lathes continue to turn up in unexpected places. Several other manufacturers also made ornamental turning equipment and it should not be assumed that Holtzapffel was the maker.

Ornamental turning patterns.

Price Guide

♦ A rare gent's lockable pine tool chest by **Holtzapffel** measuring 18" x 9½" x 9½" with one sliding container and other compartment with drawer under, also with quality brass furniture and trade label under lid with Holtzapffel & Co., 64 Charing Cross, Opposite Kings Mews, London and illustrating 18 tools etc. **£231** [DS 23/983]

♦ Two cranked end turning tools in beechwood, numbered handles by **Holtzapffel**. **£18** [DP]

♦ A brass goniostat (tool holder for sharpening to a precise angle) by **Holtzapffel**, no case. **£350** [DP]

♦ An elegant jewellers' tool handle in brass and iron with rosewood handle, the bit stamped **Holtzapffel**. **£55** [DS 23/648]

♦ A handled 1" screwbox and tap by **Holtzapffel**. **£50** [DS 22/171]

♦ A cloggers' bench knife by **Holtzapffel** with rosewood handle. **£396** [DS 22/1434]

♦ An ebony and brass thumb-screw operated mortice gauge by **Holtzapffel**. **£80** [DP]

♦ A beech jack plane by **Holtzapffel & Deyerlain**. **£24** [DP]

♦ A dado plane by **Holtzapffel & Deyerlain**. **£24** [DP]

James Chesterman & Co.

Founded in Sheffield by James Chesterman in 1820, the firm has always specialised in making products using thin steel strip and wire. In the mid-19th century these included springs, steel strips for corsets and crinolines, steel measuring tapes and land chains.

By the 1880s, whilst steel and cloth tapes remained central to the product range, machine-divided steel rules were also being made in quantity. Thereafter the firm gradually added a range of engineers' gauging and measuring devices

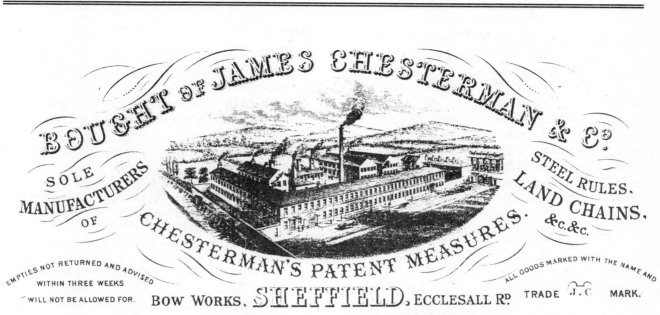

James Chesterman & Co.'s factory as shown on a billhead of the 1870s.

including combination sets and a variety of calliper and vernier calliper gauges.

In 1963 the firm amalgamated with **John Rabone & Sons Ltd.** of Birmingham to become **Rabone Chesterman.** There had, for many years, been an overlap of product range but each firm had always been predominant in some section of the market – Rabone in boxwood rules and Chesterman in precision-divided steel rules.

Price Guide

• A 2-fold steel rule marked **Chesterman** Patent measuring 16 Russian inches (28"). N **£13** [DS 22/859]

• Two cased steel tapes by **Chesterman**: 66ft in leather case, No. 155, and 25ft. steel case, No. 504, both in orig. boxes. **£39** [DS 23/551]

• An unused 30 metre surveyors' land chain by **Rabone Chesterman** with brass tallies and handles. F **£22** [DS 23/418]

• A No. 1694 boiler plate gauge with folding tongue by **Chesterman**. F **£24** [DP]

• A steel sliding calliper gauge, No. 1433, for lead pipe and sheet by **Chesterman**. **£22** [DP]

• An all steel pocket knife with one blade and two folding leaves to make a 12" rule by **Chesterman**. G **£14** [DP]

• Four different patternmakers' contraction rules by **Chesterman**. G **£18** [DP]

When looking at an 1880 catalogue one is struck by the range of foreign measurements that were available on tapes and rules. Most must have gone abroad and in small numbers, for any item marked with the different European inches is quite a rarity today and will command a premium price from collectors.

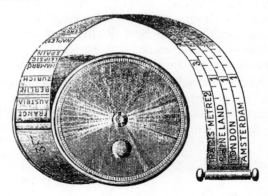

Chesterman's Patent Spring Measure with four measurements on one side and the ells of various countries on the other.

Spear & Jackson

This firm had considerable success at the great trade exhibitions of the late 19th century, earning a clutch of medals which were proudly displayed on its rather florid letterheads.

The origins of the firm go back to at least 1787 when Love & Spear are listed in the Sheffield trade directory. By 1825 John Spear and Nicholas Jackson were in partnership together. They prospered and the name of Spear and Jackson has continued in use until the present day.

An engraving of the firm's works, made in 1879, shows the conical chimneys of cementation furnaces for making (converting) the steel from which they then made the saws and agricultural and horticultural tools (heavy edge tools) that were, and indeed still are, the principal product lines. Like most of the larger Sheffield firms they also factored the products of other smaller firms.

Price Guide

♦ A brass backed cast steel 14" tenon saw and a brass backed cast steel 9" dovetail saw, both marked "Extra Cast Steel" by **Spear & Jackson**. **£35** [AP]

♦ A cast iron setting anvil and hammer (head loose) by **Spear & Jackson**. **£18** [DP]

♦ A beech plough with ebony stem wedges by **Spear & Jackson** with a set of **Spear & Jackson** irons. G++ **£55** [DS 23/1426]

♦ A **Spear & Jackson** 10 t.p.i. panel saw with rosewood handle, made to celebrate 200 years of sawmaking. G++ **£28** [DS 22/718]

Ward & Payne

The famous trade mark of this firm, the crossed hammers over an anvil, is to be found on numerous chisels, gouges and carving tools. These are generally of the best quality and are today sought by the discerning.

Founded by David Ward in 1803, the firm became **Ward & Payne** in 1845 when Henry Payne became a partner. He had died by 1850, and thereafter, the firm was managed by the Ward family but continued under the combined name.

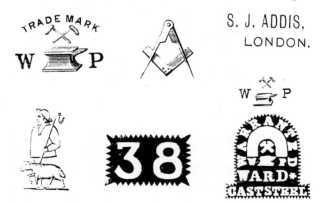

Some of the trade marks illustrated in the Ward & Payne 1911 Catalogue.

Price Guide

♦ A 1¼" bevel edge paring chisel by **Ward & Payne** with beech handle. **£35** [DS 23/287]

♦ Three little-used London pattern turnscrews with polished beech oval handles by **Ward & Payne**. **£20** [DP]

♦ A joiners' steel brace, American type, by **Ward & Payne**. **£8** [DP]

♦ Ten chisels by **Ward & Payne**. Seven ½" bevel edge with beech handles, two 1" bevel edge with boxwood handles and one ⁵⁄₁₆" firmer with beech handle, with trade labels. F **£65** [DS 21/222]

♦ An oak wall-hanging tool cabinet with two doors opening to show two drawers and racks for tools. No tools but with **Ward & Payne** label. G++ **£45** [DP]

In 1911, at the peak period of the Sheffield hand tool industry, the firm published what is probably the most comprehensive (501 page) illustrated tool catalogue ever produced by a Sheffield manufacturer. World War I, which wreaked such change on the social and trading fabric of Britain, marked the beginning of a process of rationalisation in hand tools which has, for various reasons, continued ever since.

By 1970 the firm had ceased trading – unlike many famous Sheffield names the trademark does not seem to have been taken up by anybody else.

The well-known Preston trade mark was included in many of their designs – in this case, the 5" x ⁵⁄₈" nickel plated shoulder plane [Preston, 1909].

Edward Preston

Although today best remembered for its successful range of metal planes and shaves, the firm started as wooden planemakers.

Founded by Edward Preston senior in 1825, this Birmingham-based enterprise first added rules and levels to its product range and then, by the 1880s, branched into a selection of cast iron-based products, including a wide range of shaves, routers, reeding tools and the sought-after bull nose and shoulder planes. Preston tools were always of fine quality. The firm ceased trading in 1932 but the metal plane designs were continued by **Record** (C. & J. Hampton Ltd).

A more comprehensive history of the firm and its products can be found in the introduction to the reprint of the firm's 1909 illustrated catalogue.

Price Guide

♦ A **Preston** No. 14 smooth plane. Most of original trade label remains. G++ **£55** [DP]

♦ A **Preston** No. 1393P patent reeder complete with three fences and seven cutters. **£50** [DP]

♦ A **Preston** No. 1363 narrow bullnose plane, 3¾" x ⅝". **£88** [DS 23/717]

♦ A rosewood and brass mount-cutters' knife by **Preston**. G++ **£20** [DP]

♦ A **Preston** No. 1394 adjustable circular quirk router (vertical stock, "donkey ear" handle type). **£120** [DP]

♦ A 2 ft. 4-fold boxwood rule with disc joint and level in leg by **Preston**. **£36** [DP]

Alexander Mathieson

This firm has a special place in the hearts of many for two reasons – first, it was a Glasgow firm (indeed it was the only large Scottish toolmaker) and second it was almost certainly, the most prolific British wooden planemaker of all time. That is not to say that the planes are anything but good quality. Being a Scottish firm, there are certain features in their planes – a propensity for multi-iron types, handled ploughs and fillisters – that are of interest to collectors.

Although the firm later claimed to have been founded in 1792 this is, at best, a half-truth. Alex. Mathieson, when he started in 1822, took over the business of John Manners, planemaker, which had been started in 1792. It was, however, only in the 1860s that the firm really began to expand under the management of Thomas Mathieson, Alex's son. It gradually took over the other Scottish

Best quality   *Second quality*   *Third quality*

Alex. Mathieson's trade marks from a catalogue dated circa 1895.

Price Guide

♦ A large beech bow saw by **Mathieson,** some stains. **£50** [DS 23/327]

♦ A 12" brass topped rosewood level by **Mathieson** No. 7C with trade label. **£36** [DP]

♦ A rare 2¼" triple iron quirk ogee and bead m.p. by **Mathieson**. **£190** [DS 22/1347]

♦ A d/t steel jointer by **Mathieson**, 22½" x 3⅛" with rosewood infill and handle. This plane is without the usual protruding toe and heel. 90% of the original **Mathieson** iron. F **£1200** [DS 22/1377]

♦ A **Mathieson** cast iron adjustable plane of Bailey No. 4 size and form. **£50** [DP]

♦ A bright steel Scotch brace with rosewood head. **£30** [DP]

♦ Four pairs of tongue and groove planes by **Mathieson**, some with original transfer trade labels. F **£60** [DP]

---

planemakers and widened the product range to include heavy and light edge tools, augers, etc. The dovetailed steel smoothing, jointer and shoulder planes of Mathieson's manufacture are of fine quality and should not be disregarded by collectors in favour of other makes.

Like many hand tool manufacturers, the company seems to have started a long decline in the 1920s; wooden planes were becoming obsolete, shipbuilding had almost entirely changed to steel and machine woodworking had become common. In 1957 what remained of the firm was bought by William Ridgeway and transferred to Sheffield.

Charles Nurse

A firm with an unusual history. Having started in Maidstone, Kent in 1841 (hence the Kentish horse trademark), under the direction of the next generation it moved to Walworth Road in south-east London.

Here the business grew, becoming a prolific maker of wooden planes, and continuing until the 1930s when it was the last significant London planemaker. During the period 1895-1920 there were also retail branches in the City (see *Frontispiece*).

For the purchaser who wants moulding planes for use, Nurse-made planes in general do not command high prices and can readily be found in good condition, whilst, for the collector, there are a surprising number of unusual planes – Nurse advertised that they would make planes to special order.

Charles Nurse & Co. 182 & 184 Walworth Road, London S.E. from a catalogue circa 1905.

Price Guide

♦ A 5" beech shipwrights' spar plane by **Nurse**. **£30** [AP]

♦ A pair of probably unused beech smoother and jack planes by **Nurse**. F **£33** [DS 24/1133]

♦ A set of 8 plough irons by **Nurse** in baize roll. **£31** [DS 23/1040]

♦ A brass backed 9" cast steel dovetail saw marked **C. Nurse & Co.** G++ **£24** [DP].

♦ A rare twin iron double d/t boxed airtight casemakers' hook joint plane with full length brass fence by **Nurse. £85** [DP]

♦ A 1¼" bevel edge paring chisel by **Nurse** with hexagonal boxwood handle. G++ **£50** [DS 20/1239]

♦ An ebony oval head cutting gauge with full brass face by **Nurse**. G++ **£36** [DP]

♦ A pair of planes consisting of a beech smoother and a jack plane by **Nurse** both with Nurse Patent Regulators and Nurse irons. The regulator gives a vertical and lateral adjustment. Planes fitted with the regulating mechanism are rare. G++ **£418** [DS 23/1283]

♦ A beech stop chamfer plane marked **C. Nurse**, Sole Inventor. **£40** [DP]

*Remember - condition is **G+** unless otherwise shown.*

The Buck family of London

Two firms still operating today (**Buck & Ryan** and **Buck & Hickman**) incorporate the Buck name, and the student of tools, particularly if searching in the south-east, will find many items bearing these and other Buck names.

Matthew Buck, a Clerkenwell saw and file maker, had two sons and a daughter all of whom started separate plane and toolmaking businesses in the early years of the 19th century. The daughter married a Mr John Hickman and traded under their joint names. This firm has a particular place in the tool world as the publisher of comprehensive hard-back catalogues. The print run must have been large and the recipients of these fine burgundy-coloured books reluctant to throw them away for they survive in some numbers.

A more detailed outline of the three businesses is included in *British Planemakers from 1700*.

Price Guide

♦ A 1½" bevel edge paring chisel marked **Buck & Ryan** with octagonal boxwood handle. **£26** [DP]

♦ A miniature boxwood rebate plane 3½" x ½" by **Buck. £61** [DS 23/922]

♦ A unusually small brass shoulder plane, 7⅜" x ¾", by **J. Buck**, London with rosewood infill. **£70** [DP]

♦ A large beech handled coopers' timber scribe by **Buck & Hickman. G £335** [DS 22/462]

A view of the premises at 281 Whitechapel Road in the mid-19th century.

Christopher Gabriel

A prominent London planemaker (and tool dealer) of the late 18th century. The firm worked from 1770 to 1822. Due to the unique survival of an account and inventory book, more is known about the Gabriel business than any other 18th/19th century planemaker. He was the supplier of the tools in the Seaton Chest (see *Further Reading*, p.18).

Price Guide

♦ A rare 18th C. beech coachbuilders' plough plane by **Gabriel** with brass top, single thick stem holding the fence and original fluted iron. **£2100** [DP]

♦ A pair of No. 10 H&R planes by **Gabriel. £15** [DP]

♦ A pair of side rebate planes by **Gabriel. G £72** [DS 23/109]

♦ Two wide m.p.s, 2¼" and 2⅞", by **Gabriel** (marks both G). **G £50** [DS 23/146]

♦ An brass button pad beech brace by **Gabriel** (incuse mark). **G £160** [DS 23/1391]

♦ A beech sash fillister by **Gabriel** (mark G). **£72** [DS 22/627]

♦ A rare try plane by **Gabriel** with offset handle (mark G). **G £44** [DS 21/1289]

Gabriel was a successful trader as were his sons; his grandson became Lord Mayor of London. Recent research has shown that, like some other planemakers, the firm also made other products from what was their raw material – beechwood. Seen from this perspective, the family's change to being timber merchants is less surprising.

Planes made by the Gabriels are quite common but it is difficult to determine whether they were made early or late in the period of operations.

Some of the Gabriel marks found on their planes.

The English tool making industry was a leader in the marketing developments generated by the industrial revolution. The location of production where it could be undertaken most efficiently but at sites remote from markets required not only transport but a marketing system capable of telling customers what was available. The best answer is a price list – better still an illustrated catalogue.

The earliest form of promotion was trade cards. with some examples dating from before 1700 being known. Some 18th century cards showed illustrations of the type of goods sold whilst other included long lists of tools and hardware.

Trade card of Marshall, furnishing ironmonger, c.1810.

The remoteness of the Lancashire clock tool industry from its markets resulted in the earliest known illustrated tool catalogue, produced between 1758 and 1770, by John Wyke. Others followed and by 1800 several illustrated catalogues of Lancashire and Birmingham tools had been produced. Sheffield was a little behind – not until (probably) 1816 was the first illustrated list produced – Joseph Smith's *Key or Explanation to the Various Manufactories of Sheffield.*

An advertisement in Building World, 1903. *By this date, retailers were producing their own extensive catalogues. Manufacturers' catalogues were generally distributed only to retailers and wholesalers and for this reason are rarer than retailers' catalogues.*

Copper plate

All these catalogues were printed from engraved copper plates – the lines are engraved onto the plate using a burin, each plate printing a whole page. The system is costly and inflexible – alterations are almost impossible. The result is bold and wonderfully detailed. Catalogues of this style and date, important milestones in industrial and marketing history, are very rare. Fortunately for collectors, the Wyke and Richard Timmins catalogues and Smith's *Key* have all been reprinted in recent years.

Catalogues and more catalogues

Around 1820 steel began to replace copper for the plate. Steel engravings are generally finer and less bold but they do not seem to have found widespread use in catalogues.

The real explosion in the production of catalogues came with the development of stereography and electrotyping. By these techniques metal copies of wooden blocks can be made. In both techniques a mould is made from the wood block and then a copy taken from this either by pouring in type metal (stereotype) or by electrolytic deposition of copper, backed by poured metal (electrotype). By the 1880s these techniques had become widespread. Now, the tool manufacturer could have a wood engraving prepared

A point-of-sale display card for steel pocket rules [Millard, c.1910].

from which many copy blocks were made. These could also be made available to wholesalers and retailers who wanted to issue their own catalogues.

Today, catalogue illustration blocks – both original wood blocks and electrotype copy blocks – can be found in the antiques market.

In the 1890s photographic illustrations started to appear in trade catalogues but such was the clarity of hand engraved blocks that the changeover was very slow – I have some blocks given to me by the man who made them in the late 1950s.

The retailer

Trade cards of the early 18th century show that the retailer was by then a well-established part of the tool selling system. Many retailers played some part in finishing; for example, tools were acquired without handles which were made/fitted by the retailer. All this serves to contradict the popular misconception that, until this century, tradesmen made many of their own tools or bought them from the village smith.

By the 1880s manufacturers were seeking to promote their branded goods and the customer was more likely to buy a brand name rather than be guided by the retailer.

The point of sale promotion had arrived – in this area lie some interesting items for the collector. Best are manufacturers' display cases complete with contents and oversize tools for window display. Display cards, advertising paper weights, ash trays, and thermometers are but a few of the many promotional items.

Display case for Forstner auger bits [Harding, 1903]. Some cases of this type were intended to stand on the counter and incorporated a cupboard at the rear where the stock was stored. Others were wall hanging displays intended merely as attention-getters.

> **Collecting tips**
>
> ➤ *Trade catalogues are today sought by tool collectors for the information they contain and also by those who see them as works of artistic merit. See* Old Books & Catalogues, *pp. 216– 218.*

The marks on tools – patent numbers, the registered design diamond, registered design numbers and trade marks – are all helpful both for dating a tool and, in some instances, identifying the manufacturer.

Registered design symbol

The 1842 *Copyright of Design Act* created thirteen classes of ornamental design for manufactured goods. This Act initiated the use of the diamond registration mark which remained in use until 1883, although the coding was changed in 1868. Tool collectors are most likely to encounter Classes 1 and 2, being objects of metal and wood. Registration in these classes gave three years' protection to the design.

The letters found in the diamond mark identify the class and date of registration. Below is a key to the meaning of the letters which, as can be seen, changed in 1868. The letters were not used in sequence for either year or month, so it is necessary to look these up.

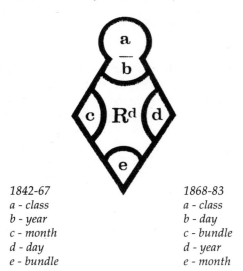

1842-67	1868-83
a - class	a - class
b - year	b - day
c - month	c - bundle
d - day	d - year
e - bundle	e - month

Registered design numbers

From 1884, the diamond symbol was discarded and consecutive numbers were used, starting with number one in 1884. These numbers are nearly always prefixed with Rd. or Rd. No. The numbers are in a single sequence and the year of registration can quickly be established by the numbers issued in any year. The design registration records from 1839 to 1910 are kept in the Public Records Office at Kew in Surrey and are open to the public. The records from 1910 until the present day are kept in the Patents Office Library in London and are not easily accessible.

Patent numbers

Many tools of more original design are marked with patent numbers. These patents can be accessed at the Patents Office Library. Before the 1852 Patent Law Amendment Act it was both difficult and expensive to obtain a patent and there are few before this date that relate to tools.

After 1852, volumes of abridgements were published in which each patent has a few lines of text and one illustration. These are listed numerically by year within a class, for example, hand tools. So if there is a patent number marked on a tool it can simply be looked up. Once an item has been found in the abridgement volume, then a full copy of the patent can be obtained by request.

Some tools are marked purely by the patentee's name. These are more difficult. They can be looked up in the Name Index; however as these are published annually, unless a date can be accurately estimated this can be a long job. Alternatively, a search can be made through the volumes of abridgements for tools of that class. For example, planes are class 61 so if you are looking for a plane, you need only look through that class.

Collecting tips

➢ *American patents, to be valid, had to state the day, month and year whereas English patents did not. If the patent in question has this information it will almost certainly be American.*

➢ *Some of the Sheffield trade directories, including the three 18th century ones (1774, 1787 and 1794), show cutlers' marks and can therefore be very useful in establishing the makers of early edge-tools.*

Trade marks

Before 1876, the registration of trade marks was haphazard and devolved to various bodies (the Company of Cutlers in Sheffield, for instance) and the citation of a "trade mark" on a tool usually indicates a date later than this.

However, a word of warning in respect of Sheffield tool makers. Many of the cutlers' marks and later trade marks were passed from company to company and often, if a name was well known, its use would be continued long after the original owner had ceased to trade. So these can be misleading as a means of dating a tool.

Iron and steel

The earliest iron smelters produced bloomery iron, a comparatively pure form of iron that needed hammering to consolidate the metal and expel slag as the furnaces did not achieve high enough temperatures to make the metal liquid. This pure iron cannot be hardened by quenching.

By 1600 higher temperatures were being achieved (around 1200° C) at which the iron melted but at the same time absorbed carbon from the charcoal charge. This liquid iron could be tapped off from the furnace and cast into ingots. The metal thus produced is cast iron – it has a carbon content of about 4%– but it is brittle and whilst it can be cast into a variety of useful objects it is of no use for making tools with cutting edges. By reheating with air blast the carbon can be burnt out but as the carbon content reduces, the melting temperature rises, and with the furnace technology available in the 17th century, the 1500°C+ needed to melt the lower carbon content material could not be achieved and the output was again pasty iron that had to be worked with a heavy water-powered hammer. The resulting product was wrought iron which had a low carbon content of 0.1%.

Steel is iron or, more strictly, an alloy of iron with a carbon content of between 0.5 and 1.5%. Today, small quantities of other metals may be added to improve the product and make it suitable for special purposes but the fundamental characteristic of steel is that, when quenched from bright heat, it becomes much harder than before. This is the real value of steel.

It took mankind thousands of years – from the start of the iron age until the 18th century – before the process was mastered and steel of some consistent quality was made. It was not until the Bessemer process revolutionised steel-making in the 1860s that steel in large quantity became available.

By the second half of the 17th century steel was being made by the cementation process in the Forest of Dean and elsewhere. This technique is similar to surface hardening. Iron is heated in close association with carbon (charcoal) and over a period some of the carbon diffuses into the metal. The problem is that the carbon will burn away but the cementation process overcame this problem by enclosing the iron bars in a brick chest with the charcoal and then excluding the air by capping with spoil. The whole mass was then heated and held at red heat for around seven days. The carbon diffused into the iron bars producing blister steel.

This was not a homogeneous material; the outside might have a high carbon content and the inside little. Some degree of regularity could be achieved by cutting up the bars and then forging

The Albion Steel Works, Sheffield from an advertisement of 1868 for J.R. Spencer & Son. The conical chimneys over the cementation furnaces are visible at the rear of the works. Like many Sheffield manufacturers, this firm not only made steel but were also manufacturers of files, edge tools, saws, cutlery, razors and shovels.

The crucible steel melting shop at the Fitzalan Steel and File Works, Sheffield from The Penny Magazine, *30 March 1844.*

them together in bundles – this was shear steel – and the process could be repeated to make double shear steel but the metal still had layers of differing carbon content.

Cast steel

The need for a more homogeneous steel was keenly felt by clock-makers who wanted to make springs of absolute regularity and it is not therefore surprising that Benjamin Huntsman was originally a clockmaker. He reasoned that, if he could melt blister steel, a superior steel could be made. The problem was to achieve high enough temperatures (1600°C) and to make pots (crucibles) that could be fired to this temperature.

A considerable folk lore surrounds his early days with stories of spies attempting to learn his methods and finally succeeding. By the 1750s he was regularly making small quantities of cast steel; each pot contained only six or eight pounds of metal and a decade later total production was only ten tons per year. Production of crucible steel continued in Sheffield until the 1950s.

The superior quality of cast steel for edge-tools was quickly realised and the material became a selling point, much trumpeted in catalogues, and often marked on the tools themselves. The high cost of steel, particularly cast steel, caused many tools to be made using only a small piece of steel fire-welded to a body of lower quality metal. The steel is usually noticeable, particularly on chisels and axes and is a useful indicator of an early tool.

Brass and gunmetal

The best quality joiners' tools of the late 19th century were exotic combinations of brass and dark woods – ebony, rosewood and others. What is perhaps surprising is that virtually all the brass fittings to be found on joiners' tools, including the flat plates which would seem to lend themselves to stamping from sheet, were made from castings which were filed to a bright finish. If you don't believe me remove the facing from a quality mortice gauge and look at the back! Only in the 20th century did sheet brass replace castings on this type of tool.

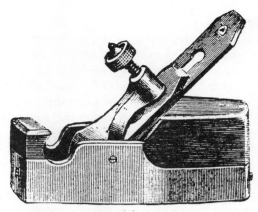

Gunmetal smoothing plane [Melhuish, 1899]. The metal is soft and better quality planes had steel soles.

In the 18th century brass was expensive and was used sparingly. Whilst brass fittings, depth stops, thumb screws etc. became usual around 1775, brass ferrules for chisels were uncommon until well into the 19th century.

Brass is an alloy of copper and zinc which can be in widely variable proportions – all of which are called brass! When some tin is added it then becomes gunmetal by name. There are those who maintain that they can tell old brass or gunmetal by the colour. One suspects that they are, in reality, being influenced more by the surface oxidation and the patina of old metal than by the constituents. Old metals acquire a surface patination of minute dents and scratches which hold surface oxides even after cleaning and this affects the colour. Incidentally, such patination is almost impossible to fake and is a good indicator of age. Brass can be almost white when first cut but quickly yellows; gunmetal shows more of the copper colour of its major constituent and bronze is similar in appearance but is more difficult to cut or file.

On joiners' tools it was usual to lacquer or French polish brass fittings – this was, in general, the same finish as that applied to the wooden parts. With scientific instruments – sighting levels, theodolites, etc. which are usually kept in cases – the lacquer will last for a century or more. The

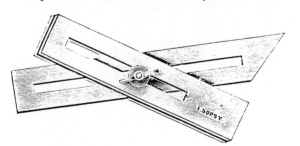

Masons' brass shift stock [Turner Naylor, 1928].

brass will become golden under the lacquer but in areas where the lacquer has worn off, it will go black and the object acquire a piebald look.

The ruthless striping of old lacquer and the crude refinishing of many brass tools – levels and braces seem to be particularly at risk – has, in recent years, reached epidemic proportions. Tools in original condition with sanded or filed finish just visible beneath a mature lacquer deserve to be more highly valued, and one day they will be, so if in doubt, leave alone!

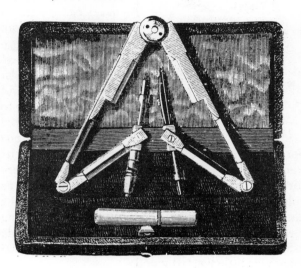

Best German silver Napier pattern compasses, cased [Melhuish, 1904].

Nickel silver

Also known as German silver and, in drawing instruments, electrum, this is an alloy containing nickel, copper and zinc. By the 1860s better quality drawing instruments were being made from it. Its resistance to tarnishing was an improvement on the brass previously used, and in fitted cases it presented a very pleasing appearance when contrasted with the usual blue linings.

The metal mounts on ivory rules were normally nickel silver, and a small proportion of boxwood rules also have nickel silver mounts, an indication that the rule was for the "gent's" trade.

Silver

Real silver tarnishes to completely black and this characteristic is often helpful in identifying the **very** occasional silver item that turns up in old boxes of tools long untouched.

Silver was a favoured material for the construction of some medical items – catheters, etc. and also as decoration on tweezers, lancets and the like. Some of these items are hallmarked.

Beech

In Britain beechwood was always the choice for wooden planes. The only exception is boxwood, used for a few premium products, for example,

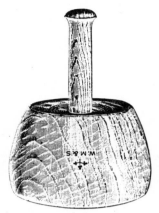

Masons' beech mallet [Marples, 1909].

miniature "thumb" planes. Thus for every boxwood plane there are hundreds, if not thousands, made in beechwood.

The beech tree, *Fagus sylvatica*, grows through- out Britain, developing into a large tree with a good straight trunk. Whilst the wood is hard and not particularly easy to split, it is not durable in exterior situations nor does it have an interesting grain so its traditional uses were generally limited to country furniture, carcasing and small products such as culinary spoons and boards where its hard-wearing properties and resistance to splitting were needed. Some planemakers also made a wide range of other beechwood products ensuring that there was little wasted wood.

Trade uses

Amongst the trade-related items made in beech are the following: brushes, saw handles, bench screws, screwboxes, sash templates, bow saws, the cheaper forms of marking gauges, the tops of joiners' benches, and handles for a wide range of tools including chisels.

In Germany beechwood planes are still being made today some being sold in both Britain and America. In Britain, beechwood is still widely used to make such items as mitre boxes, marking gauges the cheaper plumber's dressing sticks and many tool handles.

Appearance

Beechwood can be recognised by its tight grain, and its small flecked medullary rays which, on the end grain, are thin but bright whilst the growth rings are not prominent. Without finish beech ages to a grey, uninteresting tone but the application of varnish or French polish will keep the wood light for an extended period. Conversely, in the workshop situation, the application of linseed oil – and this was normal practice – will cause a plane to darken.

How fast this happens is dependent on how often the oil is applied and the conditions in the workshop. Recently we examined a group of planes from the earliest period of English planemaking which, although now 300 years old, are still light in colour; but we have also seen planes which date from between the two world wars which are already completely black! Colour is not a reliable guide to the age of a beechwood plane or tool.

It was not normal practice for the planemaker to apply a finish; some did state that the planes could be had oiled and a few (including **Alex. Mathieson**) offered a varnish finish. The occasional planes with this finish – often ploughs or sash fillisters – are to be found, still light in colour.

A workshop trick was to dam the mouth of a new bench plane with putty and then fill it with linseed oil. The toe and heel would also be stood in a tin lid of oil. A raw plane will take up a surprising quantity and the density of the stock will be noticeably increased. Incidentally, the oil inside the wood never seems to dry and in hot

Improved pattern cabinet and carpenters' bench [Tyzack, c. 1910]. The whole of this bench is beechwood. In some catalogues, this style of bench is called a German pattern as it has a wooden tail vice – more common on the continent.

weather, even decades later, it will ooze from the end grain – in the process obscuring the maker's marks which all collectors want to see!

Over-oiled planes have a sticky and often black coating which conceals the beauty of the wood (this needs to be remedied – see *Care and Cleaning*, p. 21). However many beechwood planes will have an attractive mid-brown colour and patina that needs little attention.

Improved beechwood mitre box [Turner Naylor, 1928].

Boxwood

In a world without plastics, boxwood was an important material. It is hard, close grained and light in colour and each of these characteristics is of importance. If it has a fault, it is that it can be brittle and its use for chisel handles is therefore perhaps surprising.

Boxwood has never been a cheap wood; the English box tree, *Buxus sempervirens*, is really no more than a bush, slow growing and slow to develop thickness in the trunk and branches. As it will grow in light shade and the leaves are evergreen, it was widely planted in the 19th century to provide cover for game and the English boxwood available today often comes from these plantings. Sadly, much of the wood cleared is burnt, even though there is a good market for it amongst turners. Woodsmen seem unable to appreciate that even logs of only 3" (75mm) diameter are usable, any boxwood of 4" (100mm) diameter is premium material and 6" (150mm) diameter wood is practically unobtainable.

English boxwood is difficult to season. It must be taken under cover immediately after cutting and

Plumbers' boxwood bending stick [Melhuish, 1904].

the initial drying must be quick or it will very soon be shot through with black mould. Thereafter the drying should proceed slowly or the wood will develop nasty shakes. Some older books suggest burying in sand or waxing the cut ends to control the rate of drying but in our experience this does not make much difference except to extend an already lengthy procedure and that some shakes seem inevitable.

Trade uses

A principal trade use was rule making. For rules no wood could compete with it – only ivory makes a clearer rule. Boxwood has given its name to the hard-wearing strips (boxing) inserted into moulding planes at points of particular wear. It was also the choice for small thumb planes.

It was used for making blocks for wood engraving – the cuts that illustrate this book were originally engraved on the end grain of highly polished blocks of boxwood – and it was used for a wide range of turned items which today would be made in plastic. As this is being written there are, sitting on the desk, two nesting sets of watch-maker's movement stands (used to support watch movements while they are under repair). The one consists

Boxwood ring size stick [Isaacs, c.1890].

of thin rings of boxwood with an everted lip, not dissimilar to pastry cutters and often sold as such; the other is plastic!

Working boxwood

Boxwood is not easy to work with normal woodworking tools. Moxon writing in 1678 said

> If it be very hard wood you are to Plane upon, as *Box*... It [the plane iron] is set to 80 Degrees, and sometimes quite upright...

A toothing plane is effective, and files and sandpaper are often needed to finish flat surfaces.

Above all, boxwood is a fine wood for turning: chisel or awl handles represent simple items; oboes and clarinets, items of some size and complication.

The English box also grows in France and some other European countries. Related species of *Buxus* grow in the Balkans and Turkey. Christopher Gabriel (the London plane and tool seller) listed in his c. 1800 inventory "14 cwt 3 qutrs Turkey box".

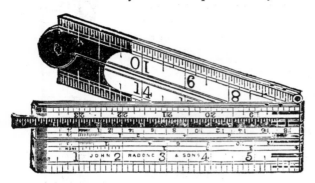

Arch joint, 2ft. 4-fold boxwood rule with brass slide [Tyzack, c. 1910].

By the early years of the 19th century, Maracaibo boxwood, *Gossypiospermum praecox*, was being imported from South America and the West Indies. This tree is unrelated to the English box but is available in larger sizes, 20" (500mm) diameter logs being not uncommon. The wood is slightly lighter and softer but this is difficult to distinguish when finished and patinated by the passage of time. This was the material used by the rule makers of the 19th century and by the commercial turners of larger items such as bottle cases, musical instruments, etc.

Newly cut boxwood has a pale, almost lemon colour which initially looses colour before very gradually darkening with the passage of time. Old boxwood, aided by polish or oil finish, becomes progressively more "golden", eventually becoming a deep gold in colour. To artificially colour boxwood is not easy and attempts can be detected by the experienced observer.

Collecting tips

➤ *Boxwood tools command a considerable premium and there are a considerable number of recently made objects now circulating in the market place. So beware!*

Lignum vitae

Lignum vitae, *Guaiacum officinale*, has several notable characteristics. It is very heavy, will sink in water, is tough with an interlocking grain and is heavily impregnated with resin. But most noticeable is that the heartwood is dark brown whilst the sapwood is a light yellow. It turns well and its popularity for the production of turned domestic wares is confirmed by the many mortars, coffee grinders and large wassail bowls dating from the 17th century onwards. Lawn bowls in their thousands were made from lignum vitae and these can be a source of good wood for the turners of today.

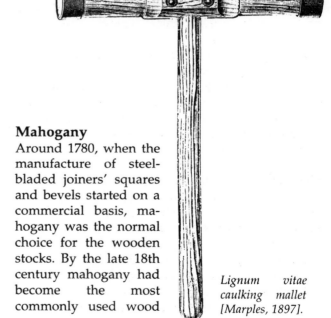

Lignum vitae caulking mallet [Marples, 1897].

Mahogany

Around 1780, when the manufacture of steel-bladed joiners' squares and bevels started on a commercial basis, mahogany was the normal choice for the wooden stocks. By the late 18th century mahogany had become the most commonly used wood

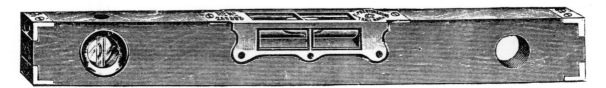

Mahogany plumb and spirit level [Preston 1909]. This level could be had in rosewood, mahogany or ebony.

A large log of Honduras mahogany being sold at Birkenhead. From "The Illustrated London News" April 6, 1850.

for furniture of any quality. Not surprising, for mahogany, whether it be hard Spanish mahogany or the softer Honduras variety, is easy to work and is also very stable. This is almost certainly why mahogany was chosen for stands and lathe-beds by the makers of the best quality lathes.

However by around 1830 rosewood replaced mahogany as the material of first choice for squares etc. The larger sizes of builders' levels have always been made in mahogany; that is until recently, when the extruded aluminium level has taken over the market – and very good they are too, accurate and impervious to wet and warp!

The collector will find many craftsman-made tools in mahogany. As we have said, the wood was widely used in the joinery trade, so when the impecunious apprentice needed to make a pair of winding sticks, a panel gauge or a mitre paring guide, his hand, as likely as not, would light upon an off-cut of mahogany and some very good and collectable tools originated in these circumstances.

Ebony

Used decoratively in furniture making in Europe since the 16th century or before, ebony, a wood hard to work, had no place in commercial tool making until the second half of the 19th century.

The family of trees to which ebony belongs, *Diospyros*, grows in many parts of the world including India, Sri Lanka and Africa and there are quite a number of trees of that genus that produce

Best ebony saw pad [Turner Naylor, 1928].

black or stripy black wood (which is usually described as Madagascar ebony). The exact species need not concern the tool collector; it is the beauty of the tool, glowingly black or finely striped, that is important.

From the last decade of the 19th century until 1940 ebony was used for the finest quality squares, bevels and mortice and marking gauges. The marketing strategy of English tool manufacturers always seemed to be directed more to producing fine quality tools made of much brass and exotic woods than to innovation of design and construction. Ebony certainly makes an impressive looking tool but it is brittle and the collector should inspect potential purchases with great care as many of these items are chipped or cracked.

Brass faced, oval ebony head and stem mortice gauge [Melhuish, 1905].

Rosewood

Rosewood, so named because when worked it gives off a distinctive sweet smell, comes from a number of the species of the genus *Dalbergia*. This characteristic is useful when trying to identify whether a tool is rosewood as, unless totally overwhelmed by polish, a faint aroma can often be picked up.

The original supplies came from India, but nowadays the timber comes from sources as far apart as Madagascar, Honduras and Brazil. The wood is dark brown striped with black lines, perhaps with a hint of red when newly cut. It

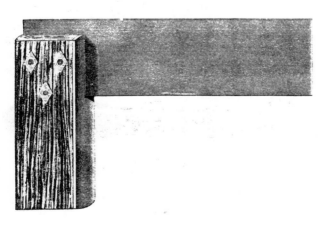

Best plated rosewood square [Wingfield Rowbotham, 1903]. The "best" indicates that it has a three rivet, one piece escutcheon.

became fashionable for cabinet making in the first decades of the 19th century and retained some popularity throughout the century.

Rosewood is the normal choice for the stuffing of English metal planes, and Stanley also used it for handles and knobs. It may be no coincidence that the first commercially made metal planes appeared around the time of the first fashion for rosewood.

Cormier

Why French planemakers should have preferred to use cormier, *Sorbus terminalis,* and to have ignored beech which was more widely available and in much larger size seems a mystery of technology not yet satisfactorily answered. The wood is in many respects similar to fruitwood, mid-brown in colour with a hint of red and close grained with little figure. For the collector the identification of cormier in a tool is a strong, almost conclusive, indicator of continental origin.

Padauk

Identifiable by its bright red colour when first cut and by its heavy weight, padauk, *Petrocarpus sp.,* has been used in France for commercial planemaking, for which purpose it is well suited. Occasionally craftsmen-made English planes made in this wood also appear. With time the wood becomes dark reddish-brown and thus contrasts well with light coloured handles.

Fruitwoods

Prior to the modern fruit-growing use of miniaturised trees, pears and apples grew into medium sized trees. These woods are usually lumped together although the wood most commonly encountered is pear, *Pyrus communis*. This is a pink-brown wood of fine texture. Railway curves and other draftsmen's templates are usually

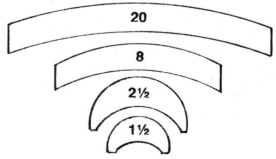

Railway curves were sold in cased sets of twenty-five (1½" to 30" radius), fifty (1½" to 120" radius or 3cm. to 3m) and one hundred (1½" to 240" radius).

made of pear and the wood was also used for wood-cut blocks. When faded and impregnated with oils or polish, small pieces are not easy to identify.

Yew

Beloved of country furniture makers, as it turns well and can be bent, yew handles will occasionally be found on craftsman-made tools of country origin. With time the wood fades to a mid-brown colour but the darker and lighter patches that are characteristic of the fresh wood – and that make it a very attractive wood – can still be seen. Yew takes a fine finish and many say that it has a unique silky and satisfying feel in the hand.

Ash

Traditionally a wagon and coachmakers' wood as it is both reasonably stiff and resistant to shock. The principal use in toolmaking is for handles for agricultural and garden tools. When human labour was the main source of energy, it was most important to maximise production by having exactly the right tool. The right handle is every bit as important as the right shaped and sized fork, scythe or spade attached to it. **Swindell & Co. Ltd.** of Dudley, Worcestershire, offered a choice of eight styles of handle on each of the spades and forks that they made. Long handles, straight except for a

slight waisting near the top, graced hay forks and hoes; adzes required carefully angled dog-leg shapes whilst that aristocrat of agricultural tools, the scythe, was fitted with a serpentine handle that had to be steam bent.

Inside the workshop ash found fewer uses: handles for hammers and mallets and, traditionally, for the heavy weight chisels used by the wagon and railway carriage builders.

Ash grows just about everywhere in Britain; in some western areas it is almost a weed. For this reason, the use of ash for chisel handles seems to have become more common during the Second World War and the years of austerity that followed.

Other woods

Craftsman making for themselves tended to use whatever was to hand. Planes in southern Europe were often made from evergreen oak – both heavier and more resistant to splitting than English oak which is seldom used in toolmaking. In central Europe, where carved decoration is frequently a feature of planes, hornbeam and even some quite soft woods like lime were used. In America the planemakers used the most suitable of the native hardwoods in their area, including beech (not the same as European beech), cherry, maple, hickory and birch.

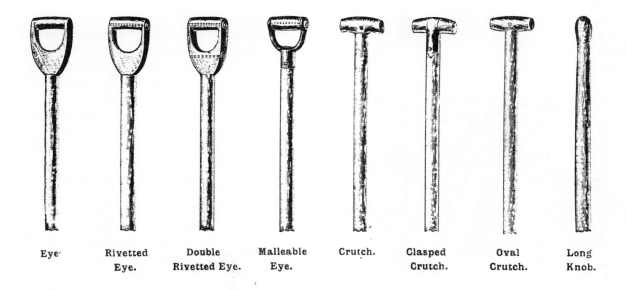

| Eye. | Rivetted Eye. | Double Rivetted Eye. | Malleable Eye. | Crutch. | Clasped Crutch. | Oval Crutch. | Long Knob. |

The eight handle styles available for shovels, spades, draining tools and forks [Swindell & Co. Ltd., 1903]. Ash was the preferred wood for the hafts (handles) of builders', horticultural and agricultural tools.

Ivory and bone

In this ecologically concerned age no one can fail to be shocked by those Victorian photographs of heaps of ivory tusks in the London dock warehouses but, in a pre-plastic age, ivory was an important material. The decorative possibilities are obvious; it contrasts well with all dark woods, it can be carved and turned and the surface can be engraved and filled. Its also had useful properties: as an electrical insulator in scientific equipment; as a heat insulator in handles for silver teapots and, because it was easily cleaned and durable, as handles for medical and dental instruments and untold numbers of table knives and penknives.

Cutting ivory handles with a circular saw from "A Day at the Sheffield Cutlery Works", The Penny Magazine, *27 April 1844.*

The principal use in tools was for rules of all types. Except for the ivory rings in Ultimatum braces, ivory never found any decorative use in English commercial tools. It was occasionally used by craftsmen making for their own use but this was infrequent. The collector should therefore approach any tool made from or decorated with ivory or bone with considerable caution. The greater proportion of such tools in the market today are of recent manufacture.

Price Guide

◆ Two ivory handled diamond tipped glass cutters by **Chater & Sons**, London and **Hoe & Sons**, London, crack to one handle. **£66** [DS 22/381]

◆ Prisoner of war bone alphabet set in bone screw top box. G **£121** [DS 21/1739]

◆ A 10" ivory double slide excise slide rule by **Aston & Mander**. **£200** [DP]

◆ A 12" ivory Gunter's rule, c. 1800. G **£125** [DP]

◆ An ivory draughtsman's triangular square, 10½" x 3½" by **Troughton & Simms**. G **£65** [DS 18/349]

◆ An attractive ivory gavel, 7½" long x 2¾" dia., shaped as a masons' maul. **£180** [DP]

◆ Two different sized gun turnscrews with horn handles by **Holland**. F **£55** [DP]

◆ A surgeons' steel bow metacarpal saw with octagonal horn handle. G++ **£75** [DP]

◆ A three blade fleam by **Arnold** with horn scales. F **£30** [DP]

*Remember - condition is **G+** unless otherwise shown.*

IVORY & BONE

To differentiate between bone and ivory is not always easy. Ivory has a little grain but no open channels whereas bone, which usually comes from the shinbone of cattle, has open channels for the blood vessels. Commercially made items will almost certainly be ivory – one sees a few "zig-zag" rules of continental origin that are bone but these are the only likely exceptions to this rule.

By 1900 casein-based ivory substitutes, with trade names such as Xylonite, were being used in Sheffield for knife handles etc. Later versions of these materials included an artificial grain and these can be extremely difficult to distinguish from the real thing. If there is any failing in them, it is that the induced lines are too straight.

Ivory is tricky stuff. Like wood, it needs to be seasoned before use and failure to do so results in later warping. Quite a few rules suffer from this and it is impossible to correct permanently. It also stains easily – old inks, which contained iron, and iron pins in rules, if allowed to get wet, will produce stains which are impossible to remove.

In the past few years, legislation to save elephants has resulted in some difficulty in exporting ivory objects between countries and this has, in some measure, affected the market. To the best of our knowledge, no ivory rules have been made since 1940 and very few since 1914. Whatever were the wrongs of the past, it seems pointless to discriminate against objects made in the time of our grandparents. Follow this logic to its conclusion and we would smash all English china made before 1850 – after all, it was fired by coal extracted by children working in the mines!

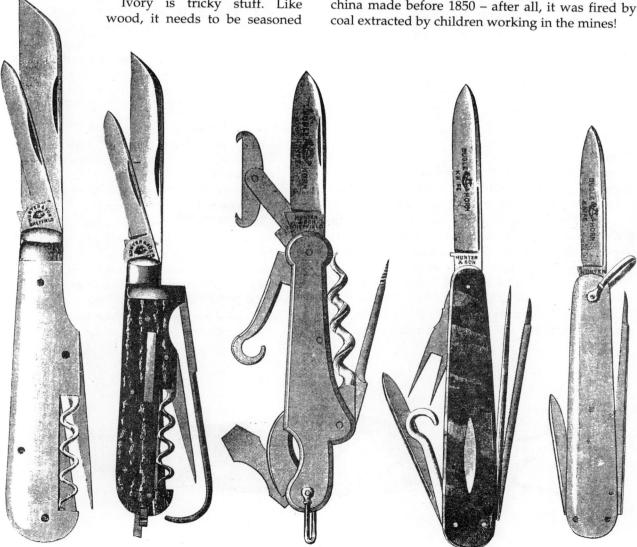

From left: bone handled sportsmen's knife; staghorn handled sportsmen's knife; nickel silver handled shooting knife; tortoiseshell handled smokers' knife; ivory handled gent's knife [Hunter, 1916].

Horn

Farmers today prefer breeds of cattle that have the smallest possible horns, and even then, many cattle are de-horned. Older breeds had impressive horns – what would the average cattle stampede be without the Texas longhorn? – and mankind has used this material for thousands of years, complete as a container or drinking vessel or as the raw material for a host of products.

When heated in boiling water horn softens so it can be split and flattened or pressed into moulds. Thicker material can be built up with a number of layers, the large pieces needed to stuff the rare horn-filled Ultimatum braces must have been made thus. The colour of horn is dependent on the breed from which it comes and varies from near black to light buff. Mostly it is somewhere in the middle, often with a strongly streaked colouration.

Of interest are butchers' steels with buffalo horn handles (illustrated as translucently marbled) shown in the 1914 catalogue of **M. Hunter & Son**, a leading Sheffield cutler. The most common use of all types of horn, and this includes deer antler, was as scales on cutlers' wares of every type – sportsmen's knifes, fleams, razors.

Tortoiseshell

Prized for its distinctive translucent red-brown mottled colour, tortoise shell was used it the cutlery trade for scales on quality products such as protective scales on lancets, razor cases and the scales on the finer types of pocket knifes. There were, of course, many other uses outside the area covered in this book – veneers, frames for spectacles and combs are but a few. Like horn tortoiseshell is a keratin-based material and can be heated and pressed to shape. Until the advent of plastics it was the finest textured material of this type available. It will take a fine polish and it is a form of flattery that some plastics were designed as copies of tortoiseshell.

Whalebone

Not the stuff in ladies' corsets, which is baleen – thin strips that formed part of the whale's plankton straining system – but bony material from the whale's skeleton. Although originally white, it has almost always become stained or discoloured brown. The texture is coarser than bone with large open channels. Sailors on whaling ships made all sorts of things that they needed from this available material – fids and knife handles seem to have been the most common. Whalebone is not easy for the average faker to get hold of so whalebone items are usually of some age. Sperm whales also yielded teeth, 4"–5" long, of pointed form with an open root, much used by sailors for scrimshaw work and folk art pictures, often of nautical subjects – alas, so much faked that even original material is called into question.

Price Guide

• Three good quality butchers' steels, all with swivel loops for hanging and different pattern horn handles. **£39** [DP]

• A unique 9" whalebone spirit level with brass top and shaped ends. **£370** [DS 22/403]

• An early and ornate 5" sailmakers' whalebone seam rubber with nicely carved decoration. **£165** [DS 21/1741]

• Two early whalebone bodkins with turned tops and a small 6" walrus tusk fid. **£60** [DS 21/1743]

• A fine 14" whalebone fid, the round body tapering to an octagon at the top. **£320** [DP]

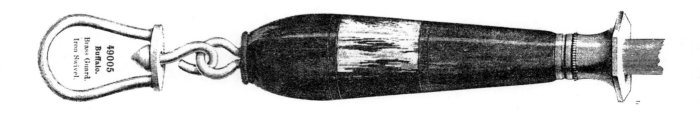

Buffalo horn handled butchers' steel with brass guard and iron swivel [Hunter, 1916].

Working wood

AXES & ADZES

Museums everywhere overflow with axes dating from the Stone Age to the 19th century and the archaeological literature would fill a library.

Axes

Until the 19th century the making of axes was a local trade carried on by smiths who made their tools to local patterns developed over time. Then, as with most tools, the trade gradually became concentrated into fewer hands. Axes are "heavy edge-tools" (as are billhooks, forks, spades, shovels, etc.). In this trade the Sheffield manufacturers never achieved the virtual monopoly that they gained in other fields. The West Midlands was a principal centre of manufacture but other provincial makers continued into the 20th century.

British pattern: Typically with flaring shoulders (the Kent axe) or flat topped with a straight flaring underside.

American pattern: More compact but thicker cheeked.

Continental patterns: These appear archaic to Anglo-Saxon eyes – the shapes are often flared, the poles are round, the handles straight. Bearded and goose wing shapes, which were still being made in the 20th century, are relatively common.

*A wheelers' axe and a broad axe [Sheffield List, 1862]. Both these axes differ significantly from the **Brades Co.** patterns with the same names on page 59 – the collector should be aware there are few absolutes regarding axes.*

Construction

Before steel became cheap, axes and adzes were made "bitted" – that is, the main body was of iron and only the working edge was of steel. The usual method was to place the steel between the iron that had been folded round, thus forming the eye for the handle, but they were also made the other way, with the steel outside. If an axe is bright it is possible to see the difference in colour and grain of the two metals and, if the tool is far gone, the difference in corrosion pattern will show. All-steel construction became the norm after 1900.

Fully axed

Broad (side) axe: Intended for chopping or even paring away to create a flat surface and therefore bevelled on one side only. The largest sizes, with long handles, are the true broad axes but hand versions were used by coopers, coachmakers, cricket bat makers and in ship building trades.

Felling axe: Long handled axes for felling and cleaning up felled timber. The English pattern has now given way to the rounded cheek American patterns. A characteristic is a heavy pole.

Carpenters' axe (hatchet): In the past, a basic tool of the carpenter and used much more than is generally supposed today.

Hatchets: Small axes intended for use with one hand are often called hatchets. But there is an implication in the name hatchet that it is a tool for a less skilled trade or chopping up firewood, whilst the word axe implies more skill. Lathing and shingling hatchets, both of which incorporate a hammer and a nail removing notch, are multi-purpose tools and this may be the reason that they are called hatchets rather than axes.

Mortice axe: Narrow long bladed axes were used in Britain to chop out large mortices but their use never seems to have been common. In France a development of this type of axe, the twybill, was extensively used.

Twybill (French): Typically up to 5ft. long with a chisel-like blade at each end and a short metal handle in the middle, supported by the shoulder, it was, until recently, used for cleaning out large mortices.

Price Guide

Axes

◆ A 4¼ lb. felling axe by **Wm. Swift** (with large WS mark). G **£60** [DP]

◆ A 3¼ lb. Suffolk axe head by **Ward & Payne**. G+ **£45** [DP].

◆ Kent pattern broad axe head by **Thomas Turton & Sons**. **£65** [DP]

◆ A Kent pattern r/h side axe with 4" edge by **Marples**. **£35** [DS 24/933]

◆ A wheelwrights' side axe by **Gilpin**. G **£50** [DP]

◆ A wheelwrights' r/h side axe with 8" edge by **Spear & Jackson**. G **£110** [DS 25/1060]

◆ A chairmakers' r/h side axe with 7¼" edge, smith's mark and beech handle. G **£66** [DS 24/965]

◆ A double ended brick cleaning axe by **Marples**. G **£16** [DP]

Hatchets

◆ A shingling hatchet by **Gilpin**. G **£20** [DP]

◆ A No. 2 polled Kent hatchet by **Sorby**. **£10** [DP]

◆ Morticing axe by **Brades Co**. G++ **£120** [DP]

Twybills

◆ A hurdlemakers' twybill with the usual spear point end and an ash handle. **£24** [DP]

◆ A 42" French twybill with smith's marks, light pitting. **£72** [DS 24/913]

See also Coopers' Tools, *p.221 and* Shipwrights' Tools, *p.224.*

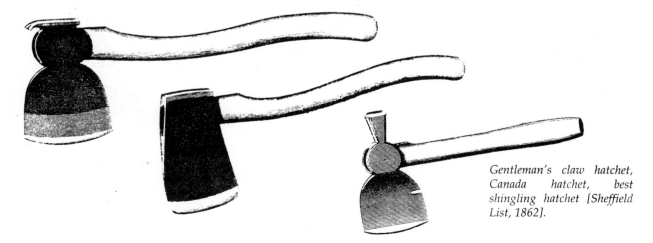

Gentleman's claw hatchet, Canada hatchet, best shingling hatchet [Sheffield List, 1862].

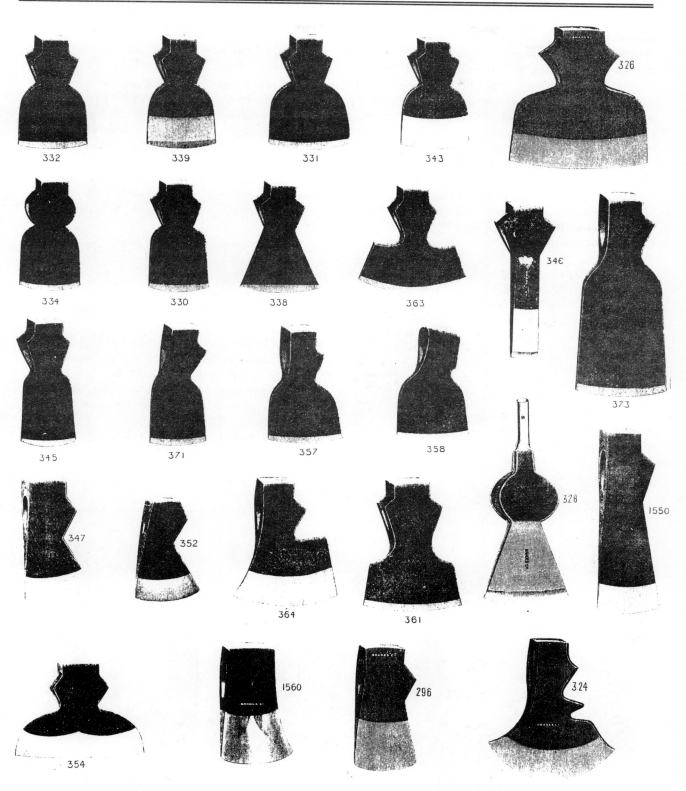

*Some types of axe available in **Brades Co.**'s 1905 list: Nos. 332, 339: Kent axes; 331: broad Kent axe; 343: Kent side axe; 326: hop pole side axe; 334: Scotch axe; 330: Suffolk axe; 338: broad axe; 363: crate axe; 346: morticing axe; 373: Newcastle ship axe; 345: hedging axe; 371: Yorkshire hedging axe; 357: topping axe; 358: topping axe, round poll; 347: felling axe; 352: Sussex felling axe; 364: coach side axe; 361: wheeler's axe; 328: butchers' poll axe; 1550: Welsh miners' axe; 354: broad axe; 1560: Yankee axe; 296: Kentucky axe; 324: Dutch side axe.*

Adzes

The function is the same as the side axe – to chop or pare away leaving a reasonably smooth surface. The blade being at right angles to the handle, it will conveniently work areas that are difficult with an axe. Used by many trades including shipwrights, chairmakers, railway tracklayers, wheelers and carpenters on heavy work; there are numerous patterns, not always easy to identify. A folklore of accidents to feet and toes surrounds its use.

Coopers used several forms of hand-sized adzes all of which also incorporate a hammer head. With the exception of the strap adze, other forms of hand adze are rare.

Strap adze: These are often the subject of some discussion as they appear to be primitive tools of some age. This however is not the case. They were made by most British edge-tool makers in large numbers for export. They are often called Brazil adzes, as they were particularly popular in South American countries, but are also known by other names. The handles were made and fitted locally.

Brazil strap (slot) adze [Sheffield List, 1862].

Price Guide

Adzes

• A No. 1 adze by **James Cam**, 3½" edge with shaped ash handle, pitted. G **£50** [DS 24/907]

• An adze with hammer head poll (improved Scotch pattern) on shaped ash handle. **£35** [DP]

• An adze head by **Brades Co.** with trade label. G++ **£18** [DP]

• A wheelers' (flat poll) adze by **Marples** on shaped ash handle. **£22** [AP]

• A guttering (or spout) adze by **Ward & Payne**. G **£55** [DP]

• An unusually large adze with hammer poll, possibly shipwrights', razor sharp with cover. G++ **£36** [AP]

Hand adzes & other types

• A coopers' nailing hand adze by **Sorby**. **£24** [DP]

• A No. 2 hand adze by **Greaves**. **£16** [DS 24/1082]

• A large French Bordeaux pattern coopers' (bowl shaped) hand adze. G **£60** [DP]

• A French socketed double ended adze, the large edge 3¾", the small edge 1½", with maker's mark. **£53** [DS 25/1103]

• Two unused stirrup adze heads, 4½" and 4¾" by **Isaac Greaves** in original paper packing with trade labels. G++ **£55** [DS 24/897]

• A sculptors' hand adze, marked **Tiranti**. **£25** [DP]

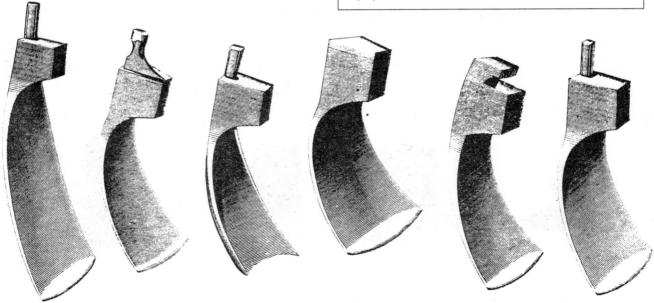

A selection of adzes [Sheffield List, 1862]. From the left: Ship, improved Scotch, carpenters', wheelers', improved wheelers' and spout.

The modern hand saw, cutting on the push stroke, requires stiffness that can only come from properly tempered steel: too hard and the saw would be brittle and impossible to sharpen with a file, too soft and it would bend under the thrusting force.

Until the end of the 17th century it would seem that the necessary skills did not exist for, or at least were not directed into, sawmaking. Until then, many saws were made as frame saws in which the blade is held under tension – in such saws even an unhardened iron blade will work satisfactorily – or were much smaller than the modern hand saw with its 26" blade. Pit saws (of the un-framed variety) and large cross-cut saws were made in mediaeval times but it should be remembered that both these types of saw are pulled on the cutting stroke and in this vital respect differ from the joiners' hand saw.

The development of the saw from earliest times has been traced by W. L. Goodman in *The History of Woodworking Tools*.

London made

The development of the modern hand saw, or at least a leap forward in its development, seems to have taken place in London around 1700 with the adoption of the split wooden handle fixed onto the blade with rivets or screws. 17th century saws generally have a pistol-grip handle which does not have the firmness of attachment or control of the newer form. At the same time, the technology to make bigger and better plates seems to have arrived and the hand saw grew to its modern size of 26" or even larger.

Collecting tips
➤ *In all saws, the quality of the handle is a good indicator of the quality of the blade.*
➤ *Famous makers to look for: **Disston**, **Spear & Jackson** and **J.V. Hill**, who also marked his saws **Hill Late Howel**.*

Below: an English patent hand saw. The holes through the blade were intended to reduce friction and to provide guidance when re-sharpening [Melhuish, 1899].

Something of the pistol-grip origins of saw handles can be seen in the open handles that are still fixed to the smaller sizes of back saws. These were, until recently, set at a higher angle than the closed handles that developed when larger and better blades permitted the use of greater force.

Collecting
Collecting interest in saws is, in general, confined to limited areas.

"Old saws": Used regularly, a saw will quite quickly disappear from sharpening and, when set aside, will rust and become useless. For these reasons, in contrast to most other types of tools, almost no 18th century saws have survived. Of the few that have come down to us most are back saws. The rarity of demonstrably early saws has led to some very high prices.

Fine hand saws: Many discerning users seek out older hand saws as they cannot today buy anything of the same quality. The deterioration in the quality of handle on modern saws is perhaps more noticeable than in the blade. Users today are more at home with the skew backed form – the older style of saw with straight back seems clumsy to those not brought up with it.

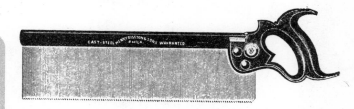

*Brass back saws: Above top: a Sheffield made dovetail saw; above bottom: brass back saw by **Disston & Sons**, with applewood handle [Melhuish, 1905].*

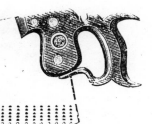

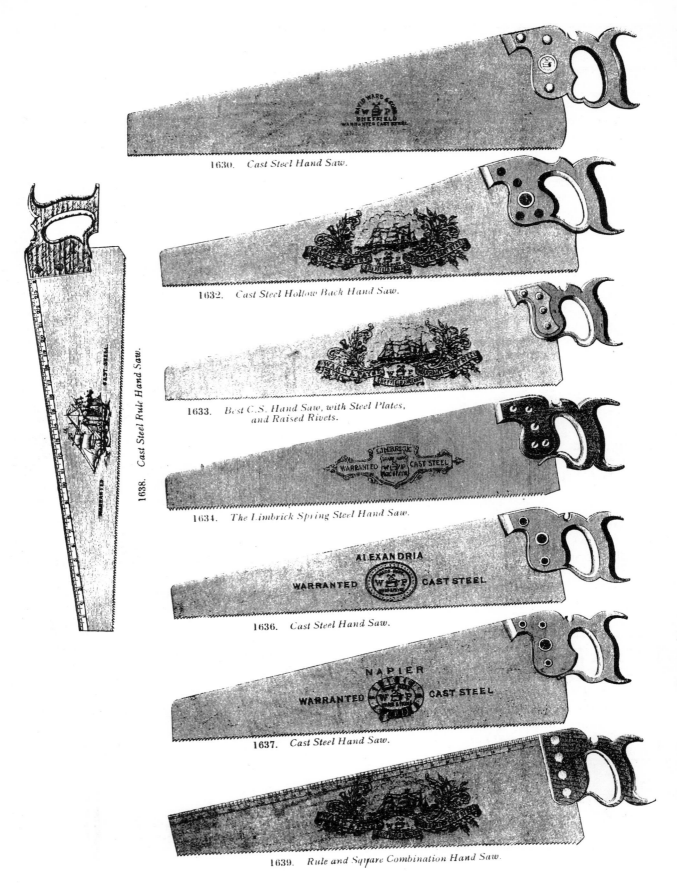

1630. *Cast Steel Hand Saw.*

1632. *Cast Steel Hollow Back Hand Saw.*

1633. *Best C.S. Hand Saw, with Steel Plates, and Raised Rivets.*

1634. *The Limbrick Spring Steel Hand Saw.*

1636. *Cast Steel Hand Saw.*

1637. *Cast Steel Hand Saw.*

1638. *Cast Steel Rule Hand Saw.*

1639. *Rule and Square Combination Hand Saw.*

Hand saws. Note the depth of plate of the top saw, No. 1630, and the handle which is set well outside the plate. This is the traditional form of English hand saw that was already old-fashioned by 1911. It was superseded by the pattern shown in No. 1632, the hollow or skew-back saw [Ward & Payne, 1911].

Price Guide

Brass back tenon saws in good usable condition (not short). **£12–20.** Skew-backed handsaws, 7–10 teeth per inch (but not ripsaws) in good bright usable condition. **£8–16.** Ditto but U.S.A.-made **Disston** saws + 50%.

♦ A rare named 18th C. brass backed dovetail saw by **Shepley & Brain** (1798) The 13 t.p.i. blade is well worn but the handle is undamaged. G **£308** [DS 25/540]

♦ A combination saw, square and level by **Henry Disston** with integral screwdriver and apple-wood handle. **£550** [DS 25/574]

♦ A 7" German silver backed dovetail saw by **Matthews,** rosewood handle. Blade badly pitted. G **£42** [DS 25/616]

♦ A 24" steel back saw by **B. & J. Wilcock,** Sheffield, for a mitre frame. F **£32** [DP]

♦ A 13" **Disston** saw marked "made in USA for **Robert Kelly,** Liverpool" with two sizes of teeth. Five teeth missing, chip to handle spur. **£72** [DS 24/602]

♦ A brass back dovetail saw by **J.V. Hill** Late Howel, London. **£36** [DP]

♦ A 28" saw by **Taylor Bros.,** Sheffield, with perforated blade with incremental teeth. Patent No. 2579. F **£350** [DP]

♦ A well-shaped beech staircase saw, the closed handle of normal saw shape with peaked front tote. **£48** [DP]

♦ A 72" pit saw without box or tiller. Bright condition. **£70** [DP]

*Remember - condition is **G+** unless otherwise shown.*

Pad saws: The best are fine examples of turning in boxwood or ebony. The slot for the blade that passes right through the handle seems difficult to make – how was it done?

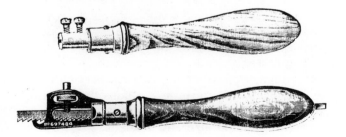

A beech handled pad saw and a "lock tooth" beechwood pad saw [Turner Naylor, 1928].

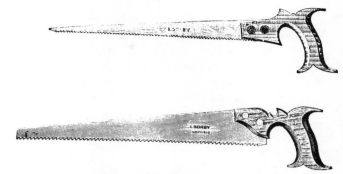

Top: compass saw; bottom: table saw [Turner Naylor, 1928]. Often misidentified as pruning saws, these were fine saws for craftsman and are under appreciated today.

Price Guide

Pad saws, ebony handled **£8–12,** rosewood or boxwood handled **£6–10,** beech handled **£3–5.**

♦ A rosewood handled pad saw by **Alfred Ridge.** F **£22** [DS 25/543]

♦ A rosewood and brass tool handle with a 6¾" saw blade and a boxwood pad saw all by **Holtzapffel.** G++ **£149** [DS 24/679]

♦ A 24" compass saw with finely worked beech handle by **Moseley. £12** [DP]

♦ A large (26") table saw by **Hill** Late Howel. **£24** [AP]

Medical saws: With the advent of antiseptic procedures many surgeons' saws with ebony handles made their way into workshops. The acceptance of antiseptic principles did not happen overnight – the first nickel plated instruments date from the 1870s but ebony handled items were still available in 1900.

Price Guide

♦ A surgeons' brass backed amputation saw by **Evans,** London with removable ebony handle. Minor pitting. **£110** [DS 25/570]

♦ A late 18th C. surgeons' amputation saw of hacksaw form, well shaped frame with bead decoration on frame, decorative tightening screw and plain horn pistol handle. **£400** [DP]

♦ A "Doctor Butcher's" surgeons bone saw. Steel frame with tensioning turnbuckle and ebony handle. **£95** [DP]

♦ A nickel plated open handle amputation saw by **Allen & Hanbury** – takes apart for cleaning. **£24** [DP]

Frame saws: There are many varieties of woodworkers' frame saws from the giant 10ft. pit saw to the small bow saws made to take 6" blades. Almost all are of interest to collectors and the decorative examples, particularly if boxwood, sell for considerable sums.

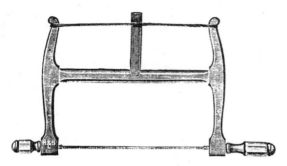

A beech bow or turning saw, with octagon boxwood handles. [Harding, 1903].

Price Guide

Bow saws to take 10"–12" blades. Beechwood frames, craftsman or commercially-made, handles commercially-made **£8–12**, more if smaller. In boxwood or rosewood **£20–30**, more if small or artistically pleasing.

♦ A little used beech bow saw by **Flather** and another by **Sorby**. G++ **£57** [DS 25/550]

♦ A small craftsman-made ebony bow saw. G **£72** [DS 24/19]

♦ An 11" sycamore bow saw, ebony handles and decorative toggle. G++ **£143** [DS 24/650]

♦ A small well-made bow saw (8" blade) in rosewood with octagon box handles. **£55** [DP]

♦ A beech fret saw with brass fittings and screw tensioner in boxwood handle by **Buck**, 245 Tottenham Ct. Rd., some original finish. G++ **£132** [DS 24/612]

♦ A 48" box-framed pit or veneer saw with slight decoration to ends of cross pieces. **£85** [DP]

♦ A coopers' beech frame (buck) saw, 20" fixed blade. **£30** [DP]

♦ A wheelwrights' felloe saw, a heavy beech frame turning saw with handle on an extended leg with 22" blade. **£45** [DP]

Pit saws & cross-cuts: These appeal to more collectors than you might think. Impressive in size, and, in the case of the cross-cuts, with some very interestingly named tooth designs. These innovative shapes were intended to reduce the energy needed (how about "lance perforated" or "cedar savage").

Lancashire pattern hacksaws: The severe but elegant Lancashire pattern hacksaw was made in a virtually unaltered form from the late 18th until well into the 20th century. The smaller sizes are particularly attractive.

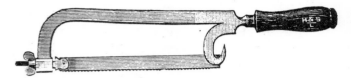

A Lancashire pattern hacksaw [Harding, 1903]. Made to hold 4" and upwards blades.

Early metalworkers' saws: To call these hacksaws is an insult; the best are finely ornamented, and some bear 17th century dates.

Price Guide

Lancashire pattern hacksaws, 8"–10" **£8–16**. 4"–6" sizes **£20–30**. 15" or over **£25–35**. Jewellers' adjustable piercing saws, late 19th/early 20th C. **£15–20**.

♦ A small Lancashire pattern hacksaw with mahogany handle and 5" blade. G **£40** [DS 25/549;

♦ An 18th C. armourers' saw, the metal frame of twisted iron flattened and decorated, the turned ebony handle with brass ferrule, some original gilding remains. G **£110** [DS 25/557]

♦ An early 19th C. miniature hacksaw with decorative brass wing nut, 2½" blade and turned rosewood handle, 5½" overall. G++ **£150** [DS 25/594]

♦ A beech open handled Aubin's Patent hacksaw made by **Tyzack Turner**. G **£10** [DS 24/668]

♦ A large iron bow saw with 22" blade riveted to the tensioner with long mahogany handle, for cutting ivory. **£85** [DP]

♦ A **Hobbies** treadle fret sawing machine in good working order. **£60** [DP]

Saw sharpening

When sharpening a saw it must be held close to the teeth. The larger hand saws are too deep to be accommodated in normal vices so the special saw sharpeners' vice is called for.

The first stage is to level down the tops of the teeth – small pieces of file in a special holder are available for this purpose – before the teeth are filed. Filing one hundred and fifty teeth to the correct angle and size and with the proper cross bevel is skilful and a bit boring. Judging by many of the saws around, it is not a skill that is often mastered!

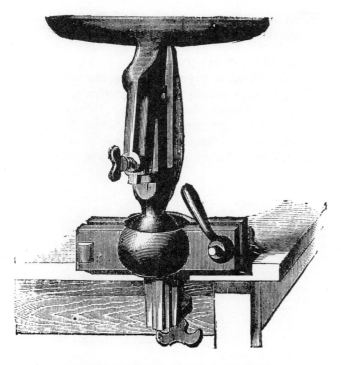

A ball and socket saw filers' vice [Melhuish 1905].

Setting

After sharpening, the teeth must be set. Today they are set individually, bending alternate teeth to the left and right. However, in the 17th century, Moxon's description indicates that the wrest was twisted sideways which must have given a waved set more like today's hacksaw blades.

Setting hammer and anvil: A block of steel around 7" long with a bevel to one edge. The saw is laid on the anvil and the teeth are bent using the thin edged hammer.

Setting punch: A development of the anvil and hammer. The tooth is bent by a triangular-ended punch guided in the stock. Several manufacturers including **Ed. Preston** produced versions of this tool but all have the drawback that the sharpened teeth rub against the stops and blunt themselves.

Saw wrest: A simple slotted tool for bending the teeth. Some also have a sliding gauge or guard. There is a wide range of quality, age, and design.

Pliers type: The Eclipse 77, Morrill's patent and a German-made light pattern with finely decorated castings are all commonly found. There are also others.

*Right: Saw sets and saw setting tools: a **Morrill's** "Perfect" patent saw set [Melhuish, 1905]; a saw wrest with sliding gauge; a saw setting hammer and plate (anvil) [Ward & Payne, 1911].*

Price Guide

Saw sets, late 19th/early 20th C. Pliers type, £5–8, saw wrests in beech handles £4–8, with brass stop **£6–10**.

◆ A saw setters' anvil and hammer. **£18** [DP]

◆ An **Eclipse** No. 79 saw set for cross-cut saws with instructions in original box. **£18** [DP]

◆ Two boxwood handled saw wrests with brass guards, one by **Marples**. **£28** [DS 23/265]

◆ A large saw wrest for circular saws with rosewood handle by **Marples**. **£12** [DP]

◆ Three saw sets by **Stanley, Morrill's** and **Disston**. **£10** [DS 22/686]

◆ A well-shaped and patinated craftsman-made mahogany saw sharpening vice. **£48** [DP]

◆ A cast iron **Preston** Patent punch saw set, bronze painted. G++ **£24** [DP]

◆ An iron spring arm form of punch saw set mounted on a mahogany block. **£28** [DP]

Collecting tips

➤ *Saw setting seems to have interested the inventors and there are numerous different patent or original saw sets, most of which do not fetch large money – a good area for the impecunious collector.*

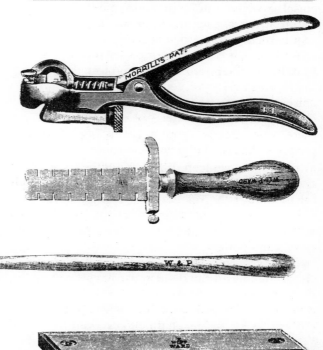

History

Chisels must be amongst the earliest woodworking tools to have been developed: they were in use in Egypt by 1500 BC and probably earlier. Medieval and Renaissance period illustrations and carvings show many chisels and gouges. Surprisingly, most of these seem to have tanged blades although the majority of chisels recovered from archaeological sites are of socketed design.

Even until recently the heavy duty members of the chisel family were usually of socketed construction. The exception to this general rule is the full mortice chisel – at first sight the mightiest of the chisel family but in reality not used for the roughest work.

The production of a successful tanged chisel requires greater smithing ability than the production of a socket chisel. The bolster must be shaped onto the tang, without which the blade will just drive into the handle and better skills in hardening and tempering are needed – too soft and the neck will bend, too hard and it will snap.

Design and variety

During the 18th and 19th centuries the chisel evolved into more and more different types, each made to suit the requirements of a particular work.

As early as 1825 the price list of **Marshes & Shepherd** contains no less than sixteen series of general purpose chisels (tanged), five types of mortice chisels and seventeen types of socket chisels. Then there were gouges and a sprinkling of special purpose chisels The general purpose chisels were available in seventeen different widths (⅛" to 2⅛") and the mortice chisels in twelve sizes. As a rough calculation, if the retailer of the time had bought only one of each of these chisels, he would have taken delivery of about three hundred and fifty chisels and gouges.

The conclusion to be drawn is that commercial chisel making in Sheffield must have been taking place for many years – a product range of such diversity and extent, made to designs published in pattern books, had taken time to evolve.

Of significance to the collector is the subtle change of shape that took place over the years. The basic progression is that from having little or no shoulder, the chisel gradually acquired progressively more pronounced shoulders.

A chisel for every purpose

By 1911, **Ward & Payne,** a leading Sheffield maker of the time, listed no less than forty-nine different ranges of chisels: twenty-three of tanged type, seventeen socket and nine varieties of mortice chisel which included both tanged and socketed types. Not included in these figures are those that

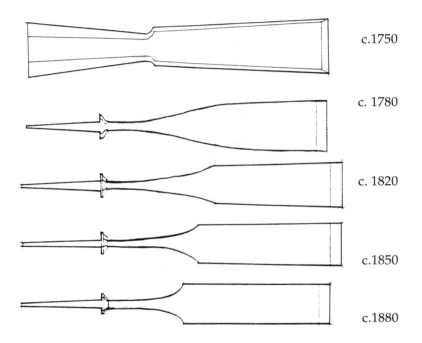

c.1750

c. 1780

c. 1820

c.1850

c.1880

The changes in shape of chisels from the mid-18th to the end of the 19th century.

had integral metal handles (such as the solid steel wagon builder's chisels) and the wide variety of special purpose chisels – which ranged from lock bolt chisels to button hole and roller coverer's chisels.

Cast steel

By the 1760s Benjamin Huntsman's crucible (cast) steel was still only being made in limited quantities. This improved steel was expensive and initially was only used for the cutting edge on the

Gunstockers' chisel [Ward & Payne, 1911].

292. *Light Socket Bevel-edge Firmer Chisel, Ash Handled.*

295. *Strong Socket Firmer Chisel, with Iron-hooped Handle.*

296. *Strong Socket Bevel-edge Firmer Chisel, with Leather-tipped Handle.*

271. *Boatbuilders' Chisel.*

298. *C.S. Socket Framing Chisel*

244. *C.S. Long Thin Paring Chisel, Round Beech or Ash Handled.*

269B. *C.S. Short Strong Wagon Builders' Chisel, Ash Handled.*

A selection of chisels [Ward & Payne, 1911].

larger sizes of chisels, the general body of the tool being made of steel of lesser quality or even wrought iron.

In 1791, **Christopher Gabriel & Sons,** the prominent London planemaker and tool seller, listed his stock of 1,292 chisels. Of these, about one-third were cast steel. In 1796, when Joseph Seaton bought a comprehensive kit of tools for his cabinet-maker son, Benjamin, he acquired two sets of apparently identical chisels – but one set was described as cast steel. In this set, the larger sizes are faced whilst the smaller are wholly of cast steel.

The tools are not marked "CAST STEEL" and it seems that not until perhaps as late as the 1840s did such marking become usual. Thereafter the wording was widely applied until World War I.

The gradual reduction in the price of cast steel meant that the production of tools made entirely from this material became steadily more economic. Indeed, by 1911 **Ward & Payne** was listing nothing but cast steel chisels for general purposes and it is likely that, with the exception of a few of the largest types – mortice chisels, lock mortice chisels etc. – welded (faced) manufacture had ceased.

Collecting chisels

Years ago whilst still a tyro tool collector I asked a leading tool dealer if he had any collector's chisels. The shop was well stocked with a glowing range of tools, many priced in hundreds, if not thousands, of pounds – this must be the place to find something special. "There's no such thing as a collector's chisel – they're bought for use, not to hang on walls". He was, of course, right.

A month or two later I attended a sale where a single long paring chisel, the metal glowingly black with the oxide which is the sign of metal of age that has never been allowed to become rusty or been over-cleaned, fetched the then very high price of over £70. I still remember it: a beautiful object, with a generously proportioned ebony handle that came straight out of *Plumier*, satisfying to hold in the hand, to be placed in a drawer and got out occasionally to be stroked. Its appeal was certainly more than just visual – tactile, yes – the coolness of the steel, the comfort of the handle and the imagination of what might be accomplished with it.

But all but one in a thousand chisels and gouges are bought with at least the *intention* of use.

Collecting tips:

➤ *Sets of chisels from the 19th century are rare.*
➤ *Original octagon boxwood handles add significantly to the value.*
➤ *Early names to watch for: Robert Moore; John Green; Green; P. Law; T. Shaw.*
➤ *Special purpose chisels are collectable.*
➤ *Never divide matched sets.*
➤ *Bevel edged chisels were not made until around 1870.*
➤ *Good makers include Ward & Payne, Wm. Marples, Sorby Punch Brand, Thos. Ibbotson.*

Price Guide

Most old chisels command prices of **£5–10**, much the same as new chisels. But the quality available today is not what it was and discerning users seek fine examples. Paring chisels by good makers in boxwood handles, not short **£10–15**. Ditto, bevel edged **£12–20**.

• A mid-18th C. chisel by **Stidman** with octagonal socket and flaring blade. Excellent condition for age. **£220** [DP]

• Two early mortice chisels, short from use, by **James Cam** and **John Green. £32** [AP]

• A 2¾" paring chisel by **W.C. Earl**. The blade wide but thin with flare from shoulder. Octagonal mahogany handle. **£28** [DP]

• A 1" firmer chisel by **P. Law** with plain octagon beech handle. **£35** [DP]

• A 1½" bevel edge chisel by **Sorby** with octagonal box handle. **£33** [DS 22/37]

• A 1¼" bevel edge paring chisel by **Marples** with octagonal box handle. G++ **£18** [DS 22/44]

• Three paring chisels by **Ward & Payne**, ½", ⅝" and ¾", the boxwood barrel handles scratch engraved with erotic decoration. **£48** [DP]

• Little used set of ten bevel edged chisels by **Marples**, ⅛" to ¾" with boxwood handles and trade labels. G++ **£143** [DS 22/1]

• Set of seven Kangaroo Brand bevel edge chisels by **Sorby** with trade labels on octagonal boxwood handles. F **£75** [DP]

• Seven long paring gouges by **Marples** with some original trade labels. **£85** [DP]

• A set of six full mortice chisels by **Thos. Wales** as new. **£75** [DP]

• Five heavy socket chisels from a wheelwrights' kit, makers include **Brades**, **Woodcock** and **Ward**, the largest 1½" wide. **£40** [DP]

• Seven boxwood paring chisels by **Bedford**, Sheffield. **£105** [DP]

• A fine 2" bevel edge chisel by **J. Buck** with octagon box handle, the blade and bevels lightly flaring. **£66** [DS 23/289]

• Five sash mortice chisels (not matching) by **Marples. £35** [DP]

Cast steel bevelled long thin paring chisel, best London octagon box handled [Ward & Payne, 1911].

By the 17th century, and probably long before, a tapered octagon was the normal form of handle for chisels, gouges and carving tools. This style, which has no ferrule, can be made by any competent woodworker from a scrap of wood, and many tradesmen continued to make and fit these handles until well into the 20th century. The numerous chisels in the Seaton chest (bought in 1796) are all fitted with tapered octagon handles whilst only the awl (or file?) handles were turned and had brass ferrules.

The octagon handle has a pleasing continuity with the octagon of the bolster and although by the middle of the 19th century turned handles were available and were supplanting the older form, the manufacturers continued to make all but a few types of chisels with octagonal bolsters.

Handle it right

After 1850, with increasing industrialisation and competition a wider range of handles became available. Whereas beech or ash had hitherto been the usual material, boxwood was now used for the best quality handles for chisel types such as paring chisels and carving tools which were not intended to be pounded with a mallet. Some preference for ash was apparent in the choice for heavier duty tools. Almost all handles for tanged chisels were now made with ferrules, brass on the general ranges and iron on the heavy duty series.

The vast variety of handles that eventually became available can be illustrated by referring to **Ward & Payne's** catalogue of 1913 in which twenty-six types of chisel handles for tanged chisels were listed – each being available in eight sizes. In addition there were handles for socket chisels, turning tools, carving tools and mortice chisels.

We have recorded at least five different registered designs or patents for chisel handles. One of the more interesting is Wilfin's patent, No. 221871, of 1923. The boxwood handle is strengthened by a pewter ring which is cast onto the handle being secured by invisible lugs.

Collecting tips
➤ *Early chisels and gouges are most often found with plain beech octagon handles without ferrules.*
➤ *Look out for registered designs and patents.*

Best Improved Round Beech Chisel Handle, London pattern.

Best Taper Round Beech Chisel Handle.

Best Carving Pattern Beech Chisel Handle.

Best Carving Pattern Box Chisel Handle.

Plain Beech Octagon Chisel Handle.

Common Octagon Box Chisel Handle.

Best Taper Octagon Box Chisel Handle.

Double-hooped Ash Ship Chisel Handle.

Bright Double Iron Hooped Ash Registered Chisel Handle.

*A selection of the eighteen different handles offered by **Ward & Payne** in 1911.*

SPECIAL PURPOSE CHISELS

Sash mortice chisel: Light-weight mortice chisels, originally used in sash window making, for cutting the small mortices needed. As full mortice chisels are no longer produced, these are the only type of mortice chisel now available new.

Mortice chisel (full mortice chisel): The cutting out of deep and true mortices for work such as framed doors is surprisingly easy with a full mortice chisel. The depth of the blade keeps the tool square in the work and the width of the mortice is defined by the width of the chisel. For this reason the correct size chisel is needed for each width of mortice.

Lock mortice chisel: in the days when joiners really were joiners and doors had good brass knobs and locks which were six inches long, the mortice to accommodate them extended right through the door jamb into the mid-rail. To cut away the wood from the bottom of such mortices required cutting across the grain. The heel of the lock mortice chisel bearing against the inside of the mortice allows the joiner to exert great force. Only the London pattern, the lightest of the variety of all the patterns once available, is now made.

Sash pocket chisel: This is a **very** special purpose chisel about which there is much misunderstanding. The blade is quite thin and sharpened from both sides, like a knife. Its purpose is to cut through those parts of the timber that cannot be reached with a tenon saw when cutting into the side piece of the sash box to form the sash pocket – that small but indispensable, removable piece of wood which allows access to the sash weights. The chisel is **not** for opening up sash boxes when re-cording – the blade is far too thin for levering. This explanation is however clouded by the "improved" sash pocket chisel shown in a number of manufacturer's catalogues. This has a thicker and shorter blade than the normal sash pocket chisel and would be less suitable for straight cutting in.

An interesting variation is the interchangeable blade sash pocket chisel made by **Ed. Preston & Sons.**

Butt chisels: Short chisels for cutting out for hinges (butts). **Ward & Payne** list these chisels from ½" to 2". Note the leather end to the handle so this chisel could be used on site with a hammer.

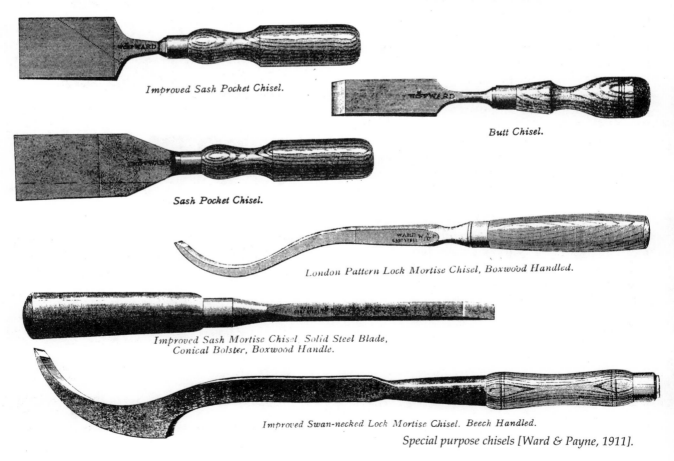

Improved Sash Pocket Chisel.

Butt Chisel.

Sash Pocket Chisel.

London Pattern Lock Mortise Chisel, Boxwood Handled.

Improved Sash Mortise Chisel Solid Steel Blade, Conical Bolster, Boxwood Handle.

Improved Swan-necked Lock Mortise Chisel. Beech Handled.

Special purpose chisels [Ward & Payne, 1911].

Price Guide

Lock mortice chisels: London pattern (light), boxwood handled **£12–16**. Improved pattern (heavy with heel), socketed **£14–20**. All metal **£8–10**.

- A pair of drawer lock chisels by **Ward** in boxwood handles. F **£45** [DP]

- A double ended drawer lock chisel by **Marples**. **£9** [DP]

- Three mortice lock chisels ⁷⁄₁₆", ½" and ⅝" by **Sorby** and **Marples** with matching boxwood handles. G++ **£77** [DS 22/18]

- Seven full mortice chisels by **Greaves**, Nos. 1 to 8 (No. 7 missing). **£85** [DS 22/697]

- A fine and unusual set of three sash pocket chisels by **W. Butcher**, 1" 1½" and 2", in boxwood barrel shaped handles. F **£60** [DP]

- An unused 2½" sash pocket chisel by **Sorby** with beech handle. F **£35** [DS 22/22]

- A 1¼" sash pocket chisel by **Sorby** with beech handle. **£18** [DS 22/24]

- A rare interchangeable blade sash pocket chisel in boxwood handle by **Preston**. **£80** [DP]

- Three cranked paring chisels by **Marples**, ¾", 1" and 1¼". Circa 1960, as new. **£50** [DP]

- Wheelwrights' bruzz (corner chisel) by **Sorby**. Ash socket handle. **£15** [DP]

- Three bruzzes by **Marples, Ward** and **Ibbotson**. **£36** [DP]

- A group of six printers' chisels, used for trimming electrotype, all below ³⁄₁₆" wide. **£48** [DP]

- Three butt chisels by **Sorby**, 1", 1¼" and 1½" with correct short boxwood handles. **£36** [DP]

- A beech handled button hole chisel, with parallel blade by **Ward & Payne**, the handle similar to a file handle. **£22** [DP]

Drawer lock chisels: Anyone who has tried to cut out the recess needed to house a lock inside a drawer will immediately appreciate the need for these chisels. The 18th century name for this tool was *bolting iron*; it had two cutting edges at one end with a handle at the other. Modern versions are all metal with the two edges at opposite ends. 19th century catalogues show two separate tools with normal chisel handles but these seem to be surprisingly rare.

Button hole chisel: Early examples and those made for fancy sewing kits can have decorative shanks but later examples are workaday and could be mistaken for turnscrews.

(See also *Shipwrights' Tools*, pp. 224–226)

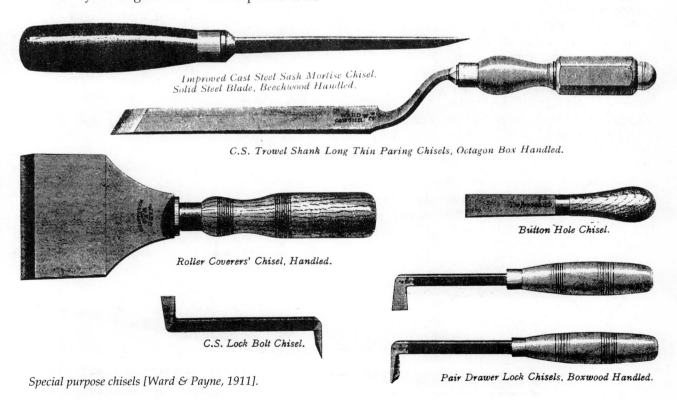

Improved Cast Steel Sash Mortise Chisel. Solid Steel Blade, Beechwood Handled.

C.S. Trowel Shank Long Thin Paring Chisels, Octagon Box Handled.

Roller Coverers' Chisel, Handled.

Button Hole Chisel.

C.S. Lock Bolt Chisel.

Special purpose chisels [Ward & Payne, 1911].

Pair Drawer Lock Chisels, Boxwood Handled.

There are two basic types of gouge: in-cannel, which has the sharpening bevel on the inside of the blade, making the tool useful for paring to the radius of the outside, and out-cannel, which has the bevel on the outside, making a tool useful for gouging out depressions of a freer form.

365.   C.S. Firmer Gouge, Cannelled inside.

361.   Cast Steel Firmer Gouge.

Top: An in-cannel gouge and bottom: an out-cannel gouge [Ward & Payne, 1911].

Patternmakers' gouges

Today, long gouges, whether cranked or straight, are invariably referred to as patternmakers' gouges. However, this terminology is modern. In their 1911 catalogue, **Ward & Payne** listed long thin paring gouges and in 1928 **Turner Naylor & Co. Ltd**. listed both long thin paring gouges and paring gouges "to circles", i.e. to specified radii. This latter group, although the trade is still not specified, *is* for patternmaking.

In 1911, the only *specialist* patternmakers' gouges listed were the long strong spoon gouge, which is a handsome (and rare) long-shanked paring gouge with little or no resemblance to a spoon, and the more aptly named patternmakers' spoon bit gouge which is so tightly curved that the cutting edge is almost at right angles to the axis of the tool.

One suspects that spoon bit gouges were one of those tools which every patternmaker felt he should have "just in case", but, as unused sets of six or so regularly appear in sales, it would seem that suitable jobs didn't come along too often.

Curvatures

Smith's *Key*, published in Sheffield in 1816 shows four curvatures – these are annotated A to D. **William Marples**, 1900, illustrates five curvatures whilst **Ward & Payne**, 1911, had seven, as shown below. In all cases A is the flattest curve.

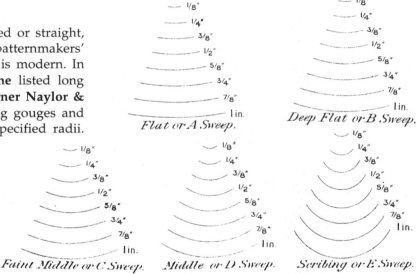

Flat or A Sweep.   *Deep Flat or B Sweep.*

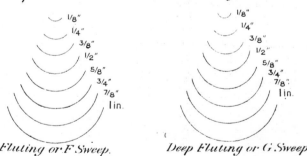

Faint Middle or C Sweep.   *Middle or D Sweep.*   *Scribing or E Sweep.*

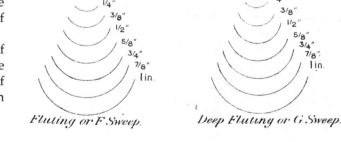

Fluting or F Sweep.   *Deep Fluting or G Sweep.*

414.   *Pattern Makers' Spoon Bit Gouge.*

415.   *Pattern Makers' Long Strong Spoon Gouge.*

Patternmakers' gouges [Ward & Payne, 1911].

GOUGES

Shoulders

In general, modern gouges, like chisels, have pronounced shoulders. This was not always the case as pre-1820 gouges have sweeping shoulders as do carvers' gouges. This feature, together with the thickness, indicates whether the tool in question is a true gouge or a carving tool. Like all rules, there are a few exceptions – such as the scribing gouge below.

Sash scribing gouge

Used to shape the ends of ovolo sash bars, the characteristic feature is a handle extended to form

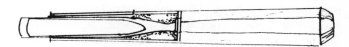

A sash scribing gouge

a stop half an inch or so back from the cutting edge. Desirably these should come with the matching sash templet with which they were used – rare without the templet – even rarer with it.

(See also *Carvers' Tools*, pp.74, 75, *Turning Tools*, p.76 and *Bookbinding Tools*, pp.228, 229)

C.S. Long Thin Paring Gouge, cannelled inside, Beech Handled.

429.   C.S. Socket Gouge, Handled.

Trowel Shank Long Thin Paring Gouge, Octagon Box Handled.

Various types of gouges [Ward & Payne, 1911].

Price Guide

Single gouges of late 19th/early 20th C. date, both in-cannel and out-cannel **£4–8**. Paring gouges (always in-cannel) **£8–14**.

• Four 18th C. gouges, three by **John Green** but probably all by the same maker. G **£110** [DS 24/615]

• A set of six boxwood handled gouges by **Ward**, ½" to 1". **£53** [DS 23/314]

• Four cranked paring gouges by **Marples**, ½" to 1¼" with boxwood handles. **£44** [DS 23/315]

• Three patternmaker's cranked gouges by **Stormont** with boxwood handles. **£25** [DP]

• A patternmakers' long strong spoon gouge by **Marples** in fine boxwood handle. **£36** [DP]

• A set of six little used spoon bit gouges by **Addis** with boxwood handles. **£72** [DP]

• A set of seven long paring gouges by **Marples** with boxwood handles, ⅛" to 1¼". **£83** [DS 22/28]

• A ⅝" sash scribing gouge by **Marsden Bros. £24** [DS 22/40]

• A fine quality 2" scribing gouge, 21" overall with faceted shank and brass ferrule on turned yew handle, not named. **£77** [DS 23/305]

• A sash paring gouge by **Howarth** with beech extended handle. **£24** [DP]

• A cranked in-cannel gouge in a handle with swelled top end by **Marples**. It is uncertain whether this pattern was intended for sash scribing or patternmaking. **£18** [DP]

• A massive shipwrights' smith-made socket gouge, 3¼" wide. **£30** [DP]

• A 1¾" deep gouge (½ circle) by **Ward** with boxwood handle. G++ **£44** [DS 21/178]

• A heavy 2½" wrought iron in-cannel socket gouges by **Ward**. G **£24** [DS 23/292]

• Three paring gouges by **Taylor**, 1" to 1¼" with beech handles. **£29** [DS 23/294]

• Five heavy in-cannel socket gouges with steel bound ash handles by **Greaves**. **£54** [DP]

As a general rule of thumb, carving tools are thinner in the body than gouges or chisels – or at least they were until the last couple of decades. Modern carving tools have become much thicker and, for most purposes, less useful so it is no surprise that the older tools are keenly sought by today's carvers.

A look at Smith's *Key* shows a plate titled carvers', print cutters' and gunstockers' tools and another, showing what are plainly carving tools, but titled Italian chisels and gouges. The Italian influence on carving in Britain is also demonstrated by the names of some of the carving tools, such as the flutaroni and macaroni.

Amateurs

Wood carving seems to have become particularly popular as a pastime around 1900. Many manufacturers sold sets of tools intended for amateur use. These are mostly small in size and were often fitted into oval or barrel pattern handles. The sets made by **Wm. Marples & Sons** had oval handles in three different woods – boxwood, rosewood and ebony – intended to ease the task of finding the right tool when working. Amateur pattern tools were often finished "straw", that is, they were polished over a greater area than the professional tool and then tempered to straw, leaving a brown tinge.

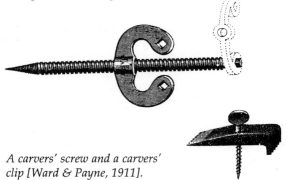

A carvers' screw and a carvers' clip [Ward & Payne, 1911].

Collecting tips

➤ *Names to look out for:* **Addis, Herring, Ward & Payne**.

➤ *Some manufacturers sold cheap ranges which did not have bolsters. It is easy to miss this when looking at the business end!*

➤ *Collectors often search for matched sets of tools – but remember, these will most likely be of amateur size and quality.*

➤ *The majority of carving tools are bought for use.*

Professionals seemed to have preferred to have a whole range of different shapes and sizes of handle, which solved the tool finding problem.

Other tools

Handled carvers are not the only tools used – carvers' knives (whittling or chip carving knives) were sold in a variety of shapes. The other tools used are carvers' vices, screws, routers, mallets, background punches and coarse rifflers.

London made

The **Addis** families made carving tools in London from the beginning of the 19th century. **Addis** was purchased by **Ward & Payne** in 1870 following the death of Mr. S.J. Addis. However, they continued to market the tools under the Addis name.

The other well-known London firm is **Herring**. Thomas Herring came from a family of Sheffield edge tool makers but around 1850 came to London and married into the Addis family before setting up in business on his own. This business continued until the 1940s.

Opposite page: London pattern carving tools [Ward & Payne, 1911]. This is a coded way of saying that these are the best quality and are for professional use.

Price Guide

Small/medium sized tools by good makers **£5–8** each. Matching sets of twelve or more **£9–12** per tool. Large tools **£12–18** each, price depending on size.

• A fine set of eighteen very little used box handled carving tools in original fitted pine box with brass **Marples** label. F **£280** [DS 25/613]

• Nine varied carving tools by **Addis** fitted to craftsman-made shaped octagon handles without ferrules. **£105** [DP]

• Three very large front bent gouges by **Ward** in beech handles. **£54** [DP]

• Six matching curved gouges by **Marples** in beech handles. **£45** [DP]

• A set of twelve amateur pattern carving tools with oval handles in three woods, a little pitted. **£75** [DP]

• A professional carvers' kit of tools in a five drawer chest with two mauls, sharpening stones etc. Eighty-nine tools, mostly by **Herring** and **Addis**. **£750** [DP]

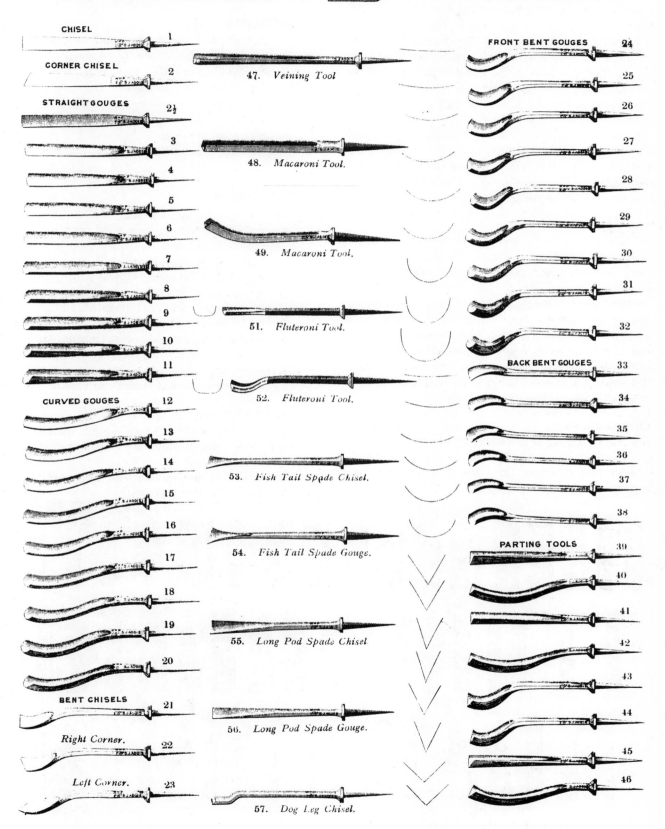

CHISEL — 1

CORNER CHISEL — 2

STRAIGHT GOUGES — 2½

3
4
5
6
7
8
9
10
11

CURVED GOUGES — 12
13
14
15
16
17
18
19
20

BENT CHISELS — 21

Right Corner. — 22

Left Corner. — 23

47. *Veining Tool*

48. *Macaroni Tool.*

49. *Macaroni Tool.*

51. *Fluteroni Tool.*

52. *Fluteroni Tool.*

53. *Fish Tail Spade Chisel.*

54. *Fish Tail Spade Gouge.*

55. *Long Pod Spade Chisel*

56. *Long Pod Spade Gouge.*

57. *Dog Leg Chisel.*

FRONT BENT GOUGES — 24
25
26
27
28
29
30
31
32

BACK BENT GOUGES — 33
34
35
36
37
38

PARTING TOOLS — 39
40
41
42
43
44
45
46

Turning tools do not have a bolster – there is no need for one as the pressure on the tool is downwards at the working tip, upwards where supported on the rest and downwards at the handle. For this reason the handles are large. The normal shape is a form swelled near the ferrule, tapering away to the end, in length, up to 15" or even more.

Cuts made on the lathe are either slicing (made with a gouge or chisel type tool) or scraping. Flat tools for slicing cuts are normally sharpened from both sides whilst scraping tools have a steep angle bevel and a flat upper surface.

Fancy tools

Collectors may come across sets of scraping tools with a variety of shaped ends. These were sold for work on both wood and the softer metals such as brass.

More extensive sets including this type of tool and many others were made for turning on ornamental lathes. The best of these sets will have matched handles in rosewood, cocobolo or ebony although beech was also an alternative. **Holtzapffel** and some other lathe makers supplied hanging cases for these tools.

Thread chasers

For hand chasing of threads onto both metal and wood, these tools must come in matched pairs, internal and external. Their use is a skill that has largely been lost today.

Collecting tips

➤ *Old lathe tools, apart from fancy ones, are rare – they are constantly re-sharpened and disappear quickly.*

➤ *The handle is important. Large tools should have appropriate size handles. Needless to say, some well shaped handles in unusual woods are to be found.*

Price Guide

◆ 144 ornamental turning tool bits in three matching boxes to fit a rosewood and brass handle. G++ **£725** [DS 25/1613]

◆ 14 fancy turning tools by **Buck** in boxwood, barrel shaped chisel handles. **£120** [AP]

◆ A diptych (two-fold cupboard) containing 54 ornamental turning tools by **Holtzapffel**, all with rosewood (?) handles. **£1150** [AP]

◆ Four pairs of internal and external thread chasing tools, no handles. **£28** [DP]

◆ Four very large gouges from a bobbin turning workshop with unmarked blades and very large beech handles believed to have been made in the workshop. **£48** [DP]

◆ 11 assorted turning tools, mostly marked **Sorby**, with variously shaped craftsman-made yew handles. **£95** [AP]

◆ Three unusual turning tools with cranked ends and rosewood handles. **£24** [DP]

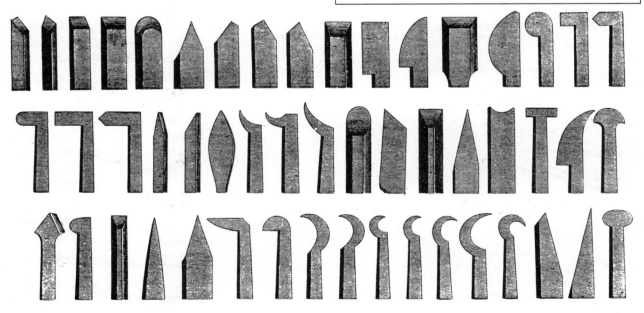

A set of turning tools [Wingfield, Rowbotham, 1904].

When you need to cut a wooden thread there is nothing but nothing that will do except a screwbox and tap of exactly the right size – so if you think you might need to cut wooden screws buy when you have the opportunity. Screwboxes are made today but the supply is fitful and needless to say they are expensive.

Unlike engineering threads there is no standardisation; indeed even the diameters are pretty nominal. English taps are cast iron, minimally finished with a file and taper a little. Our experience is that the boxes used on a suitable wood – boxwood and beech, not overdry – produce good results but the taps are always tricky and will burst the work unless supported with clamps.

Best English screw box and tap [Tyzack, c. 1910]. Sizes from ¼" to 3" are listed.

The smaller size screw boxes – up to ¾" – are usually in boxwood and unhandled. Larger sizes will usually be beech and will have handles.

During the early decades of the 20th century, **Peugeot Freres** metal bodied boxes and machined steel taps were being sold in Britain in some quantity. The taps are parallel and have a single cutting edge at the commencement of the screw. They work well.

The most useful sizes and the most demanded by users – and there aren't many collectors of screwboxes – are from ⅜" to ¾". The largest sizes in the 2" to 3" range are handsome pieces for collectors. Incidentally these seldom have matching taps, as it is quite normal to form the larger female threads using slips of boxwood, morticed in.

Really large screws were marked out by wrapping a strip of paper, the width of which corresponded to the thread pitch, around the shaft. The thread was then cut with chisels and gouges. This was probably the earliest method of marking and making screws.

A few large taps, 4" to 6" diameter, for larger presses have been recorded – these are of different construction with one or two metal teeth inserted into the body.

Price Guide

♦ A rare 1½" beech handled screwbox and tap with keeper by **Holtzapffel**. **£160** [DS 23/39]

♦ A little used ¾" boxwood and beech screwbox and tap by **Preston** with keeper. F **£18** [DS 23/707]

♦ Two handled beech screwboxes and taps, 1" and ¾" by **Marples**. G++ **£55** [DS 22/176]

♦ A 1" beech screwbox and tap by **Marples** in original box. F **£34** [DS 22/449]

♦ A ½" boxwood screwbox and tap by **J. Buck**. **£30** [DP]

♦ A 2¼" beech handled screwbox with keeper, no tap. **£40** [DP]

♦ A 1¼" boxwood handled screwbox and tap by **J. Buck**. **£35** [AP]

♦ A set of three boxwood taps, ⁵⁄₁₆", ⅜" and ½" marked **Melhuish** with taps in nicely turned handles. **£55** [DP]

♦ A 38mm **Peugeot** screw box and tap. **£45** [AP]

♦ A 20mm **Peugeot** single handle screw box and tap. **£35** [DP]

Peugeot boxes and taps [Tyzack, c. 1910]. Sizes from ⅜" to 2" are listed.

The design of an efficient and durable brace took centuries to achieve. Indeed it was only in the 1870s, with the advent of the American pattern brace with a metal crank, adjustable chuck and ratchet, that all the problems were finally overcome.

With a brace the bit is driven continuously, which is not the case with augers and gimlets, but the problems to be overcome include making a crank that will resist the high forces applied, the design of a bearing that will transfer pressure from the head, and, above all, quick and easy insertion and removal of **any** bit – not just those fitted to **that** brace.

The earliest braces had fixed bits – this type of brace was still being used by chairmakers well into the 20th century – a hang over because the trade used a limited number of sizes of hole. In most trades this was not the case and the advantage of interchangeable bits was obvious.

It is now clear that, by the second half of the 18th century, commercial manufacture of wooden braces, then called bitstocks, was widespread. Those being made were of a style that continued throughout the 19th century – a plain beechwood stock with a turned handle and heavily chamfered top and bottom webs. Interchangeable bits were held in a brass pad.

Improvements

The history of the wooden brace is a tale of the gradual replacement of wood by metal as each point of weakness was tackled. Around 1820 improved versions with brass plates to reinforce the short grain in the webs were being made and by 1840 the all-wood head began to be replaced by a composite head with brass collar and wood cap. Unfortunately for the collector, and this is general in most types of tools, even when improvements

were made, the old type continued to be made as a cheap line, so examples of simple types can be much later than might be supposed.

The Ultimatum brace

A further step towards yet more metal took place in 1848 when John Cartwright of Sheffield took out a patent for a brass-framed brace. The patent actually contains two drawings; one shows a divided brass frame with wood infill between (this was the type that was made in quantity), the other

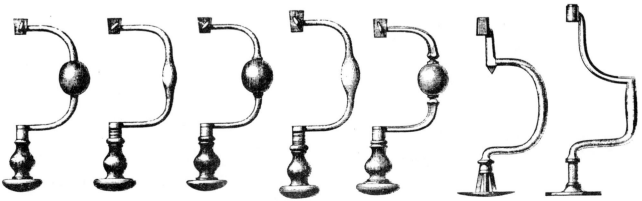

Just a few of the iron braces from the 1862 Sheffield List.

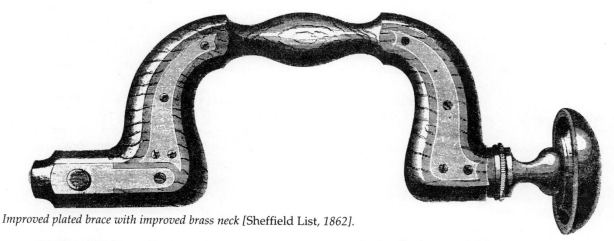

Improved plated brace with improved brass neck [Sheffield List, 1862].

a solid brass frame. The other significant advances are a rotating handle, a pad (chuck) operated by the nozzle and the suggestion that horn might be used for head and filling. The patent was purchased by **William Marples** who, by 1854, had put the "patent metallic-framed brace" into production and coined the very memorable description "Ultimatum brace". When the patent protection ran out around 1870 other Sheffield makers muscled in – eventually the brass-framed brace became such a standard product that it appeared in the *Sheffield List*. These were often supplied without trade names – the so-called "trade braces". Manufacture of ebony braces had ceased by 1905.

Collecting tips

➤ *From time to time much excitement has been generated amongst collectors when "undertakers' braces" have been auctioned for considerable sums. These are similar to the cheap sorts of iron brace except that the head can be swivelled inside the crank. Ostensibly, these are for the use of the undertaker when screwing down the lid of a coffin and were made to fold so that they could be fitted discreetly into a pocket. It's a good story but beware – the unscrupulous can alter a sixpenny brace with ease. Whist this may well be how they were originally made the ones I have seen mostly look like modern adaptions.*

➤ *The 1840s and 50s were a period of innovation in braces – even small variations are very attractive to collectors.*

Price Guide

◆ A rare heavy plated beech brace by **Pilkington, Pedigor & Co.** also stamped "Brass Mounted Brace, Sheffield and by her Majesty's Royal Letters Patent". **£1150** [DS 19/1178]

◆ A long model brass framed horn filled Ultimatum brace by **Marples**. **£715** [DS 25/1648]

◆ A "Horton" brace with rosewood handle and unusual rosewood head, the brass frame with floral decoration stamped **G. Horton**, Sheffield. Registered No. 2528 Nov. 8- 1850. **£572** [DS 25/1667]

◆ A brass plated button pad rosewood brace by **Thos. Ibbotson** with ivory button in ebony head. G++ **£682** [DS 23/1376]

◆ A brass framed rosewood brace by **Alfred Ridge** with ivory ring in rosewood head. G++ **£561** [DS 23/1394]

◆ A short sweep, long model brass framed beech Ultimatum brace by **Marples** with much original finish with a roll of 22 notched bits all marked **Marples**. G++ **£814** [DS 23/1375]

◆ A long model brass framed ebony Ultimatum brace with ivory ring by **Wm. Marples**. Circa 1860. G++ **£462** [DS 24/1629]

◆ A short sweep long model Ultimatum brass framed beech brace by **Marples**. Ebony head with ivory ring. **£390** [DS 24/1639]

◆ An early brass framed beech Ultimatum brace by **Marples** with double ring pad and "57 Spring Lane" address (1854–55). Some stains. **£330** [DS 25/1650]

◆ A brass framed ebony Ultimatum brace by **Marples**. Early model, 1853–54, together with roll of 13 mostly Marples bits. G++ **£219** [DS 25/1666]

Further reading: Eaton, *The Ultimate Brace* and Roberts, *Some 19th Century English Woodworking Tools*. (See p. 18)

Price Guide

◆ A "Ne Plus Ultra" brass framed ebony brace by **Henry Pasley** with ivory ring in **Marples** ebony head. **£220** [DS 24/1630]

◆ A brass framed ebony brace stamped **Frederick Willey** Opposite Corn Exchange Leeds. **£286** [DS 25/1668]

◆ A brass plated button pad ebony brace by **Marples** with ivory ring in head. Minor chip. **£290** [DS 24/1626]

◆ A brass plated button pad beech brace with brass head. **£77** [DS 25/1656]

◆ A brass plated button pad brace by **D. Flather** with ebony head and brass medallion with set of 14 shell bits. G++ **£143** [DS 24/1616]

◆ A brass lever pad beech brace by **Howarth** with brass emblem in rosewood head. **£55** [DS 25/1635]

◆ A little used brass button pad beech brace by **Marples** with ebony head. F **£110** [DS 25/1649]

◆ A chairmakers' brace with fixed screwdriver bit. G **£25** [DS 25/1661]

◆ A small chairmakers' beech brace with iron ferrule and spoon bit. G **£33** [DS 24/1585]

◆ An elegant ash boatsway with iron ferrule and fruitwood handle. (A shipwrights' brace, usually primitive, with a shallow sweep.) **£110** [DS 24/1587]

◆ An unused coopers' beech brace by **Mathieson**, Glasgow. F **£209** [DS 24/1590]

◆ An iron geared drill brace with beech handles. G **£57** [AP]

◆ An all brass (spark free) brace with rotating head. **£32** [DP]

◆ An iron gas fitters' brace by **Newey**. **£20** [DS 25/1624]

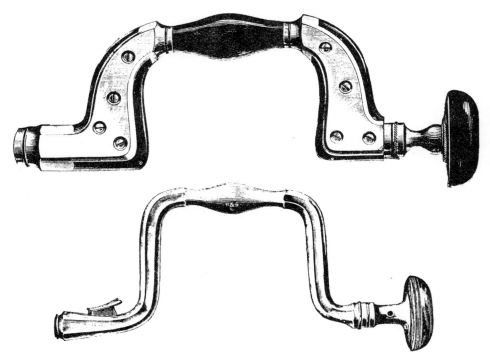

Top: A brass framed, ebony filled brace [Harding, 1903]. Often called an Ultimatum, this example, from a merchant's catalogue, is in fact a trade brace.

Bottom: A Scotch brace with ball bearing head [Harding, 1903]. Earlier versions would not have had ball bearings. Scotch braces are sometimes confused with wagon builders' braces which are similar heavy, large sweep braces but with less flat on the stock and a thumb screw chuck.

Metal

By the 1850s, when the brass-framed brace was entering the market, it was already obsolete. The Scotch brace, a strong and durable forged brace with a sprung catch, was by then widely available as were a wide range of cheaper iron braces. For all square socket braces, it is necessary to fit the tang of the bit to the socket and also to file a nick in the tang to engage the retaining dog. For this reason it was quite usual to supply braces together with a set of bits.

The quest for better methods of holding the bit resulted in many patents and claims of improvements, particularly in America. In 1865 the Barber chuck, now the most universally used type, with jaws closed by an outer ring, was patented but other forms of improved chuck, the Spoffard (split jaw form) and the Fray were amongst the many competing designs. Around the same time ratchets were developed. Perhaps the only further significant improvement to be made was the 20th century addition of ball races to take the thrust from the head.

Accessories and alternatives

The difficulty of boring holes where the crank of the brace could not be rotated was only partly solved by the addition of a ratchet. One solution for working a cramped space was simply to make the brace smaller – the charming small electricians' brace, rather overlooked by collectors, was the

result; but the mechanically inventive were also at work and the last decades of the 19th century saw corner braces, extension bit holders, corner ratchet braces and some ingenious angular boring devices being added to the carpenters' armoury

A brace for the trade

As with many tools, special braces to meet the needs of particular trades were available. We have already mentioned the rather simple chairmakers' brace, with its fixed bit, whilst coopers used similarly simple but robust braces but with means for changing bits whilst smiths, wagon builders and gas fitters all had special metal braces.

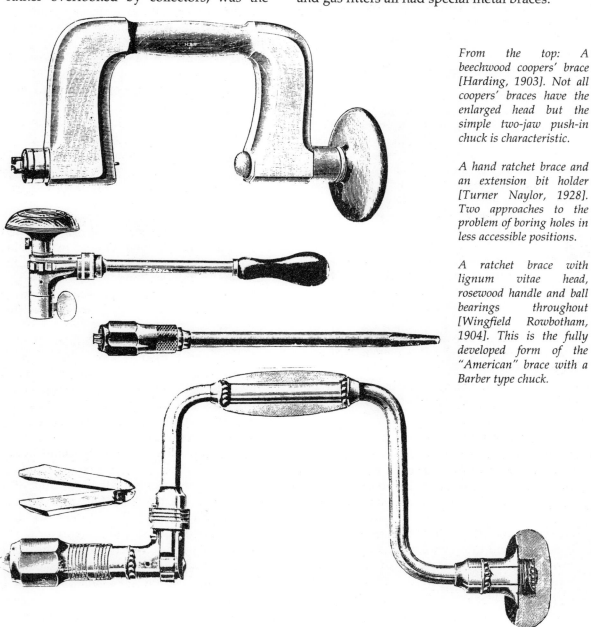

From the top: A beechwood coopers' brace [Harding, 1903]. Not all coopers' braces have the enlarged head but the simple two-jaw push-in chuck is characteristic.

A hand ratchet brace and an extension bit holder [Turner Naylor, 1928]. Two approaches to the problem of boring holes in less accessible positions.

A ratchet brace with lignum vitae head, rosewood handle and ball bearings throughout [Wingfield Rowbotham, 1904]. This is the fully developed form of the "American" brace with a Barber type chuck.

The bag of tools contained something for every eventuality except, that is, any drill bits. There was only one thing for it – make a bit. The head was cut from a 1½" nail, the pointed end hammered flat and the holes were chewed out in a minute or two. The bit I had made was a rough form of spade bit, similar to those still being sold in the 1930s with small archimedean drills.

The truth is that almost anything will form a hole in time but the search for clean working and efficient bits has taken millennia. The recent discovery of some Anglo-Saxon boat builders' tools including several shell bits that might well have passed for 19th century date suggests that, in hole boring, progress was for many centuries pretty slow – until the 19th century inventors got to work.

Price Guide

Unless in sets, most bits sell for only £1–2. Rarer types commanding up to £10 include: the larger sizes of centre bit, best with two spurs; dowel pointers and rounder, usually **Preston**; deck dowelling bits and large taper bits. Around £15: cock plug bits; patent/proprietory countersink bits and screw nose deck dowelling bits.

• A complete set of thirteen **Russell Jennings** spur auger bits in their original three tier fitted box. £100 [DP]

• An unused **Irwin** No. 22 micro dial expansive bit with instructions in original box. G++ £24 [DS 24/1577]

• A set of twenty-two brace bits by **D. Flather**. G++ £40 [DP]

• A little used set of twenty-nine notched centre countersink, shell, taper, etc. bits by **John Wilson**, Sheffield. £60 [DS 23/1038]

• A set of five shell auger bits, the largest 1½" with an ash handle, brass reinforced. £36 [DP]

Good bits – bad bits

Shell type bits: A venerable form of bit still used well into the 20th century. The simplest form has no particular cutting edge at the point apart from the edge to the shell but there are some more developed forms including the spoon (a more closed end to the shell), the nose (an ear provides faster cutting) and those with a gimlet point. Taper versions of varying sizes right up to the giant coopers' bung reamers were also made. The Achilles heel of this type of bit was always starting, particularly starting in the right place.

Flat type bits: Usually called centre bits, they are efficient and quick cutting considering their simplicity – but only for shallow holes. The modern version, the flat bit, is a poor descendant and chews its way through solely because of the power of the electric drill.

Twist type bits: A twist bit will clear the shavings from a deep hole – at least, that is the theory. Whilst bits of this general type had been made since early times, it was only in the 19th century that manufacturers, mainly in America, produced improved proprietary types.

The observant will detect that there are many variations: a simple twist, usually called an auger bit, a solid centre (often referred to as Irwin bits), or a hollow centre.

The nose may be a simple point or a gimlet point which has the advantage of drawing the bit into the work. It may have one or two cutting edges with a variety of spurs whilst the spiral may have thicker or thinner lands (the outer edge of the spiral). Thin lands supposedly gave an easy working/clearing bit; thick ones better directional stability.

We are frequently asked about bits with the cutting spurs turned back or with the spurs turned so far back that they are united with the body giving a hole through the cutting edges. These were sold as "unbreakable" bits or bits that could be used to enlarge holes. Unbreakable they may be but they don't work cleanly.

Rimmer type bits: Commonly found in bundles of bits but it is difficult to understand what they were used for: they don't drill holes and, if used to ream, they exert a huge outward pressure and, given half a chance, will split the work. They come in half-round, square, and octagon form.

Engineers' type twist bits: Considering the millions in use today it is worth remembering that these only became available after 1864 when **Stephen Morse** started in business in Massachusetts, U.S.A.

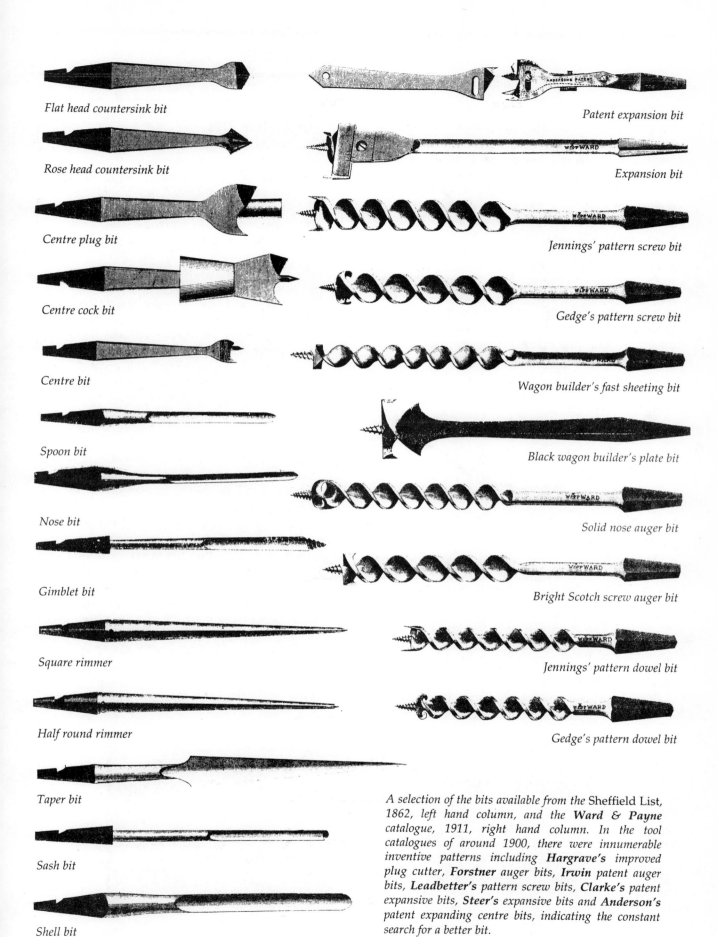

Flat head countersink bit

Rose head countersink bit

Centre plug bit

Centre cock bit

Centre bit

Spoon bit

Nose bit

Gimblet bit

Square rimmer

Half round rimmer

Taper bit

Sash bit

Shell bit

Patent expansion bit

Expansion bit

Jennings' pattern screw bit

Gedge's pattern screw bit

Wagon builder's fast sheeting bit

Black wagon builder's plate bit

Solid nose auger bit

Bright Scotch screw auger bit

Jennings' pattern dowel bit

Gedge's pattern dowel bit

A selection of the bits available from the Sheffield List, 1862, left hand column, and the **Ward & Payne** catalogue, 1911, right hand column. In the tool catalogues of around 1900, there were innumerable inventive patterns including **Hargrave's** improved plug cutter, **Forstner** auger bits, **Irwin** patent auger bits, **Leadbetter's** pattern screw bits, **Clarke's** patent expansive bits, **Steer's** expansive bits and **Anderson's** patent expanding centre bits, indicating the constant search for a better bit.

History

Once wheelwrights must have been the principal users but today – and this has been the case for generations – spokeshaves are used for a wide variety of tasks and it is doubtful if one in a thousand is ever used to shave a spoke. If you want to shape a new handle for your hammer, an oar for your boat or clean up a chair splat, this is the tool you will need.

Wooden spokeshaves

In the basic form of spokeshave, the iron is secured into the stock by two tapering tangs, friction-fitted. Adjustment is by tapping. Although simple, this arrangement seems to have been remarkably effective and the great majority of wooden spokeshaves found are of this simple form.

A patternmakers' boxwood radius shave [Turner Naylor, 1928].

The improvers were however at hand and versions adjusted by thumb turns acting on the threaded tangs were produced. **Ward & Payne** also made a "Patented Beech Holdfast Spokeshave" – this is a tanged type, locked by means of toggle head screws bearing on the tangs – probably unsuccessful and therefore rarely found.

Beech and boxwood

As with planes, beechwood is the standard material for wooden spokeshave but the smaller

Collecting tips

➤ *Early spokeshaves have less flare on the handles than later examples.*

➤ *Look out for miniature size shaves in all materials – boxwood, brass and cast iron.*

➤ *Don't buy brass miniatures unless you are certain that they are not recently made.*

➤ *Don't confuse woodworking shaves with leather working shaves – often metal with convex soles. They aren't worth much.*

➤ *Most spokeshaves can be bought for under £10 and few command a price above £30. For those with a limited pocket and space, a collection in the single genus "Spokeshaviana" would provide variety and interest.*

types could also be had in boxwood at a cost of around 50% more and a few of the smallest types – the "Extra Small Boxwood Spokeshaves for Fine Work" (4", 5", 6" or 7" overall with 1" or 1¼" irons) – were only made in boxwood. The better quality shaves had brass wear plates fitted to the face.

Spokeshaves are either flat faced or round faced, which allows the tool to be used on a tighter radius. In wooden spokeshaves, the face shape is seldom, if ever, specified in the catalogues, the purchaser altering it if he thought necessary. With the advent of the metal spokeshave, this was no longer possible and listings specified curved or flat faces.

The wooden spokeshave is no longer made – production by **W. Marples** in Sheffield ceased around 1960.

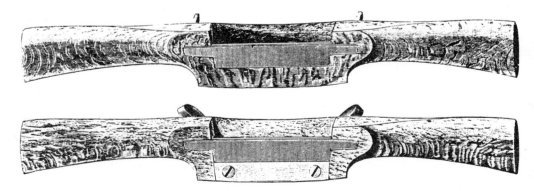

Top: Plain beechwood spokeshave. Bottom: Plated beechwood spokeshave with screwed iron [Wingfield Rowbotham, 1904].

Metal spokeshaves

By the last decades of the 19th century both British and American manufacturers were producing iron spokeshaves. These are an interesting area for the collector with examples of double acting (cuts in both directions), ingenious methods of fine adjustment and shaves to cut to chamfers and to minimal radius.

The earliest metal spokeshaves were gunmetal but the handles were wooden – fixed onto a spike extending from the metal stock. Some examples of this rare form are marked JOHN GREEN – probably the Sheffield firm of **John Green** (later **Hannah Green & Son**) 1774–c.1845.

Top: Stanley No. 55 hollow faced spokeshave. Bottom: Stanley No. 67 universal spokeshave [Melhuish, 1905].

By 1870, the **Stanley Rule & Level Co.** of New Britain, Connecticut, USA, were listing 13 types of iron spokeshaves. These included plain and adjustable models, hollow faced, combined hollow faced and straight, and extra long/heavyweight models for coopers.

The adjustable versions, No.53 (raised handle) and No.54 (straight handle), are interesting. The adjustment is made by moving the front of the mouth – there is a similarity with the earliest method of adjustment used in Stanley planes. These shaves certainly had a long run, remaining in production until 1947 and 1935 respectively.

After 1880, **Ed. Preston & Sons** of Birmingham built up a most comprehensive range of iron shaves. The firm developed the open-handled screw adjustable and transverse adjustable shaves, often with the legend "PRESTONS PATENT" in the casting. These are the handiest shaves ever made. There is also a range of small shaves.

Intended to reach work of the smallest diameter, the **Millers Falls** circular spokeshave (made 1872–present) with its cylindrical body and rosewood handles is attractive but of dubious utility.

Top left: Registered design iron spokeshave [Preston, 1909]. Top right: Patent iron spokeshave [Preston, 1909]. Bottom: Millers Falls No. 1 circular shave [Melhuish, 1925].

Edward Hoppus, writing in 1765 (*Practical Measuring Made Easy to the Meanest Capacity*), besides insulting his readers, records that "of wood screws there are thirty-one sizes which are sold from 1s 6d to 36s per gross". At this time screws were only used where absolutely essential and until the end of the century, much ironmongery, on even good quality houses, was still fixed with nails.

Get to the point

I don't know what it has to do with screwdrivers but it is worth noting that woodscrews in the 18th century did not have the gimlet point of today's screws. Only in the 1840s was the machinery to make pointed screws invented, and thereafter screws became cheap and easy to use.

old files with the teeth still evident to a greater or lesser extent. The best of these, with artistic shaping at the sides and simple, sometimes eccentric, handles, are attractive and are sought by the discerning collector.

Screwdrivers

Cabinet pattern: Used by woodworkers. Characterised by oval handles and a blade that is mostly round, waisted but flat at the ferrule. From around 1900, a closed thimble form of ferrule became normal.

Spindle pattern: Tradition holds that these were originally made from worn spindles taken from spinning machinery. The blades are round and long with one or more steps in diameter.

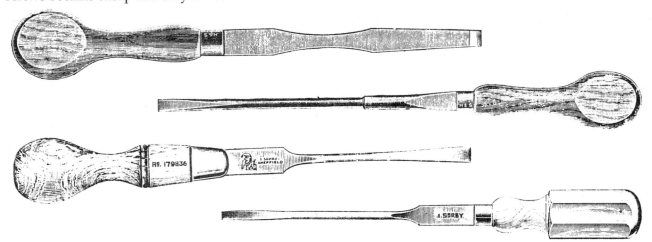

From top: London pattern; spindle pattern [Harding, 1903]; registered cabinet pattern, boxwood handle; electricians' light cabinet pattern [Turner Naylor, 1928].

The largest turnscrew I have ever seen was a monster, 4ft. long with an edge at least 1" wide, reputedly used on a naval gun.

Earlier turnscrews are most likely to have flat blades. Many were craftsman made – usually from

Collecting tips

➤ *Screwdrivers must be the most abused of all tools. Bent blades can be straightened and tips reground but if the handle is split and damaged little can be done.*

➤ *If you want to sound like a knowledgeable tool buff always refer to screwdrivers as "turnscrews".*

➤ *There's more variety here than you might think.*

Gun turnscrews: There are two types, the one intended for the sportsman's pocket and the other for inclusion in a gun case. The pocket type are only about 2" long with some versions being pairs that close into one another. The case type are also quite small, around 4", with a short blade and distinguishable by horn, ebony or rosewood handles.

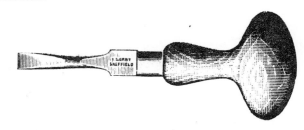

Motor turnscrew, cabinet blade [Turner Naylor, 1928].

London pattern: Waisted flat blade pattern available in a wide range of sizes from 3" to 18" (screwdriver sizes are always the blade size not overall).

Electricians' or pianoforte makers': The same pattern turnscrews were sold for both these trades – slender long bladed turnscrews with boxwood handles of either octagon or worked oval pattern.

Perfect pattern: Drop forged steel blade extends right through the handle – the wooden scales are riveted on. Made since the 1920s.

Driving home – quickly

As a child I seem to have watched a lousy class of workman. With some chortling, the "Brummagen driver" (a large hammer!) was applied to screws, until only a turn or two needed to be made with the screwdriver.

Putting in many screws is certainly hard on the wrist, work only to some degree eased by the spiral screwdriver. These were developed in America in the 1870s. and became colloquially known as "Yankee drivers" after a best selling brand produced by **North Brothers** (now part of Stanley). Growth was explosive – a Chicago tool merchant writing in 1895 records 18 different manufacturers making them. Other commonly encountered brands are **Goodell-Pratt** and **Millers Falls.** Early examples in good condition are not easy to find.

Price Guide

Few screwdrivers command prices of more than **£10.** Screwdrivers have a hard life so items in fine condition are rare and sometimes make high prices. 19th C. ebony or rosewood handled gun turnscrews (from cased sets) **£12–20.**

* A well shaped 32" ash handled screwdriver, craftsman-made from a file. **£26** [DP]

* A 23" turnscrew by **Preston** with boxwood handle. G **£49** [DS 25/1024]

* A fine 18" boxwood handled turnscrew by **Mathieson. £40** [DS 25/1517]

* A set of three screwdrivers with worked oval boxwood handles by **Sorby.** F **£36** [DP]

* A 7" rosewood handled turnscrew with spiral twist to blade, Timmins pattern. **£18** [DP]

* A spiral screwdriver with rosewood handle by **A.H. Reid** (No. 1 Lightning Brace). **£24** [DP]

* Three motor pattern turnscrews, all with boxwood handles. **£20** [AP]

* A 16" forked flat blade screwdriver with craftsman-made turned yew handle. **£12** [DP]

* A large (30") spindle pattern turnscrew with three changes of diameter and oval boxwood handle. **£24** [AP]

* A **Millers Falls** No.620 double acting spiral screwdriver in manufacturer's cardboard box. F **£22** [DP]

* Three "Perfect Pattern" screwdrivers by **Toga** of ascending size. G++ **£16** [DP]

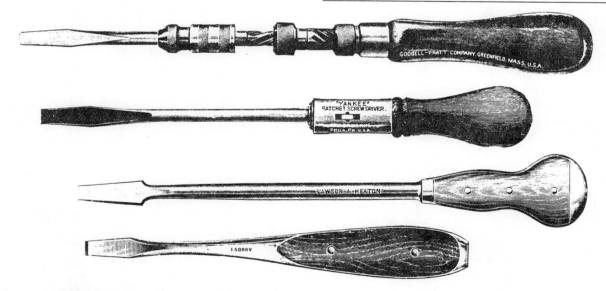

From top: Goodell-Pratt reversible automatic screwdriver – this tool has separate spirals for right and left hand twist and in this respect differs from most other pump screwdrivers where the two spirals are combined; "Yankee" ratchet screwdriver No. 797 [Melhuish, 1905]; solid forged pattern screwdriver [Lawson & Heaton, c. 1930]; Perfect Pattern turnscrew [Turner Naylor, 1928].

Bow drills

A very old form of drill. Until wheel braces became common in the 19th century, this and the pump drill were the only methods of providing a fast-turning drilling action. Some types incorporate a handle, whilst others turn in a socket worn on the chest of the operator. The bow drill was used by many trades, continuing into the 20th century. Perhaps the most collectable are the so-called piano makers' drill stocks. Most of these seem to have been made by **Buck**, Tottenham Court Road, London. They are beautifully engineered.

Piano makers' ebony handled bow drill.

Pump drills

Another ancient form of drill capable of high speed rotation and therefore useful for drilling small sized holes in hard materials. A favourite of the china repairing trade which drilled holes and inserted rivets in the days before modern resin glues.

Wheel braces

The modern form of cast iron framed wheel brace with a jaw chuck is an American development.

Price Guide

Archimedean drills of reasonable quality **£5–10**.

♦ A piano makers' ebony and brass bow drill by **Buck** – engraved name and address – without bow. **£130** [DP]

♦ A sculptors' or masons' beech handled bow drill stock with brass spool. **£90** [DP]

♦ A miniature (8") side handled archimedean drill with ebony handles. **£40** [DP]

♦ A 5" watchmakers' archimedean drill with brass knobs. **£12** [DP]

♦ A wrought steel wheelbrace with mahogany handles in good bright condition. **£40** [DP]

♦ A commercially-made pump drill with brass flywheel and beechwood bar. **£24** [DP]

Collecting tips

➤ *Bow drills: Commercially made bows are rarer than the drill stocks.*

➤ *Large cast iron breast drills are common and of little appeal to the collector.*

➤ *Archimedean drills: Quality is most important. Examples with a side handle are more desirable.*

There are many variations including changeable gear ratios, storage for drills in the stock, movable handles, etc. but all are of late 19th or 20th century date. Some apparently early looking wrought steel, bright-finished wheel braces may not be as old as they look.

Archimedean drills

The variation in quality and size (from 4" to 24" or more) is notable. Beautiful examples were made for use by dentists, watch makers, jewellers and others whilst appalling poor examples were included in fretwork sets.

Modern reciprocating drill

Half-brothers to the Yankee screwdriver, the best examples operate on both the push and pull stroke.

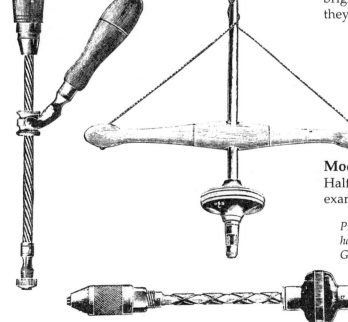

Pump drill (upright drill), Archimedean drill with side handle [Harding, 1903] and reciprocating drill by Goddell-Pratt [Melhuish, 1912].

Setting out

LEVELS

Invention of the spirit level tube has been credited to a Frenchman, Jean Thèvenot, in 1662. Initially used in scientific instruments, this "high tech" invention gradually filtered down to the man on the job. Included in the few levels listed in Smith's *Key* (1816) is a "best brass stock level" which had sliding panels to cover the top. These cost the astonishingly high price of 108s per dozen.

The level tube
All tubes are not the same. The manufacturers' descriptions varied somewhat but all offered different grades.

The cheapest were just called tubes. Next in quality were *proved* tubes, which were usually of thicker glass of larger diameter and were inspected and marked to show the central position. Both **Rabone** and **Ed. Preston** claimed to mark the proved tubes with a dot at the end. Better quality still were *ground and proved* tubes – these, believe it or not, were ground and polished on the *inside* and were marked with either two or four lines showing the central position.

The most accurate tubes were ground and polished and then *selected and graduated* to indicate the exact degree of movement. This latter quality will only be found in scientific instruments.

Spot the bubble
Always a problem with levels is the visibility of the bubble within the tube. The introduction of a bright dye (flouroscene) into the spirit helped and

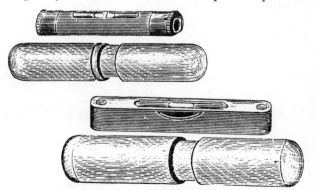

Small levels [Mathieson, c.1900]. Top: brass tube level with boxwood case. Bottom: Brass level with case.

this became common after 1914. A level made before 1730, for the 4th Earl of Orrery, now in the Museum of the History of Science, Oxford, contains a pink fluid showing that colouring the spirit, however, is nothing new.

Other visibility-improving features include the so-called mirror tubes – actually a piece of metal foil placed behind the tube; a white glass backing to the tube; and, best of all, Wood's cat's eye tube. This clever improvement consists of a green line on a white background put into the glass tube when it is made. The line is magnified by the bubble, thus making it very clear.

Levels

Level tubes of whatever quality have been installed in an extraordinary range of both wood and metal stocks as can be seen in any tool catalogue. Suffice it to say that many are fine examples of the ingenuity and art of the joiner's tool maker, whilst others are so poor as to be contemptible.

Collecting tips

➤ *Do not leave levels in full sunlight – this can break the tube.*

Strong brass plated level.

Rosewood level, brass plated and tipped round top.

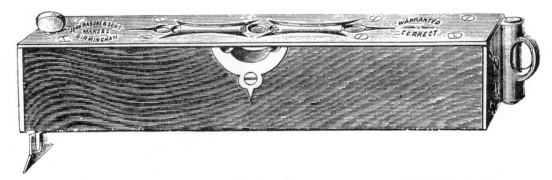

Plumb and level with field sight and graduated screw slide.

Rosewood plumb and level with tapered ends, brass tipped [Rabone, 1892].

Ebony "Protector" level with brass rotating cover to protect tube [Mathieson, c. 1900].

A few features that the collector should note:

"Adjustable": Usually found on the larger sizes of wooden level. The tube is in a metal carrier which can be adjusted by screws either visible or accessible by removing a plate.

"Protected tube": In flat-topped levels, this is by means of a sliding cover; in standing tube types, by means of a rotating cover.

Levels for falls: The Victorian obsession with drainage resulted in a number of levels intended to indicate falls for pipe laying.

Sighting levels: A longitudinal hole and cross wire allows a sight to be taken over some distance – a simple form of the surveyors' telescopic level. The most common design has an adjustable foot.

Boxwood spirit levels: These are well made and interesting level/rule hybrids with the tube in an inset brass carrier. Surprisingly for such finely-made items, they were advertised for masons. This is no doubt the reason why many examples are found in poor condition.

Gradient indicators: Comprises a long bent tube installed in a wooden stock with the top plate marked with graduations. The examples we have seen are by surveyors' instrument makers.

Bright iron level.

Extra quality small ebony level.

Boxwood rule level [Rabone, 1892].

Price Guide

Late 19th/early 20th C. rosewood or mahogany, brass plated top boat (taper end) levels, 6" to 12" **£12–18**. Ditto but with shaped brass centre plate and tips **£18–24**. Mahogany parallel sided levels, plated tops, 8" to 12" **£8–12**. Ditto, rosewood or ebony, (usually of better quality) **£15–20**. Ditto with decorative plates on sides **£24–30**. Levels of the finest quality or decorated with shaped brass inserts fetch more.

♦ A **Preston** 18" improved plumb and level in *mahogany* with brass frame centre section, Reg. Design No. 269393. **£50** [DP]

♦ A **Preston** 18" improved plumb and level in *ebony* with brass frame centre section, Reg. Design No. 269393. **£90** [DP]

♦ A **Preston** 10" ebony parallel side level with fancy brass inserts to top and sides, Reg. Design No. 390137. F **£65** [DP]

♦ A rare **Preston** 9" boat ebony level and plumb (rare in ebony). **£45** [DP]

♦ A **Preston** 8" ebony level with thick brass fittings. **£35** [DS 24/893]

♦ A 12" **Preston** No.1258FS mahogany sighting level with shaped opening for tube in brass top, screw slide foot and flip-up brass sights. **£157** [DS 24/1312]

♦ A 12" ebony level with rotating brass cover by **Buist** also stamped D. Abbot. G++ **£209** [S 24/103]

♦ A 10" ebony level by **Buist**, Edinburgh, with fancy brass inserts to top and sides. **£75** [DP]

♦ A 10" brass topped rosewood level by **Rabone**. G++ **£24** [DS 24/23]

♦ A 12" rosewood and brass tipped boat level and plumb by **Rabone**. G++ **£40** [DS 24/511]

♦ A 10" ebony and brass level by **Mackay & Co.**, Glasgow with attractive brass plates, tube empty. G++ **£28** [DS 24/512]

♦ A 12" boxwood and brass masons' plumb and level with measurement markings to sides. **£25** [DP]

♦ An 18" boxwood and brass masons' plumb and level with measurement markings to sides in original leather case. **£45** [DP]

♦ A 9" rosewood boat level and plumb with ivory and brass tips. G **£42** [DS 24/24]

Engineers' levels

There are several basic forms: a) Brass tube levels fixed to a base plate. b) Cast iron stocks – usually with a V-base. c) Tube levels with adjustable mountings on a heavy base with a curved recess on the sole. d) Cast iron parallel stock levels 12" to 24" long. The stocks are infilled with decorative latticework.

Collecting tips

➤ *Levels aren't rare, so buy only the best.*

➤ *Look out for owners' names and dates under the tube.*

➤ *Cast iron levels are brittle and many examples are chipped or cracked. Satisfactory repair is impossible so look carefully.*

Replacing a tube

Perhaps the most difficult part of renewing a tube is to find a suitable replacement. For small tubes, it is usually possible to recover one from a cheap or very worn stock but for longer tubes this can be a real problem New glass tubes are available from some dealers in antique tools.

Remove the top plate and dig out the old tube and setting material. Traditionally, level tubes were set in plaster of Paris but as this sets fast and is difficult to obtain in small quantities, the best alternative is polyfiller.

Replace the top using the original screws if at all possible. Old screws have narrower slots than modern screws and will be quite apparent to the initiated. Commercially made levels had *steel* screws which, in the best quality levels, were blued. To make blued screws is easy; polish the heads using fine wet and dry paper, oil the screws and then place on a metal sheet and heat in a gas torch until they go blue.

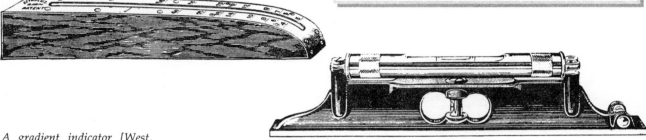

A gradient indicator [West, 1930]; an engineers' iron level [Rabone, c. 1930]; a brass tube level [Rabone, 1892].

Price Guide

◆ A 12" ebony level with brass top decorated with different engraving styles to advertise his trade, e.g. H. Sayer, Engraver 8 Denbigh Road, with engravings of a bird, lion etc. circa 1860. **£341** [DS 23/193]

◆ A 4" brass tubular level in boxwood case. **£20** [DP]

◆ A **Mathieson** No. 018C 10" ebony "Protector" level with revolving brass cover. £115 [DS 23/331]

◆ An 8" rosewood and brass parallel sided level with sliding cover to tube. **£32** [DP]

◆ A 9" brass tube level by **Rabone**. The brass tube (no cover) stands on a brass base. **£24** [DP]

◆ A 12" gradient indicator level by **Reynolds**, Birmingham, striped ebony (coromandel?) stock, in original morocco-covered case. **£120** [DP]

◆ A "Dr Bates" drain clinometer by **W.H. Harling**. The hinged underside is adjustable with a thumb wheel, the mahogany stock with nickel silver fittings. **£80** [DP]

◆ A 6" nickel plated engineers' level by **Preston** marked "Patent Cat's Eye Tube" in original leatherette-covered box. F **£35** [DP]

◆ An engineers' 12" adjustable brass level by **J. Casartelli** in fitted case. The tube, with rotating cover, stands above the heavy base. **£55** [DS 24/21]

◆ An 8" engineers' steel and chrome level by **Rabone Chesterman** No. 1388 in original box. N **£24** [DS 24/1316]

◆ A 3½" cast iron level with brass plate top with facility to fix onto a joiners' square. **£16** [DP]

*Remember - condition is **G+** unless otherwise shown.*

Squares

It is perfectly easy make a carpenter's type of square from wood and many woodworkers did so right up to modern times. But wooden squares quickly become battered in use and, if a knife is used frequently for marking out, the edges deteriorate quickly. Far more durable is a metal-bladed square.

The 18th century

From surviving dated examples it would seem that, by the middle of the 18th century, craftsmen were making squares and bevels with metal blades.

In comparison with the later commercial products, the 18th century craftsman-made squares and bevels often have eccentric features – nicely profiled ends to the blades and unusual proportions between stock and blade. Today these early items are keenly sought by collectors.

The 19th century

The earliest commercially made items are in mahogany although, like all joiners' tools by the middle of the 19th century, ebony and rosewood were being used in preference. Today, these woods are no longer available in sufficient quantities and mahogany is once again the normal material.

In general, squares and bevels have not altered much since 1800. The 19th century manufacturers produced flashier products with ebony and large brass shields on the expensive ranges but the basis of a steel tongue riveted to the stock remained.

Mitre squares, which have the blade fixed at 45°, are used, as the name indicates, for marking out mitres.

Variations on a theme

Mostly rare but certainly worth looking out for are:

Level in stock: Quite an early development – they are shown in the *Sheffield List* of 1862. Rare today, probably because the tubes invariably got broken and were almost impossible to fix.

Fully bound: The stock is a brass casting stuffed with ebony. It would seem that most of these were made by **Wm. Marples** – some bear a design registration number 3835.

Solid stock: Identifiable by an extended and ornate shield. The real distinguishing characteristic is that the steel blade is L-shaped, extending down the stock.

Roofing: Steel squares, usually with arms 18" by 24". The more recent examples are marked with a

Price Guide

Commercially-made squares, 4½" to 12" blade, rosewood, brass faced stock, diamond plated **£6–12**. Ditto ebony with shield **£10–20**. Larger sizes command higher prices. Mitre squares, 8" to 12" blade, rosewood, diamond plated **£12–18**. Ditto, ebony with shield **£18–30**.

♦ An 18th C. square, 15½" blade, with mahogany stock (not brass faced) carved with the date 1772. The steel blade with large ogee at end. **£300** [DP]

♦ An early 19th C. square by **Moon**, 4" blade, with mahogany brass plated stock. **£30** [DP]

♦ An ebony stocked square, 30" blade, with six point shield by **John Wilson**. **£125** [DP]

♦ A rosewood stocked square, 24" blade, with six diamond joint by **Marples**. F **£70** [DP]

♦ A rosewood stocked square with 18" blade, the flower shield made of alternate leaves of brass and horn. G **£38** [DS 25/437]

♦ An ebony stocked square, 15" blade. **£35** [DP]

♦ A fully brass bound ebony square, 14" blade, by **Marples** stamped with Reg. Des. No. 3835 (The stock is a brass casting with infill). G++ **£66** [AP]

♦ Three matching fully brass bound squares, 6", 7½" and 10½" blades (the stocks brass castings with infill). **£85** [AP]

♦ An ebony stocked square, 9" blade, the multi-point shield extends the length of the stock. **£90** [DP]

♦ A solid brass stocked square, 3½" blade, with owners' name and the date 1853 engraved on the stock. **£35** [DP]

♦ An ebony and brass square, 12" blade, by **Clegg** with level in stock and engraved Robert Marshall Law. Pitting to blade. **£24** [DS 24/543]

♦ A bricklayers' London pattern square rosewood stock and 7" brass blade both shaped at ends (commonly misrepresented as early squares). **£30** [DP]

♦ A 12" ebony and brass stocked mitre square by **J. Frost**, Norwich. **£42** [DS 25/436]

♦ A 12" mitre square with solid brass stock, nicely engraved C. Humphries. **£49** [DS 25/1395]

♦ A coachbuilders' square with gunmetal stock and 7½" steel blade by **Preston**. The stock rotates and is lockable with a thumb nut. **£85** [DP]

♦ A steel roofing square, 18" x 24", marked with numerous tables for setting out rafters. **£15** [DP]

Don't forget - fine condition will often radically increase the value.

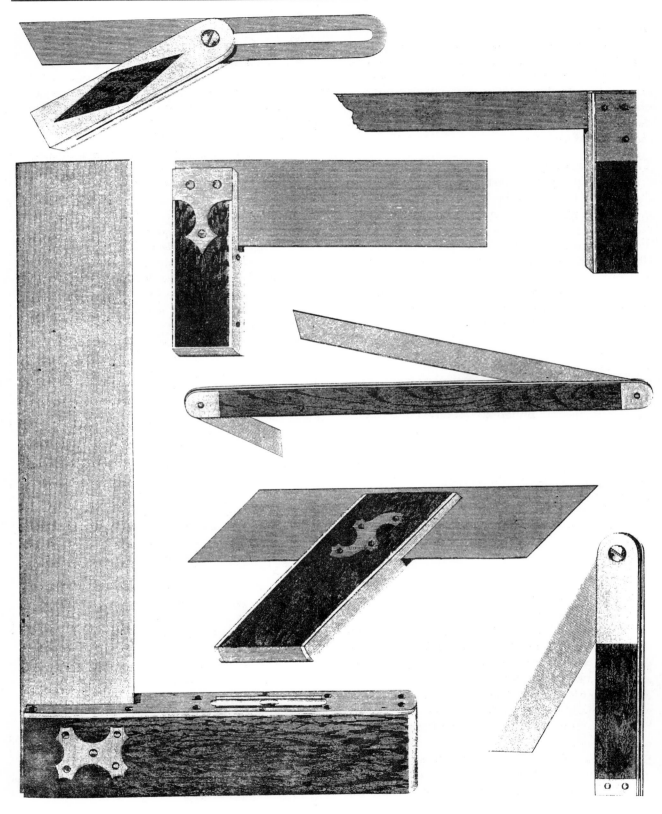

Squares and bevels [Sheffield List, 1862]. From the top: an improved metallic frame sliding bevel, a London pattern bricklayers' square, an improved brass mounted square, a boatbuilders' bevel with two brass blades, a best plated mitre square, a best London pattern bricklayers' bevel and a best plated square with level.

variety of scales intended for setting out the cuts needed for rafters and other roof timbers. Some authorities have suggested that the origin of these is the mediaeval mason's square.

Joiners': Craftsman-made squares, most often cut from a thin mahogany board and incorporating a number of curves. The best are aesthetically pleasing objects.

Coachmakers': The stock is gunmetal; the blade is rotatable on the axis of the stock and can be secured at any angle by tightening the securing nut. A late 19th century development which is surprisingly rare.

Bevels

The earliest form of bevel, a simple tongue pivoted at the end of a wooden stock, has been around since the middle ages, if not long before. By the 18th century metal bladed versions, pivoted in the middle (the T-bevel), were in the tool box but the sliding bevel is fairly recent. The **W. & S. Butcher** catalogue of 1842 does not show any sliding bevels but by 1862 the *Sheffield List* contains not only the normal plated stock form but also an example with a solid brass stock – the wood being only an inlay.

Some trades used special bevels – the boat-builders' form is brass tongued, either similar to a simple bevel or double bladed in a longer thinner stock; the masons' is all brass with a slotted blade and, in some versions, also with the stock slotted; the coachmakers used the unlikely-looking spider mortice bevel.

Price Guide

Commercially-made, plated (i.e. brass tipped on the faces) bevels with steel tongues, mahogany or rosewood of normal size £6–9, ditto ebony £8–12. This price applicable to all three types of bevel, shifting, "T" and angle.

Note: Sliding and T-bevels are sized by the length of the blade.

◆ An 18th C. boxwood and brass stocked sliding bevel, the 16" arm with shaped ends and stamped Joseph Nash Sep 10 *1754*. Some pitting to arm. G **£462** [DS 25/951]

◆ An early 19th C. mahogany stocked 11" T-bevel, the brass plates well engraved with owners' name and dated 1823. **£65** [DP]

◆ A unusual and large fully brass bound 16" (stock) angle bevel with ebony infill by **Broadhurst**, Sheffield. G++ **£110** [DS 25/1343]

◆ A brass framed sliding bevel with two ebony inserts to each face by **Flather**. **£35** [DP]

◆ A 9" brass framed bevel with a diamond shaped ebony insert by **Howarth**. **£40** [DP]

◆ A rosewood and brass two blade boat bevel, 12" overall. **£26** [DP]

◆ A coachmakers' brass and steel spider bevel. **£30** [AP]

◆ A bevel by **John Wilson**, Sheffield. The stock rosewood, brass plated, the steel blade with quadrant arm passing through the stock and locking with a screw. **£140** [DP]

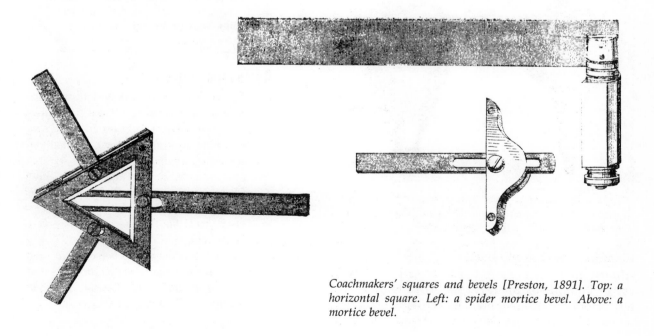

Coachmakers' squares and bevels [Preston, 1891]. Top: a horizontal square. Left: a spider mortice bevel. Above: a mortice bevel.

Down with the tube

Both the plumb bob and the mounting needed if it is to be used as a level are robust and easily made.

Although the spirit level tube had been developed by 1662, it was not until after 1850 that the plumb level was in general superseded.

Plumb bobs are still made today and they remain the simplest way of setting out the lengthy verticals needed when fixing drain pipes or papering walls.

Plumb bobs

The name derives from *plumbum*, the Latin for lead, the original material used. The egg-shaped lead bobs are however uninteresting and mostly ignored by collectors. In any case, many so-called lead plumb bobs are actually loom weights.

Brass plumb bobs and those made in exotic materials are keenly sought by collectors who have been drawn by their aesthetic appeal. Recently there has been some very strong competition to purchase examples of exceptional size or artistic merit. Interest has, in turn, spilt over into the field of technological development. There isn't much that can be done to improve the design of something as simple as a plumb bob but what little there is – such as integral reels or mercury filling – will radically increase the price.

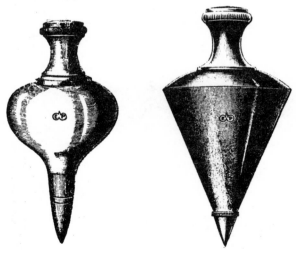

Left: a Best Quality brass plumb and right: a New Pattern plumb bob with removable point. The point can be stored inside the body. [Preston, 1914].

Commercially made bobs

Almost all commercially made brass plumb bobs are of bulbous or plain conical form with a steel tip. Modern production has crossed knurling whilst the earlier models were straight knurled.

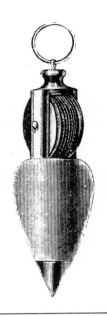

An adjustable plumb [Melhuish, 1912]. Available in bronze or brass, the reel was controlled by a friction device which could be adjusted with the centre screw.

Price Guide

Commercially-made steel tipped brass plumb bobs (not recently made) of onion or other standard shapes, weighing up to 1lb. **£12–18.**

- A 10oz. steel tipped brass plumb bob of tear drop shape with integral reel in neck of bob. **£93** [DS 24/12]

- A fine 34oz. steel tipped brass plumb bob of tear drop shape with knurled finial. G++ **£50** [DS 24/521]

- A 7" steel plumb bob of tear drop shape with elongated neck. G **£77** [DS 23/176]

- A **Preston** removable point plumb bob of conical shape. The steel point unscrews and fits into the underside of the top. **£40** [DP]

Collecting tips

➤ *The high prices for unusual plumb bobs have attracted fakers – brass bed knobs and ivory billiard balls are favoured raw materials and wholly new productions are also common.*

➤ *Look closely at the knurling – most fakers can't make or don't bother with quality knurling.*

➤ *Look inside the threaded cap hole – is it bright and new?*

➤ *In the eye of the plumb bob collector, big is beautiful – and worth paying for – but the bigger and more elaborate the item the more suspicious you should be.*

PLUMB BOBS

The most prolific maker until 1932 was **Ed. Preston & Sons Ltd.** which listed fifteen sizes numbered from 00 (1½oz.) to 12 (4lb.) The size numbers were usually stamped onto the brass.

However, most of the more artistic and larger plumb bobs are craftsman-made, sometimes from any bits and pieces that came to hand.

Steel plumb bob, No. 293 [Moore & Wright, 1938].

Plumb frames

To make a level from a plumb bob, it must be put into a mounting of some type. The commonest form is a wooden triangle with several different mounting positions for the bob, allowing it to be used either as a level or as a plumb. Also frequently encountered are T-levels. This form can only be used as a level and was the type commonly used by masons and paviors – gigantic examples up to 12' long periodically appear.

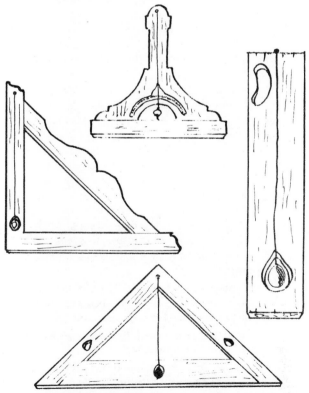

Four different styles of plumb frames: a T-level, a plumb square, a plumb board and a plumb triangle, which can be used in three attitudes.

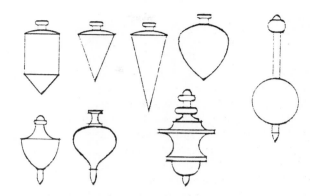

Typical plumb bob shapes. Top row: cylindrical, conical, slender conical, inverted tear drop, arrow and ball. Bottom row: urn, inverted onion (the most common commercial shape) and complex (often faked from knobs).

Price Guide

♦ A steel tipped No. 8 (1½lb.) brass plumb bob of standard onion shape by **Preston**. **£26** [DP]

♦ A **Preston** No. 1 (3½oz.) brass plumb bob of standard onion shape. F **£28** [DS 24/880]

♦ A 33oz. elongated pear shaped steel tipped brass plumb bob. **£77** [DS 23/158]

♦ A **Preston** No. 3 (6oz.) brass plumb bob of standard onion shape on an ornamental **Preston** brass reel, punched with the Preston logo. G++ **£80** [DP]

♦ A miniature 1½" silver plated onion shape brass plumb bob. **£37** [DS 21/1509]

♦ A solid steel engineers' plumb bob by **Moore & Wright** No.293, 4oz., with recessed centre portion to take cord. **£10** [DP]

♦ A 2½" all ivory plumb bob of standard onion shape, complete with screwed removable top. F **£85** [DP]

♦ A rare all-brass plumb bob, 12oz. with brass reel marked "wind up" let into the side of the bob. **£410** [AP]

♦ A giant 27½lb. brass plumb bob, 10" in height, used in the Harland and Wolf Shipyard, Belfast and engraved "H.W. Belfast" 1913. G **£1050** [AP]

♦ An early 30" pine plumb board with brass bob. **£55** [DS 24/1015]

♦ A 26" mahogany plumb triangle, with three positions for the lead bob. **£42** [DP]

♦ A 24" oak plumb square with brass bob and well shaped ogee decorations to arms. **£40** [DP]

♦ A fine Victorian carved mahogany plumb square with silver plated plaque "Presented to the Bishop of Salford on Laying of the Foundation Stone, New Priory Church. 17th Sept. 1898" with solid silver bob. G++ **£176** [DS 21/427]

Terminology here seems to have changed over time and thus become a little confusing. Strictly speaking, two trammel heads on a bar make a beam compass, but today, if a beam compass is mentioned, the assumption is that it must be a draughtsman's beam compass.

The obvious use for trammels is to scribe circles but it is probable that they were just as often used to set out repetitive dimensions.

Even though manufacturers sold nicely made brass heads, trammels seem to have been one of the items that craftsmen preferred to make for themselves, and both brass and wooden user-made examples are common.

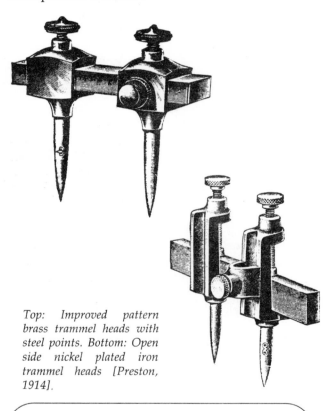

Top: Improved pattern brass trammel heads with steel points. Bottom: Open side nickel plated iron trammel heads [Preston, 1914].

Collecting tips

➤ *Screw trammels should have keeps (small slips of brass under the screws to prevent the screw digging into the bar) but these are frequently missing.*
➤ *Bigger is better.*
➤ *Who cares about the bar?*
➤ *Artistic interpretation is important.*
➤ *Well-engraved names or initials are desirable. Badly stamped names which distort the piece are not.*

In early examples the heads are secured with wedges and the simplicity of this method ensured its continuance even after the use of securing screws had become common. More sophisticated examples sometimes include some method of fine adjustment.

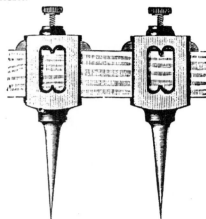

Japanned trammel points with bright fronts [Melhuish, 1899]. Steel trammels are unusual but the collector will find brass examples of a pattern similar to this.

Price Guide

✦ A set of mahogany trammels of 18th C. appearance. One head is fixed at the end of the bar (as is normal with early examples), the other moves and is secured with a boxwood wedge. **£55** [DS 23/510]

✦ A set of small (2¾") and simple early type wedged trammels in boxwood on a beech bar. One head is fixed at the end of the bar. **£40** [DP]

✦ An attractive pair of turned boxwood and brass trammels. One head, located near the end of the boxwood bar is adjustable by a screw in the end of the bar, the other is secured with a knurled brass screw. Inlaid decoration on one head. **£187** [DS 24/1293]

✦ A large pair of craftsman-made Cuban mahogany trammels of flat oval section, 9¼" high. Brass thumb screws and keeps and large steel points. **£60** [DP]

✦ Three attractively turned, 7" high, boxwood trammel heads with boxwood securing screws. Steel points fitted into brass ferrules. **£90** [DP]

✦ A pair of 4½" steel tipped brass trammels with pencil holder by **Preston**. **£24** [DP]

✦ A pair of 5½" steel tipped brass trammels complete with keeps by **Preston**. F **£35** [DP]

✦ A pair of Eclipse trammel heads in plated steel in original box. G++ **£13** [DS 24/84]

Price Guide

• A pair of plain brass, rectangular bodied trammels, 4½" high, brass keeps and securing screws. **£22** [DP]

• A fine pair of brass trammels, 6½" high. The bodies of plain rectangular shape, the fronts fretted out and engraved to form the monogram AGW. The screws with good triple lines of knurling. **£180** [DP]

• A very fine set of three brass trammels with faceted steel points on ebony bar. One head has points top and bottom, is fixed and is nicely engraved Wm. Snodgrass. **£248** [DS 23/259]

• A pair of 7" steel tipped brass trammels with steel screws and unusually, with keeps top and bottom of bar. **£55** [DS 23/1241]

• A pair of 5" nickel plated trammels by **The L.S.S. Co.**, No.50, with fine adjustment. **£31** [DS 23/212]

• A **Stanley** No. 99 rule trammel set complete with three items (two points and a pencil holder). F **£60** [DP]

• A draughtsmen's nickel silver beam compass set by **Harling** with ink and pencil heads and fine adjustment in a leatherette covered case. **£25**

• A nickel silver draughtsmen's ellipse trammel with 6" crossed channel base and 15" arm in a mahogany box, unsigned. **£220** [DP]

Trammels for ovals

One of the simplest ways of drawing an ellipse is to use sliding trammels. These consist of a scribe fixed to a beam, the movement of which is controlled by two guides sliding in channels set at right angles. By adjusting the distances between the two guides and the scribe, ellipses of varying size and elongation can be produced. Such devices are usually intended for draughtsmen but examples for workshop use can also be found.

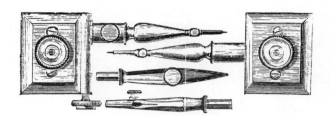

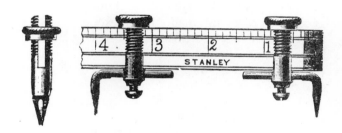

Above top: Draughtsmen's beam compasses with fine adjustment, German silver [Melhuish, 1905]. Bottom: Stanley No. 99 rule trammels with brass heads with steel points.

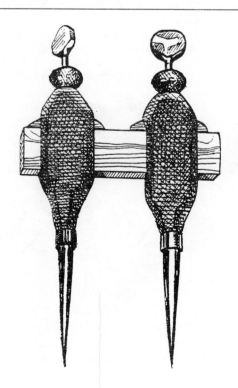

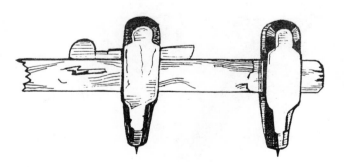

Left: Craftsman--made trammels in mahogany with steel tips and securing screws. Above: Early type wedged trammels in boxwood.

Until the 19th century improvers got to work, marking and mortice gauges were fairly simple affairs made of beech with the stock secured by a wedge or a wooden thumb screw.

By the end of the 18th century gauges of these types were being made commercially – Christopher Gabriel listed them in his stock. But the adjustable mortice gauge had not yet been developed, so a joiners' kit of the period had several mortice gauges, each with two pins set to match the width of the mortice chisels.

The cutting of a good, tight-fitting mortice and tenon joint begins with a good mortice gauge to scribe crisp lines which provide a start for both chisel and saw. The quest for a mortice gauge that was easy to set and would not move in use produced many variations and not a few registered designs and patents.

Ebony and brass

By the second half of the 19th century, the best gauges were being made from exotic woods – ebony, rosewood and occasionally mahogany. The brass fittings had become a prominent element in the design, and in some types the stem or the head was made entirely of brass.

Mortice, marking and cutting gauges are not collected as much as they should be. There are many variations to be found by the observant, and although it's true that gauges are common, ones in fine condition are not.

Price Guide

Beech marking gauges with boxwood thumb-screw **£3–6**. Rosewood mortice gauges or combination gauges, unplated or two strip brass facing and sliding adjustment of pin **£8–14**. Ditto, but turnscrew or thumbscrew adjustment of pin **£12–18**.

- A set of twelve craftsman-made birch fixed size marking gauges, ¼" to 4". **£26** [DS 21/525]
- A craftsman-made wooden screw marking gauge. The two part mahogany head nicely turned, the stem boxwood. **£20** [DP]
- A rosewood and brass mortice gauge with oval head and thumbscrew adjustment by **Marples** with part trade label. G++ **£33** [DS 22/199]
- A brass stemmed (round), brass plated oval head ebony mortice gauge by **Marples**. **£30** [DP]
- A brass stemmed (round), brass plated oval head ebony mortice gauge by **Melhuish**. Almost mint in cardboard box. **£60** [DP]
- A boxwood and brass, oval head mortice gauge by **Fenton & Marsden** with full plated face and part plated stem. **£99** [DS 21/986]
- A rosewood, long stemmed mortice gauge with oval brass head and calibrated slide. **£86** [AP]
- A boxwood and brass mortice gauge. The square head is fully brass plated and is adjusted by a longitudinal screw contained within the stem as also is the marking pin. Reg. Des. No. 1398 by **Marples**. **£140** [DP]
- A craftsman-made boxwood marking gauge, the head locks with ebony cam thumb turn. **£20** [DP]

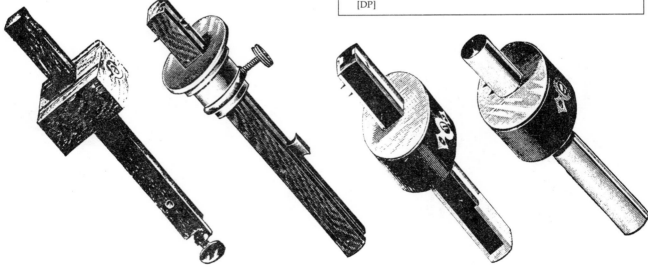

Mortice gauges [Melhuish, 1905]: From left: Rosewood thumbscrew slide; oval brass head combination mortice and marking gauge; improved brass faced oval ebony head; oval ebony head, brass stem.

Craftsman-made

Gauges are an area where many well-conceived and ingenious craftsman-made items are to be found and the collector should pay some attention to these. Common are screwed-stem examples with two-part heads that lock when tightened together. Ingenious examples with cam-operated locks were often made by pattern makers – the simplest have a cam-shaped stem which locks the head when twisted but others have cam locks installed within the head.

Beechwood panel gauge [Melhuish, 1905].

Some types of gauge

Butt gauge: Although metal (**Stanley**) versions were made, many are craftsman-made in wood. They differ from other types of gauge as they work from the face of the rebate.

Combination gauge: Two pins for mortices on one side and a single pin for marking on the other.

Panel gauge: Long gauges with wide stocks giving stability when marking wide boards etc. Mostly craftsman-made, they are often of artistic merit.

Fixed gauges: Usually in the form of a single piece of wood cut to form a number of shoulders with pins inserted at varying distances from the shoulders. Interesting, primitive and effective.

Collecting tips

➤ *Gauges are common so you don't need to buy items that are in anything but good + or fine condition.*

➤ *Brass head gauges have attracted the attention of the fakers – the originals were hollow castings, not turned from the solid.*

Price Guide

◆ Rare brass faced ebony mortice gauge by **Fenton & Marsden,** Reg. Design No. 970 adjusted by screws at each end working on a concentric slide in brass stem. **£374** [DS 20/348]

◆ A highly original mortice gauge. The ebony head and square stem of conventional form but the head is traversed by a thumb turn operating a pinion concealed in the head engaging a brass rack in the stem. **£200** [DP]

◆ Elegant boxwood butt gauge with adjustable rosewood slide. G++ **£15** [DP]

◆ Most attractive small mahogany panel gauge with satinwood strips let into each side of the stem, one piece missing. G++ **£46** [DS 21/671]

◆ Three panel gauges in ebony, rosewood and mahogany. G+ to F **£46** [DS 21/1367]

◆ 24" panel gauge with boxwood head secured with boxwood thumbscrew. **£16** [DP]

◆ Elegantly shaped, panel gauge with brass inlaid rosewood head. G++ **£45** [DP]

◆ Large beech panel gauge with handled head and roller. G **£61** [DS 22/63]

Stanley No. T340 improved butt gauge.

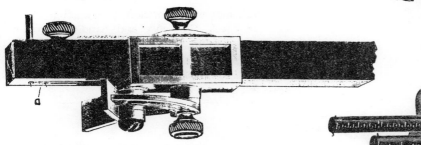

Three American marking gauges [Melhuish, 1905].

Like marking gauges, these were being made commercially by the end of the 18th century.

Cutting gauge: The cutting gauge, which has a small blade instead of a pin, is easier to sharpen than a marking gauge as the blade can be removed. It was probably used as much for marking as for cutting being less likely to deviate than a marking gauge. They were available in stocks of varied types and qualities similar to marking and mortice gauges.

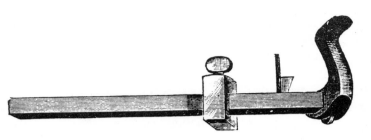

Handled slitting gauge [Melhuish, 1905].

Slitting gauges: Larger versions of the cutting gauge with longer stems, up to 18" long, and a handle near the cutting blade so that pressure can be applied. Although they were commercially made, most seem to be craftsman-made.

Cutting gauges [Melhuish, 1905]. Top: a simple beech type; below, a fine ebony and brass example. There were many qualities between.

Circular cutting gauge: Basically, a bar fitted with a cutter at one end and a rod at the other, the position of which can be adjusted. The rod is entered into a hole in the workpiece whilst the cutter is rotated. Always craftsman-made, this tool was used to cut circular holes in thin wood for items such as barometer faces and wash stands.

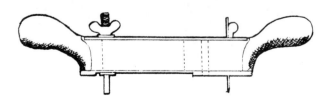

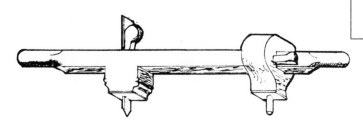

Price Guide

- A rosewood cutting gauge by **Marples** with boxwood thumb screw. **£18** [DS 24/43]
- A **Syers Regd.** cutting gauge with mahogany stem and German silver top. **£11** [DS 24/1378]
- A fine square head brass faced ebony cutting gauge with extended top brass. **£45** [DP]
- A rosewood and brass cutting gauge by **G. Preston**, Toolmaker, Sheffield G++ **£24** [DS23/571]
- A brass faced ebony long stem slitting gauge by **Moulson Bros.** with end screw fixing for cutter. **£32** [DP]
- An unusual oval head rosewood combined mortice and cutting gauge marked Patent No. 10786. The stem is round with a brass insert. The mortice pin is moved with a thumb screw from one end and the cutting iron fixed with a screw in the other end of the stem. **£60** [DP]
- A craftsman-made beech circle cutter with integral turned handles and large brass wing nut. **£24** [DP]
- An unusual combination rosewood and brass slitting gauge by **Marples** with hole for pencil, the head fixed with a boxwood thumb screw. **£36** [DP]
- A craftsman-made handled slitting gauge, the stem in beech, the fence and handle in well finished rosewood. **£48** [DP]
- A craftsman-made mahogany circle cutter with integral handles, heavily plated in brass and with brass wing nut. G++ **£65** [DP]

Two craftsman-made circular cutting gauges.

Planes

WOODEN PLANES

Not until the middle of the 19th century did metal planes appear in the joiner's tool box in any numbers. Even then, the wooden plane retained a large share of the market; indeed it was well into the 20th century before the wooden bench plane was overtaken as the tradesmen's choice. The wooden moulding plane was never seriously challenged by metal planes – after a long decline it eventually succumbed to the woodworking machine.

By the end of the 17th century, when planemaking on a commercial basis started in London, the makers had come to acknowledge that beechwood was the best available wood – and throughout the next 250 years, planemakers in Britain never used anything else, except, that is, boxwood for a very small percentage of premium products.

*Drawings of toe ends. Planes by: 1. **Thomas Granford** (1687–1713): Large round, 10½" long, 3½" deep. Note the steep drop to the shoulder and the large chamfers. 2. **Robert Wooding** (1706–39): A large ogee to work on ¾" stock, 10¼" long. The shoulder has noticeably little drop. 3. **William Madox** (1748–75): ¾" ogee, 9⅞" long. Generous flat chamfers are a Madox characteristic. This plane has a noticeably sharp arris at the shoulder.*

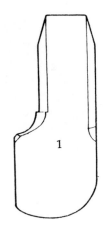 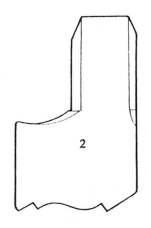 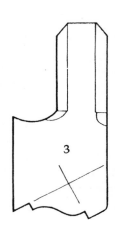

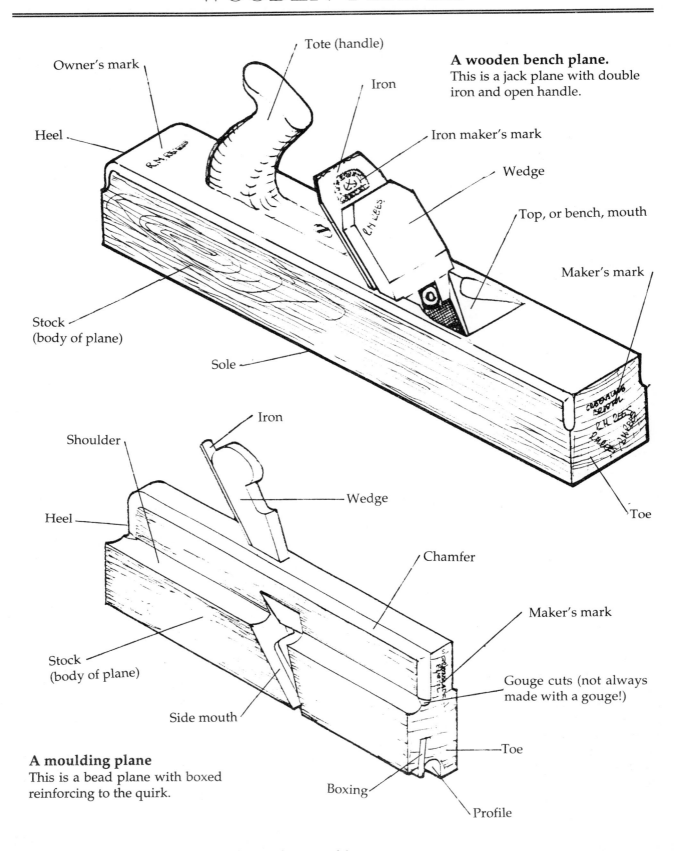

Tote (handle)

Owner's mark

Iron

A wooden bench plane.
This is a jack plane with double iron and open handle.

Heel

Iron maker's mark

Wedge

Top, or bench, mouth

Maker's mark

Stock
(body of plane)

Sole

Toe

Iron

Shoulder

Wedge

Heel

Chamfer

Maker's mark

Stock
(body of plane)

Gouge cuts (not always made with a gouge!)

Side mouth

Toe

A moulding plane
This is a bead plane with boxed reinforcing to the quirk.

Boxing

Profile

Drawings of typical 19th century planes showing the names of the parts.

Tracking 'em down

The history of British planemaking has, over recent years, attracted much attention from enthusiasts and it is now hard to believe that 30 years ago it was not possible to differentiate between planes that were 30 or 230 years old. The results of these researches have been published in successive editions of W.L. Goodman's *British Planemakers from 1700* (see p. 18). The third edition (cited in this section as BPMIII) contains the names of 1,650 planemakers and vendors most of whom were makers of wooden planes. Their locations, dates of operation and biographical details, as known, are recorded.

The beginnings

The earliest identified *commercial* planemaker is **Thomas Granford** who was working in the City of London from 1687 to 1713. Obviously there were men who made planes before this date but Granford would seem to be the first person to make a significant business from making and selling planes, and, perhaps more significantly, he marked his products. Most succeeding makers did the same so it is possible to identify the men and the planes they made.

Collecting tips

> ➤ *Does the style of wedge match the date of the plane? Many wedges have been broken and replaced over the years.*
> ➤ *If the wedge isn't fitted "tighter than a gnat's dick" it may well be a replacement.*
> ➤ *Neither colour nor condition is any sort of guide to age.*

By the early years of the 18th century **Robert Wooding,** who had been apprenticed to Granford, had taken over his master's business and had himself taken a number of apprentices who, in due course, left and established businesses of their own. By the middle of the century planemaking had become an identifiable and separate trade, with some substantial makers such as **William Madox** becoming prominent. By the 1750s or shortly thereafter, planemakers were established in most of the major provincial cities, with the rapidly industrialising town of Birmingham being a notable centre for the trade.

Price Guide

- A large hollow in exceptionally little used condition by **Thomas Granford** (1687–1713). The plane is 10¼" long with a "Granfurdeus" mark on the side of the stock. The iron is marked **Green** (ZB). **£2750** [AP]

- A 9⁹⁄₁₆" No. 8 hollow by **Thomas Granford** (1687–1713) which does not appear to have been shot. The oval mark on the side of the plane is obvious, the name is not clear but the words "This" and "Make" can be discerned, with original wedge and iron by **Hildik** (sic). Good condition for its 300 years. **£500** [DS 22/639]

- A very rare 10⅜" grooving plane (part of a tongue and groove pair) by **John Davenport** (c.1680). Snecked iron by **Aaron Hildick** and replaced wedge. G **£462** [DS 25/265]

- A 10¹⁄₁₆" x 1¾" bead and quirk by **Robert Wooding** (1706–39) (mark G). **£550** [DS 24/147]

- A 9⅞" reverse ogee and astragal, 2" wide by **Robert Wooding** (1706–39), probably shot at heel, minor worm. G **£352** [DS 23/98]

- A 10⅛" hollow by **Robert Wooding** (1706–39), replaced wedge. G **£330** [DS 22/605]

- A fine 10" No. 12 hollow by **Cogdell** (1730–52) (mark G++) **£82** [DS 25/1206]

- A 10¼" hollow by **S. Holbeck** (1730–70), the stock recut and shallow, possibly with replaced wedge, bearing the rare SAMVELL HOLBECK incuse mark. G **£440** [DS 25/264]

- A 9¾" ogee by **John Rogers** (1734–65). **£45** [DP]

- A rare 14" panel raising plane by **Jennion** (1738–78) (mark P) with offset handle, wood depth stop and **John White** cut iron. G **£209** [DS 25/1124]

- A 10" m.p. by **T. Phillipson** (1740–75) (mark G). **£121** [DS 25/252]

- A 10" long hollow by **Robert Bloxham** (1746–78) with rounded toe. **£55** [DP]

- A top mouthed handled 13" x 2" m.p. by **Madox** (1748–75) with round topped iron by **John Green** (mark G++) filled worm hole. **£352** [DS 24/108]

- A rare half set of 18 hollows and rounds, Nos. 1–17 by **Madox** (1748–75). The No. 17 is not stamped Madox but is undoubtedly the original plane. G++ **£682** [DS 24/126]

- A 2⅝" wide ogee with fillet by **Madox** (1748–75). **£55** [DP]

- A beech plough by **Madox** (1748–75) replaced wedge, damaged stems. G– **£50** [DS 21/1254]

- A fine 2⅜" wide ogee by **Small** (crown over) (1749–75). **£85** [DP]

Development

Length: Granford and Wooding made moulding planes 10½" long but thereafter the length gradually reduced until by around 1760 some makers were making to a length of 9½". By 1775 virtually all makers had adopted this length which remained standard as long as moulding planes were made. The length of a moulding plane, for planes made before 1775, is therefor a *guide* to age but there is no *direct* correlation between length and date.

Chamfers: Between 1700 and 1770 the chamfers put on both moulding and bench planes were, with few exceptions, flat. Thereafter there was a move to oval or round section chamfers. The earlier the plane, the larger the chamfers are likely to be, those of 1700 being over ½" wide whilst those of 1770 being, typically, ⁵⁄₁₆". The chamfers on a plane were however somewhat proportional to the size of the stock so again this characteristic needs to be interpreted with some care.

Wedge: The shape of moulding plane wedges evolved from the full round head and long thin neck of 1700 to the oval-headed, thicker necked style that was the standard from 1820 onwards.

A flourishing trade

By 1830 there were 25 planemaking firms working in London and 70 in the provinces. From then on, improving transport systems allowed the more efficient makers, wherever they were situated, to compete over wider and wider areas. Although the

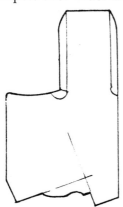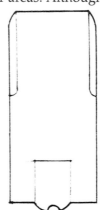

*Drawings of the toe ends of two late 19th century moulding planes. Left: a lamb's tongue sash plane (one of a pair) by **Hields, Nottingham** (1830–81). Right: a fully boxed centre bead with bevelled (coved) quirks marked **Parry, 358 Old St.** (1872–83). Note that, unlike 18th century moulding planes, the chamfers are relatively small and rounded.*

number of firms started to decline, it is likely that the production of planes continued to grow until around 1880. By this time woodworking machinery was beginning to have a significant effect on the production of mouldings.

After 1850 the strong grew stronger whilst the weak disappeared. Major survivors into the 20th century were **Chas. Nurse** and **John Moseley** in London, **Ed. Preston** and **Atkin & Son** In Birmingham, **Gleave** in Manchester, the **Varvill** business in York and **Alex Mathieson** in Glasgow. Planes made by all of these businesses are common and most people assume that the planes always date from the later periods of operation. This is often not correct.

The social convulsions of the First World War speeded the inevitable decline in the wooden

Collecting tips

➤ *The style of a plane, i.e. the length, type of chamfers, style of gouge cut and shape of wedge, is useful in determining the age **but** it is only an indicator and there are many exceptions to the rules.*

➤ *Has the plane you are considering been shortened? Early planes that were longer than the standard length of 9½" have often been reduced so that they can be stored neatly with later planes. If they have only been cut off at the heel, the makers' name will still be present but the position of the mouth may reveal the alteration.*

➤ *If the name stamped on the toe is not recorded in BPMIII, the chances are it is an owner's name. But you never know your luck, as once upon a time all makers were unrecorded!*

➤ *Worm tunnels, rather neat round holes, showing on the surface are almost always an indication that the plane has been altered – the only exception is when planes have been stored tightly packed together.*

➤ *Virtually all moulding planes were of a standard height. If you plane is less than 3⅜" high, it has been reduced. Scribing lines that do not match the shape of the mould are also a sure indication of an alteration.*

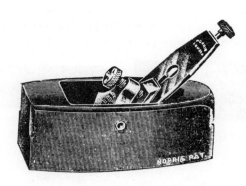

*Norris No. A70 smoothing plane [Buck & Ryan, 1930]. The reluctance of joiners to give up their wooden planes was recognised by **Stanley** and **Norris**, both of whom made wooden stocked planes with metal fittings and screw adjustable irons.*

planemaking trade whilst the upheavals of the Second War were terminal. Only at **Wm. Marples & Sons** in Sheffield did the once important trade limp on until a final demise about 1960.

Boxing

At weak points, such as the quirks in the Grecian forms of mouldings, and also in areas where a plane such as a moving fillister might be subject to heavy wear, the planemaker used boxwood for reinforcement, known as boxing.

The grain of the boxwood is set at an angle of 45°, which is approximately parallel with the iron, and at this angle it is particularly resistant to wear.

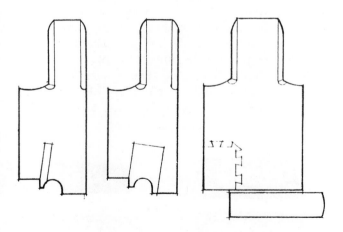

Boxing varies from the simple to the complex. From the left: slip boxing, full boxing and dovetailed boxing (on a moving fillister). The amount and complexity of boxing is a good indication of the quality of the plane.

Simple boxing is merely inserted into the appropriate groove in the stock of the plane but there are many small improvements in the form of keying and dovetailing that were utilised to keep the boxwood from moving. This work required skills of the highest order and it is surprising that it is still possible for only a few pounds to purchase a plane on which such care was lavished.

The best quality moulding planes for smaller sections, such as astragal and hollow sash bars or picture frames, were sometimes made "fully boxed", that is, the working area of the plane was reinforced for its whole width with a larger than normal piece of boxwood, often secured with dovetails.

The soles of both moving and sash fillisters of the better qualities are often fine examples of the English planemakers' skill in boxing.

Price Guide

• A moving fillister by **T. Darbey** (1765–85) with continuous lignum vitae boxing. **£95** [DP]

• A pair of hollows and rounds by **Mutter** (1766–99). **£30** [DP]

• A 24" try plane by **John Green** (York) (1768–1808) with iron by **Marsh**. Very good condition for age. **£160** [DP]

• A pair of No. 10 hollows and rounds by **Gabriel** (1770–1822) **£16** [AP]

• A fully boxed triple reed (table plane) by **John Lund** (1805–31). **£30** [DP]

• A comprehensive set of wooden planes by **Griffiths**, Norwich, (1803–1958) comprising a half set of hollows and rounds, snipe bills, side rounds, a moving fillister, three dado planes and six other moulding planes. **£360** [DP]

• A fully boxed triple reed by **B. Bown** (1718–1816). G **£22** [DS 25/1158]

• A fine triple boxed double reed by **W. Dibb** (York) (1800–60). G++ **£35** [DS 25/1190]

• A fully boxed dovetailed ⅝" bead 20th C. by **Varvill**. **£10** [DP]

• A set of five fully or double boxed, dovetailed and slipped torus beads by **C. Nurse & Co.** (1841–1937). **£28** [DP]

• A pair of ogee sash planes, marked No.1 and 2 on heel, by **Mathieson** (1822–1930s). **£18** [DP]

• A 20" closed handle badger plane by **Greenslade**, Bristol (1828–1937), the edge shoulder boxed. **£18** [DP]

London

Robert Wooding, 1706–39. Apprenticed to the father of British planemaking, Thomas Granford, Wooding, in turn, was master to several later well-known planemakers, William Cogdell, John Jennion, Thomas Phillipson and Robert Fitkin.

William Madox, 1748–75. A prolific planemaker, he worked in Westminster.

George Mutter, 1766–99. Starting in Westminster, he moved to Covent Garden, before returning to Westminster when he took over the **Madox** business.

Christopher Gabriel (& Sons), 1770–1822. Gabriel was not only a planemaker but made or sold the whole range of tools available to the 18th century carpenter. Due to the unique survival of a ledger, probably more is known about Gabriel than any other 18th century planemaker.

John Moseley (& Sons), 1778–1910. Continued the **Mutter** business and was the pre-eminent planemaker in London throughout the 19th century. His business was in turn taken over in the 1880s by **William Marples** of Sheffield. Planes with the Moseley name were available until the 1930s

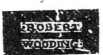

As a general rule, although there are some exceptions, 18th century makers used embossed marks (like the first five above) whilst by about 1820, most makers had changed over to incuse marks (like the last two above). Addresses were added from 1840 onwards but usually indicate manufacture later in the century.

The Buck family, 1826–present. It is important to distinguish the different branches of this family, which traded as three separate entities. Numerous wood and metal planes can be found bearing the Buck names. **Buck & Hickman** still listed wooden planes in their 1964 catalogue.

Charles Nurse, 1841–1937. Started in Maidstone but moved to London in 1887. Numerous planes made by this firm dating from the latter part of the 19th and the early part of the 20th century are to be found. In 1914, they stated that special planes could be made up the same day as ordered.

Birmingham

George Darbey, 1750–93. The earliest planemaker working in Birmingham, he was followed by his son Daniel.

William Moss, 1775–1843. Made not only planes but a range of tools needed by carpenters and joiners. The name continued to be used after the firm was taken over by **Atkin & Sons**.

George Davis, 1821–76. A prolific maker with a considerable export market. His planes are sometimes marked LONDON – more identifiable than Birmingham in the mid-19th century!

Atkin & Sons, 1839–1966. Did not start making planes until they took over William Moss in 1845. They also sold planes with the **Sims**, **Ames**, **Davis**, and **Moss** marks, so take care dating of planes with these marks.

York

John Green, 1768–1808. Perhaps the most prolific of all the 18th century makers. After his death in 1799, the business was continued by his son. In 1808, it was taken over by his nephew and his foreman, who continued until 1829. Planes of every type can be found with this mark.

Michael Varvill (& Sons), 1793–1904. With the demise of the Green business, the Varvills became the biggest planemakers in York, with a countrywide business. The firm was finally taken over by **Greenslade** of Bristol.

Hull

King & Co., 1744–1907. Started by Henry King, the firm was run by several generations of the King family in partnership with others and traded under a variety of names.

Norwich

John Griffiths, 1803–1958. Although situated in the provincial town of Norwich, Griffiths developed a significant business, distributing mainly in the south and east of England. Around the 1880s, they were selling about 6,000 planes a year but by the 1930s, the firm was moribund.

Manchester

Joseph Gleave (& Son), 1832–present. Rapidly became the principal Manchester planemakers but later also developed as a retail tool seller. Planemaking had ceased by 1926, although the firm still trades today.

Bristol

William Greenslade & Co., 1822–1937. One of the last makers of wooden planes, their output in the 20th century was high.

George Gardner, 1843–1939. Before he set up on his own, he worked for Holbrook as he marked his planes "Late Holbrook" even though that maker was still in business. From 1880 the firm was run by his son, Thomas.

Glasgow

Alexander Mathieson, 1822-1966. The most prolific of all British planemakers, he started in Glasgow. There were, at different times, branches in Edinburgh, Dundee and Liverpool. Planemaking finished with World War II

Perth

David Malloch, 1850–1932. For a company that only made wooden planes, the premises they occupied were extensive – a good indication of the number of planes being made. Taken over by **Alex. Mathieson** in 1913, the name continued in use for another 20 years.

About 1,650 makers and sellers of planes have now been identified. These are listed, together with representations of their marks, in *British Planemakers from 1700*. See *Further Reading* p.18 and *Famous Names*, pp.35–41.

Wooden bench planes are what most people would just call planes, that is, planes that are intended to produce a flat surface. The adjective "bench" hardly adds clarity to the description as there aren't many planes that are not used on a bench! A defining feature is that bench planes have bench or top mouths discharging the shavings through an aperture in the top of the stock.

Bench planes, little and large

With time the names of the planes have altered somewhat but the following are today's usage.

Jointer plane: The longest of the bench planes, from 28" to 30" or occasionally even longer. Originally intended to plane very straight edges onto boards that were to be joined edge to edge.

Try plane: A general workhorse plane, 22" to 26" long, with a 2½" wide iron. The **Ed. Preston** catalogue of 1901 lists trying and *long* trying planes.

Jack plane: Jack planes can be any length from about 12" up to 20" but are typically about 16" long with a 2¼" iron. There are some variants, such as technical jack planes (the handle is set lower into a cut away section at the rear) and gent's jack planes (small – maybe 12" long with a 2" iron).

Whilst jointers and trys have closed handles, jacks have open handles, although closed were available as an alternative from some makers.

Smoothing plane: Typically 9" long with a 2⅛" iron, the stocks mostly have curved sides and are frequently referred to as coffin or boat planes. Again there are a number of variants – some examples are straight sided and an improved and adjustable mouth made with a cast iron front shoe is quite common. User adaptations in the form of added handles are also to be found.

Smoothing plane with iron front [Harding, 1903].

Collecting tips

➤ *Late 19th and 20th century bench planes are mainly bought for use and must therefore be in good condition.*

➤ *Early planes can be speedily identified by their offset handle. By the 1830s the handle had become central.*

➤ *18th century bench planes are surprisingly rare.*

➤ *Just a few smoothing planes were made in boxwood.*

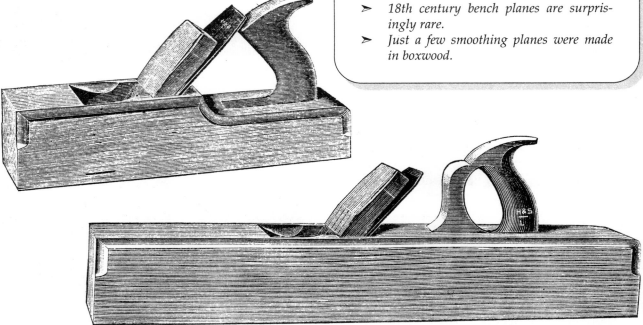

Technical jack plane and trying plane [Harding, 1903].

Thumb plane: Small smoothing-type planes. Commercially made from 3½" upwards but craftsman-made can be smaller. Sometimes also known as block planes.

Thumb plane [Harding, 1903]. Most manufacturers made these in both boxwood and beechwood.

Changing names

The term "block plane" is a good illustration of the changes in terminology that have taken place. Originally a block plane was a wooden *[strike] block* plane, parallel-sided, around 14" long, used to shoot mitres.

With the development of specialist mitre planes, it lost this use but hung on as a sort of small shooting plane. In the late 19th century, the small metal planes around 5" long acquired the name and, in turn, it has been transferred back to similar sized wooden planes.

Our advice is not to get too bogged down with the exact name of any tool; far more important is to understand its function, when it was made and something of the trade in which it was used.

Mitre – shooting – compass

These are bench planes but each has a specific use.
Mitre plane: To cut mitres or shoot end grain, planes need to have a tight mouth. Around 1780 unhandled wooden planes with a boxwood end-grain plug, immediately in front of the mouth, were developed. This is the point of maximum wear in any plane – in this design the plug could be knocked down and trued up as wear took place, so keeping a fine mouth. The development of the metal mitre plane, and these were available in 1800, made such complexities obsolete. Wooden mitre planes are quite rare and are under-regarded by today's collectors.

Price Guide

Later bench planes are bought mainly for use and, as bench planes are by no means rare, it is difficult to sell any that are not in at least G+ condition. Late 19th/20th C. bench planes in good usable condition: jointers **£20–30**; try planes **£10–15**; jack planes **£5–10**; coffin smoothing planes **£4–8**; with iron shoe **£6–12**; thumb/block planes, 4" and smaller, beech **£15–25**, boxwood **£30–40**; toothing planes **£10–16**.

◆ A 30" beech jointer by **John Green**, York, (mark G–) (1768–1808) with wide flat chamfers and round topped iron by **Mitchell**. G **£130** [DS 22/617]

◆ An early 30" jointer with offset handle by **Wm. Moss** (mark G). **£60** [DS 21/1297]

◆ A fine and rare 30" beech jointer by **Madox** with very offset handle and round topped iron by **John Green**. **£330** [DS 20/1009]

◆ An 18th C. 22" try plane by **I. Cox** (mark G+) with chip to offset handle, re-mouthed with boxwood, some worm. G **£45** [DS 24/137]

◆ A 22" try plane with offset handle by **John Green**, York, (1768–1808) **£80** [DP]

◆ An almost unused **Marples** 24" try plane and a smoothing plane, both with original trade labels and from the same tool kit. F **£45** [DP]

◆ A gent's jack plane, 14" x 2¼", in solid walnut. **£14** [DS 24/1138]

◆ A boxwood smoothing plane by **T. Turner** with scrolled wedge. G **£39** [DP]

◆ A fine beech mitre plane, 11½" x 3¼", by **John Green**, York, (mark G) with boxwood mouth adjustment and scrolled wedge. **£462** [DS 24/129]

◆ A beech mitre plane, 13" x 3½", by **Hathersich** with boxwood mouth adjustment and scrolled wedge. **£126** [DS 24/1119]

Shooting (shuteing) plane: Wooden versions are large bodied, parallel sided planes, usually with a

Shooting plane [Harding, 1903].

tote (handle) fixed to the side. They were used on the side for shooting end grain or truing up edges.

Compass plane: If the surface that is to be planed is curved in its length, it is "compassed" and a plane intended to work on such a curve is a compass plane. There are two types. Adjustable versions are to be found more frequently than the fixed type which seem mainly to have been used by the rougher trades – such as the boat builder and wheelwright.

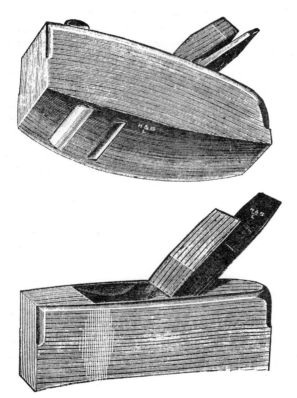

Screw stop adjustable compass plane and a concave compass plane [Harding, 1903]. Concave planes are much rarer than convex planes.

Toothing plane: With a vertical or nearly vertical iron in a stock of smoothing plane type, toothing planes are distinctive. The irons are cut with closely spaced serrations forming fine teeth when sharpened with a bevel. The planes are used to smooth woods with difficult grain and to provide a key when veneering onto very smooth grounds.

Irons

Around 1760, and this was a British invention, it was discovered that if a second or cap iron was fitted above the iron, the performance of a plane was enhanced.

Price Guide

• A beech mitre plane, 11" x 3¼", by **S. Lunt,** Liverpool, with scrolled wedge and boxwood mouth adjustment. G++ **£132** [DS 23/775]

• A late 18th C. compass smoothing plane by **S. Tomkinson** (mark G) with round topped slotted iron. G– **£35** [DS 22/631]

• A beech compass smoothing plane by **Nelson,** London with adjustable foot worked with a brass thumb screw. G++ **£32** [DP]

• A beech compass smoothing plane with boxwood toe adjustment by **Gleave,** Manchester. G++ **£26** [DP]

• A rare concave sole, beech smoothing plane, 6" long by **Malloch,** Perth. **£22** [DP]

• An adjustable compass smoothing plane by **Cowell & Chapman** with boxwood mouth plug and toe adjuster. G **£44** [DS 24/1095]

• A fine boxwood round both ways compass plane, 5" long. G++ **£40** [DP]

• A little used toothing plane by **Crow,** Canterbury, (mark G++) with **James Cam** iron. G++ **£28** [DS 24/1140]

• A very little used beech toothing plane by **Marples** with spare iron in original box. F **£42** [DS 21/534]

• A 26" shooting plane with hooked side tote by **Varvill & Sons,** York. **£42** [DP]

• A miniature boxwood smoothing plane, 3¼" x 1¼", by **Musgrave.** **£60** [DP]

• A pair of small beech flat and compass smoothing planes, 5" x 2⅛². **£42** [DP]

• A pair of miniature boxwood flat and compass smoothing planes, 1⅝" x ⅝". G++ **£170** [DS 22/1485]

• A miniature boxwood smoothing plane, 2¼" x 1⅛", iron by **Ward.** **£65** [DP]

Toothing plane [Harding, 1903].

Price Guide

• A rare small beech smoothing plane, 4" x 1½" by **Norris**, York Road, Lambeth, mark faint. **£94** [DS 24/1720]

• A 13" beech strike block plane by **Armour**. Iron by **Ward**. **£25** [DP]

• A **Norris** No. A72 adjustable beech jointer, handle spur shortened, 30% original early Norris iron remains. **£148** [DS 24/1718]

• A **Silcock** Patent 22" beech try plane (c.1845). Iron secured to body of plane with a large screw. Handle spur repaired. **£242** [DS 25/511]

• A closed handle jack plane by **Nurse**, fitted with the Nurse Patent Regulator, giving set and lateral adjustment. F **£200** [DP]

• A 12" German (Bismark) pattern jack plane, English made. **£22** [DP]

• A 17" jack plane by **Kimberley** with patent cast iron insert. **£40** [DP]

• A **Norris** No.A71 adjustable beech jack plane, 70% original iron remains. **£132** [DS 24/1719]

• An unusual "Elbydee" patent beech smoothing plane by **Gleave** with shaped hand grips to rear of iron and at mouth. G **£77** [DS 24/734]

• A beech smoothing plane by **Kimberley** with patent cast iron insert, somewhat displaced due to shrinkage but otherwise G++ **£32** [DP]

• A beech smoothing plane with double dove-tailed boxed sole by **Kimberley**. **£55** [DS 21/478]

• A beech smoothing plane by **Nurse** fitted with the Nurse Patent Regulator (adjuster). F **£150** [DP]

• An 8" smoothing plane by **Gleave**, the fine mouth closed by a boxwood plug similar to a mitre plane. **£45** [DP]

• A **Stanley** No. 24 composite wood/metal smoothing plane. **£36** [DP]

• A **Stanley** No. 32 composite wood/metal jointer plane. **£36** [DP]

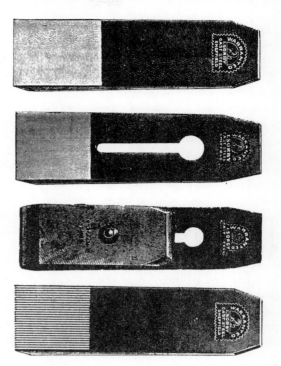

Sorby "Punch" brand plane irons [Turner, Naylor, 1928]. From top: Cast steel uncut iron; cast steel cut iron; cast steel double iron with cap held by late style brass nut; toothing plane iron.

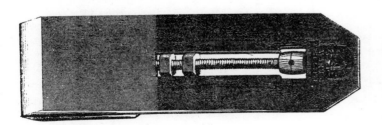

Cast steel long screw double iron [Ward & Payne, 1911]. This type of iron had been obsolete for a century by this date!

The earliest method of holding the cap is the long screw system, where the cap is loose but is prevented from slipping forwards by a screw axial with the iron. Very quickly the simple method of a slotted iron with a large head screw inserted into the cap iron was developed.

Cap irons are, however, too thin to hold a good thread and it was necessary to improve the grip. In the earliest examples the hole in the cap iron is punched upward, so distorting the metal near the hole and thus getting more threads. By the 1830s it was normal to weld a rectangular piece to the cap but, by the 1860s, the normal method for quality products was to rivet in a brass cone-shaped piece.

Unfortunately, like many tool developments when an improvement was made, the old form was continued on the cheaper lines, so the construction of the cap nut is only a guide to age.

The cap iron provided significant improvement in performance on hardworking planes and speedily became standard on try planes and jointers. Eventually most jack and smoothing planes were fitted with them but the smaller the plane the less likely was this provision.

The well-found joiner of the 19th century had a generous-sized, fitted chest to contain his tools. The planes were stored in the bottom level of the chest, there normally being space at the rear for a double row of moulding planes. As the chests were usually about 3ft. wide, this allowed about 6ft. run, accommodating around 45 moulding planes.

For the well-found journeyman joiner (an employed man), the contents were likely to have been a half set of hollows and rounds (seven or eight pairs); a pair each of snipe bills, side rounds, side snipes and side rebates; three or four rebate planes; three or four dado planes; a set of seven bead planes; three or four ovolo planes and a couple of small ogees.

With the so-called simple planes, that is the hollows and rounds, snipe bills etc., he would have been able to make almost any moulding. With the addition of the other small moulding planes (the beads, ovolos and the ogees) he would have been able to make the same mouldings, but quicker.

Hollows and rounds

The 17th and early 18th century writers all mention hollows and rounds in terms that suggest joiners had numbers of these planes, and this is confirmed by several matching planes, originally part of a set, by **Thomas Granford** that have recently been reported.

Planes made after the middle of the 18th century are numbered, on the heel. In most cases the sets numbered from 1 to 16 though 18s, 20s and very occasionally even larger numbers were made. It is clear from surviving sets that the vast majority of purchasers bought only the half sets, seven or eight pairs numbered odd or even.

Collecting tips

> Most purchasers did not buy all the snipe bills, side snipes, side rounds and side rebates so matching groups of all four pairs are rare.

> Forkstaff and the other top mouthed hollow planes are often altered bench planes resulting in a mouth that will not be the same width across the iron.

> Odd numbered sets of hollows and rounds are less common than even number sets but the actual size may well be the same as there was no standardisation of size between makers. It is therefore illogical to pay more for odd sizes.

Snipes and the like

Accompanying the hollows and rounds were the snipe bills, side snipes, side rounds and side rebate planes. The snipe bill and the side snipe, which were always supplied in handed pairs, were used to cut or clean up quirks and the side rounds, which surprisingly only seem to have been supplied in one size, will reach the hollow sections of a mould.

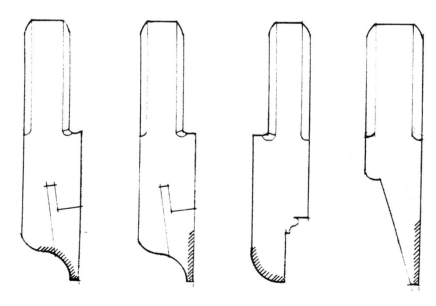

From the left: a snipe bill, a side snipe (both boxed), a side round and a side rebate. The position and extent of the cutting edge of these planes is shown hatched. All these planes were sold in handed pairs, only one of which is shown here. Single planes are of little appeal to collectors unless of mid-18th century date or earlier. These planes all bear the mark **Birch, 77 Weedington Rd., Kentish Town.** *The firm operated from 1875 to 1922.*

The snipe bill is an ancient form of plane and is known to have been made since the 16th century.

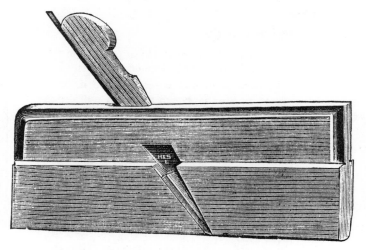

A round plane [Harding, 1903]. Planes could be had either square or skew mouthed and with irons at higher angles of 60° or more, if the planes were intended for use on hardwood.

A half set of hollows and rounds as illustrated in the **Ward & Payne** *catalogue, 1911. This is a long set of nine pairs. Many sets are shorter, seven or eight pairs. Full sets, i.e. sixteen or more pairs, including odd and even numbers, are rare.*

Price Guide

Hollows and rounds dating from post 1850: pairs of Nos. 14 and below **£6–9**; pairs of Nos. 16 & 18 **£9–12**; complete and matching half set (8 no. pairs) by single maker **£80–120**; set with one or two planes made in and not matching, less 25%.

◆ A half set of eight pairs of H&Rs, Nos. 2–16 by **Frogatt** (1760–90), the irons at cabinet pitch. **£280** [DP]

◆ A half set of nine pairs of H&Rs, Nos. 2–18 by **Mathieson. £160** [DS 20/976]

◆ A half set of nine pairs of skew mouthed H&Rs, Nos. 2–18 by **Hields**, Nottingham. G++ **£286** [DS 24/1088]

◆ A half set of nine pairs of skew mouthed H&Rs by **Greenslade**. F **£125** [DP]

◆ A set of nine "short" rounds by **Varvill**. The beech stocks are only 7" long. Several similar sets, some with hollows, are known but the intended use is a mystery. **£95** [AP]

◆ A fine set of 24 planes by **Varvill & Son** including a set of skew mouthed H&Rs and pairs of snipe bills, side rabbets and side rounds, all with same owner's mark, one plane with minor damage to toe otherwise G++. **£200** [DS 21/1486]

◆ A pair of small H&Rs by **Phillipson** (1740–75). **£40** [DP]

◆ A pair of No. 14 H&Rs by **Bloxham** (1746–78) (mark G–), with rounded tops and ends. G **£90** [DS 24/290]

◆ A pair of 9⅜" No. 2 H&Rs by **Wm. Moss** (mark G–). G **£18** [DS 24/144]

◆ A pair of No. 18 H&Rs by **Mutter. £35** [DP]

◆ A pair of H&Rs, marked **H & D** (Holtzapffel & Deyerlain). **£25** [DP]

◆ A pair of (miniature) No. 10 H&Rs, 3½" long by **Moseley**, beech with boxwood wedges. **£48** [DP]

◆ A rare 9¾" hollow by **John Grabham** (mark G) with no chamfers, with **Robert Moore** iron. G **£231** [DS 24/291]

◆ A pair of snipe bills by **Hambleton** (1823–34) (mark G+), G **£33** [DS 24/117]

◆ A pair of boxed snipe bills, a pair of boxed side rebate planes and a pair of side rounds, all by **Moseley. £65** [DP]

◆ A pair of boxed snipe bills and a pair of side rounds (unboxed) by **Griffiths**, Norwich together since new. G++ **£52** [DP]

◆ An unusual combined left and right snipe bill by **Arthington. £65** [DS 21/468]

◆ A pair of side snipes by **Stothert**, Bath (1784–1841). **£28** [DP]

◆ A pair of side rounds by **W. Dibb** (1800–43). G++ **£18** [DS 22/336]

◆ A 7" forkstaff plane by **Fitkin** (1750–78) (mark P) with iron by **James Cam**. G **£66** [DS 24/297]

◆ A small (6") forkstaff plane by **Holbrook. £26** [DP]

◆ A hollow of try plane format, 24" x 3½", by **Buck**, Edgware road (believed to have been used for planing columns). **£120** [DP]

◆ A spar plane by **Lund**, London, with round topped iron by **Greave. £22** {DS 21/1455]

Hollows – larger and larger

If you want to make something long and round don't use a lathe – use a plane. Planes to make many round products from fishing rods to columns are to be found. These planes pass under a variety of names but they are all **hollow** planes – but, unlike their humble relations the *ordinary* hollows and rounds, most have top mouths.

Forkstaff plane: Usually of smoothing plane size, either straight or curved side. Used to make tool handles, shafts for horse drawn vehicles, etc.

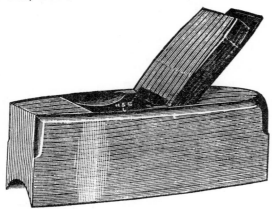

Forkstaff plane [Harding, 1903]. These planes are also called forkshaft, forkstail or even rounding plane.

Spar plane: Not really distinguishable from a forkstaff plane except that it was used for marine trades making spars, oars and the like. Some have boat uppers and parallel lower sides.

Spar or mast plane [Mathieson, c.1900].

Handrail capping plane: The use is clear – to shape the top of a handrail. For this reason the plane has a lightly incurved sole. The planes that we have seen that fit into this category have all been straight sided. Again a plane that one can never be totally certain as to identification, with context being important.

Price Guide

♦ A 14" x 2½" gutter plane of jack plane format (not an adaptation) by **Greenslade**. **£35** [DP]

♦ An almost matching set of three fishing rod planes by **Stokoe**, bench mouthed with irons set at 80°. **£140** [DP]

♦ A patternmakers' gunmetal stocked, interchangeable sole plane. Mahogany stuffing with extended handle at rear. Nine round soles and irons, each marked with a radius. **£60** [DP]

♦ A rare set of commercially made patternmakers' round planes by **Moseley**. The stocks are top mouthed, only 1⅝" high and marked with radii. **£120** [DP]

♦ A patternmakers' interchangeable sole plane by **Kimberley**. The plane is 14" long of jack plane form with six round bases and matching double irons. **£100** [DP]

Fishing rod plane: For planing along the length of fishing rods – the sole is deeply curved, giving most or all of a half circle. The iron is set at a high angle. The rod maker would have had three or four planes of ascending size. (See also *Rounders*, p. 137.)

Billiard cue plane: In almost all respects similar to fishing rod planes. The only maker to have listed planes for this purpose is **Joseph Gleave** of Manchester.

For **nosing planes** and **cock bead planes** see *Beads and Reeds*, pp.122–123.

Collecting tips

➤ *Makers' mark or owners' mark? Planemakers usually marked their planes high up on the front. If they did not, this left a nice space for the owners' mark. Most joiners had their own name stamps as they could not claim from their union or friendly society for lost tools if unnamed. Confusions therefore arise – collectors want planes by identified makers but some planes are marked only with owners' names in the position where the makers' name might be expected.*

➤ *If the mark is not high up on the toe, it is probably an owners' mark.*

➤ *If a mark has been struck on the plane more than once, it is certainly an owners' mark.*

Pricing your moulding planes

This is a not too serious system of calculation to estimate the value of a moulding plane.

We start with the premise that **"All moulding planes are worth £4."**

Add to £4 in sequence the percentage increases or decreases that, in your judgement, are applicable. You may choose as many categories as are appropriate, even from the same box. You may choose values between those shown if you feel this is more suitable for the plane in question.

This system will be more accurate for planes that are not in any way exceptional. Conversely, the rarer the item, the more likely it is to be wrong. It will not work for planes dating from before 1750.

Examples

You are the proud owner of two small ogee planes by **Varvill** [ZB], circa 1800, G+ condition with ten worm holes.

£4	+ 40% for date	£ 5.60
	± 0% for mark	£ 5.60
	+ 50% for number	£ 8.20
	+ 50% for size of plane	£ 12.30
	− 0·5% for worm	£ 11.70

Say, for the two planes **£ 24.00**

You are the proud owner of a 2½" wide ovolo m.p. by **Shepley**, circa 1800, G+ condition:

£4	+ 40% for date	£ 5.60
	+ 25% for crisp mark	£ 7.00
	+ 300% for width	£ 28.00

So, for this plane **£ 28.00**

Boxing

Boxing of exceptional quality will improve the value of a plane. The quoted prices assume slip boxing where this is normal. If boxing is full, keyed, dovetailed etc. this can add 100% to 300% to the value.

Note: The calculated price is for a single plane; therefore for pairs or sets, don't forget to multiply by the number in the set!

Pricing matrix

Condition

G or below	− 50%
G+	± 0
G++	+ 50%
F	+ 100%

Date

Clearly 20th C.	− 20%
1850–1900	± 0
circa 1825	+ 20%
circa 1800	+ 40%
circa 1775	+ 100%
circa 1750	+ 200%

Maker's mark

No mark	− 30%
Readable (identified) mark	± 0
Crisp (identified mark)*	± 25%
Rare maker	+ 50%
Very rare maker	+ 100%

*applicable to pre-1850 marks only

Number

One only of a pair **		− 25%
Set of beads	3–5 in set	+ 25%
	6–11 in set	+ 50%
Moulding planes	2–4 in set	+ 50%
	5–8 in set	+ 100%
Dado planes	3 or more	+ 100%

** e.g. hollows and rounds, sash planes etc.

Type of plane

A pair of snipe bills, side rnds. etc.	+ 50%
Rare mould shape	+ 100%
Complex mould shape	+ 100%
Dado plane	+ 300%

Size of plane

Moulding plane under 1"		+ 50%
Ditto	1 – 1½"	+ 75%
	1½" – 2"	+ 100%
	2" – 2½"	+ 200%
	2½" – 3"	+ 300%
Moulding plane over 3"		+ 400%
Hollow & round, Nos. 16 –18		+ 30%

Worm

Per worm hole	− 0.05%
50+ worm holes	− 40%
100+ worm holes	− 60%

Mouldings are in architecture what letters are in writing. –By the various dispositions and combinations of mouldings may be made an infinite number of different profiles for all sorts of orders, and compositions, regular and irregular; and yet all the kind of mouldings may be reduced to three; viz. square, round and mixt [sic] i. e. composed of the other two. *Cyclopaedia or an Universal Dictionary of Arts and Sciences, 1743.*

Joinery before the last quarter of the 18th century was based on few simple shapes – the ovolo, the ogee and the bead. Thereafter, the "rediscovery" of the antique architecture of Rome and Athens, which rediscovery was several times repeated, brought new designs for mouldings. By the 1790s multiple reads, cluster beads and the Grecian ogee were common. The Grecian mouldings, although flatter than earlier forms, are visually well defined, as they incorporate a quirk. Using less wood, they were great favourites of the speculative builder.

Moulding planes

The quick way to cut mouldings was to use a plane that produced the required profile without resort to the hollows, rounds and the other planes that can be used in combination to shape a moulding.

For small and simple shaped moulds, this is quite easy and I have worked many planes myself. Even today, this is a quick and cheap way of producing a few feet of moulding.

Larger sizes are altogether a different matter and, from the little used condition of many of the largest (3" and over), I have a suspicion that they were seldom taken from the shelf. Today large and complex moulding planes are eagerly sought by collectors, so perhaps we should not complain. Did the makers receive complaints from the original buyers?

Profiles

Moulding planes to cut the following profiles will be amongst the most frequently found.

Ovolo (Roman): The name suggests egg-shaped but in reality is a quarter-round moulding widely used on framing around panels and commonplace in early Georgian joinery. Planes to cut ovolos will be found dating from 1700 onwards.

Grecian (or quirk) ovolo: Wider and flatter in section than the Roman ovolo, the top edge being defined with a quirk. In use from around 1800 onwards.

Ogee: A favourite moulding of the joiner. Little else is to be found on the (single or double) architraves of the mid-Georgian period.

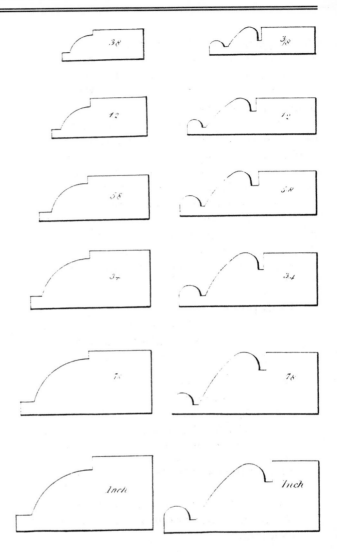

Left: a set of ovolos. Right: a set of quirk ovolo beads [Varvill, 1879]. A deep groove in a moulding is known as a quirk. The naming of mouldings can be quite variable – the right-hand set might also be described as a Grecian ovolo and bead. These illustrations show that the size of planes is specified by the thickness of the wood on which they work.

Collecting tips

➤ *Sets, by which I mean three or more planes cutting the same mould in ascending size, are rarer than one might suppose.*

➤ *Indications of altered profiles are under-depth stocks and multiple scribing lines (setting out lines on the toe) or lines that don't match the profile.*

➤ *Is the mould correct for the period of the plane?*

Common ogee moulding plane [Harding, 1903]

Price Guide

18th century profiled moulding planes

• A 9⅞" reverse ogee and astragal by **Wooding** (1706–39) (mark G), probably shot at the heel, minor worm. G **£352** [DS 23/98]

• A 10" m.p. by **Jennion** (1738–78) cutting a reverse ogee and astragal. **£90** [DP]

• A 10" x 2⅝" ogee by **Phillipson** (1740–75) (mark G++). **£231** [DS 24/148]

• Three m.p.s all with different chamfers and different marks by **Robert Bloxham** (1746–78) (marks G+), 10" (2) and 9¹¹⁄₁₆". **£160** [DS 24/315]

• A small ogee by **Okines**. **£16** [DP]

• A 9⅞" scotia m.p. by **Madox** (1748–75). **£22** [DP]

• A 10⅜" ogee by **Nelson** (1750–83) (mark G++). **£140** [DS 24/136]

• Two wide m.p.s, 2¼" and 2¾", by **Mutter** (1766–99). **£45** [DP]

• A 2¼" m.p. by **Gabriel** (1770–1822) to cut an ovolo. **£60** [DP]

• A set of three ogee planes by **Gabriel**, the widest stock 1¾". G++ **£36** [DP]

• A torus mould by **Lourie**, Edinburgh (1774–70). **£16** [DP]

• A 1⅝" wide moulding plane by **Shepley & Brain** (1798–99) to cut a cove and bead. **£22** [DP]

19th/20th century moulding planes

• A 2½" reverse ogee and bead by **Griffin**, Leicester (1830–62) (mark G+). **£44** [DS 24/120]

• A little used 2⅞" profile m.p. by **Moseley**. G++ **£82** [DS 24/1645]

• A 3¼" wide double boxed Grecian ovolo and astragal by **Ed. Preston** (1825–1924). **£75** [DS 23/1646]

• A 1⅞" wide triple boxed, very complex m.p. by **Tucker** (1833–80). **£45** [DP]

Grecian ogee: The ogee is more deeply S-shaped and finishes at the top with a quirk. This shape was so widely used for architraves that it is often called an architrave mould. Naming mouldings and planes by use rather than shape is common – but not helpful to those not in the know.

Cove and bead: The naming of hollow curves seems to have been particularly variable – cove, cavetto and scotia were applied with little logic. Cove or cove and bead were used for architraves in the mid 18th century.

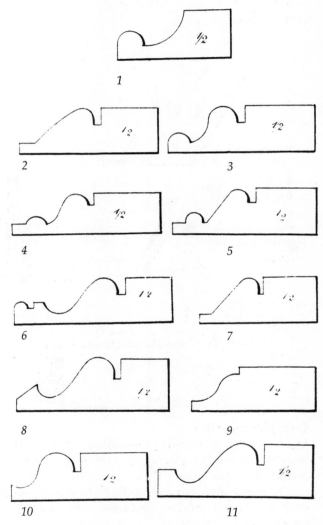

*Moulding plane profiles from **Varvill & Sons** catalogue of 1879, by which time, the Grecian or quirked moulding profiles had almost entirely replaced the simple early forms. 1. Cove and bead; 2. Quirk ovolo; 3. Quirk ogee and bead; 4. Quirk ogee and astragal; 5. Quirk ovolo and astragal; 6. Grecian ogee with bead and square; 7. Quirk ovolo; 8. Grecian ogee with raking square; 9. Ogee 10. Quirk ogee; 11. Grecian ogee with square.*

Sash planes

Sash making was an important part of the joiners' work and many of the profiled moulding planes that are to be found are "sash planes" for cutting the mouldings onto window bars and frames. Apart from the moulding shape, of which there were a good selection, an identifying characteristic of sash planes, from the beginning of the 19th century onwards, is that they were supplied in pairs, invariably marked 1 and 2 on the heel. Sash planes should be accompanied by one or more matching templets which were used for scribing or mitring at the junction of the bars.

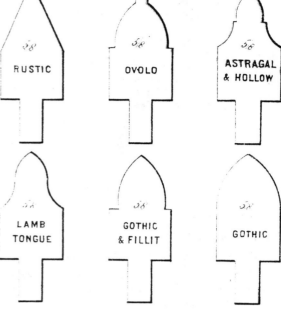

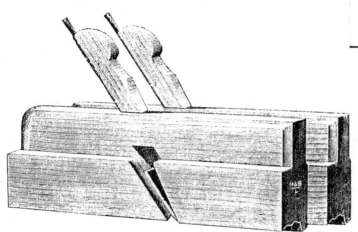

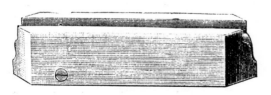

Above: Sash bar profiles [Varvill, 1879]. Left and below: A pair of lamb's tongue sash planes and mitre templets [Harding, 1903]. Other shapes of sash bar were often scribed together and the templets for this form of joint have brass ends.

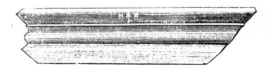

Complex profiles

More complex mouldings combining several of the basic moulding shapes were part of the palette available to designers. These might be cut using two or more of the basic moulding planes but some surprisingly complicated profiles were made as single planes.

Two – three – four irons

The difficulty of sharpening the iron, or more precisely the difficulty of keeping the iron to profile when sharpening, is the reason why many complex moulding planes have two or more irons. The Scots embraced the multi-iron principal wholeheartedly and at an early date. A reported four-iron plane to cut an astragal and hollow sash bar by **John Manners,** Glasgow, 1792–1822, is evidence of this.

Cornice planes

Wide planes to cut the moulding used for cornices. These mouldings are wide but the depth of cut is comparatively light as the moulding is struck onto the face of a plank that is then fixed at 45°, thus appearing to have been made from a much larger piece of wood. They differ from most moulding planes in being of bench plane format – that is with a top mouth. Today the term is used somewhat incorrectly to indicate any large moulding plane of bench format.

In Britain, cornice planes are rare but are more common in America.

Bolection moulding

A bolection moulding is one that stands up above the general adjacent surface. The term, bolection mould, does not describe any specific shape, only its general format and use. The commonest place to find such mouldings is on panelled doors of the late 19th century. But not all upstanding mouldings are bolection moulds – for example, a dado rail moulding is not.

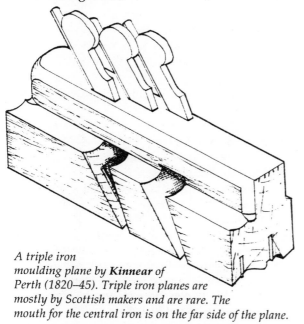

*A triple iron moulding plane by **Kinnear** of Perth (1820–45). Triple iron planes are mostly by Scottish makers and are rare. The mouth for the central iron is on the far side of the plane.*

Price Guide

19th/20th century moulding planes, cont.

♦ A 2¾" wide m.p. by **Moon** (1831–51) to cut a quirked architrave mould. **£32** [DP]

♦ A 2¼" wide complex m.p. by **Mathieson**. **£40** [DS 24/1762]

♦ A 3⅝" wide m.p. by **Elsworth** (1845–94) to cut a shallow reverse ogee suitable for part of a cornice. **£75** [DP]

Sash planes

♦ A pair of ⅝" ovolo sash planes by **Sims**, London. **£15** [DP]

♦ A pair of ovolo sash planes by **G. Berry** with matching saddle and templets. **£26** [DP]

♦ A 7½" long compassed astragal and hollow sash plane by **Bewley**, Leeds. **£45** [DP]

♦ A sash scribing plane for ovolo bar by **Malloch**, Perth. **£35** [DP]

♦ A shop front plane – to cut a large lamb's tongue – one only of a pair by **Barnes**, Worcester. 2⅛" wide. **£24** [DP]

Price Guide

Two, three and four-iron moulding planes.

♦ A pair of twin iron astragal and hollow sash planes by **Malloch**. **£44** [DP]

♦ A twin iron m.p. by **Moir & Co.**, Glasgow, to cut a ⅝" Gothic sash bar from the edge. **£80** [DP]

♦ A handled stick and rebate sash plane by **Watkins**, Bradford, 13½" long, with two irons, the one cutting the glazing rebate and the other, an ogee sash mould. **£70** [AP]

♦ A 3¾" wide twin iron double shoulder m.p. by **Gleave** to cut a complex moulding, probably for a dado rail. **£90** [DP]

♦ A 3½" wide handled twin iron m.p. by **Stewart**, Edinburgh, to cut a Grecian ogee and fillet for an architrave mould. **£110** [DP]

♦ A twin iron handled 3¾" wide boxed Grecian ogee by **McGlashan**, Perth (1827–49), with rope hole. **£70** [DS 24/1642]

♦ A pair of twin iron boxed Grecian ogees, 2¼" and 2¾" wide, Nos. 5 & 7, by **Gleave**, Manchester (1832–present). G++ **£88** [DS 24/1647]

♦ A twin iron 3" wide ogee and fillet by **Hoffman**, Edinburgh (1842–72). **£93** [DS 23/1276]

♦ A 2⅝" wide twin iron double boxed m.p. by **Cowell & Chapman**, Newcastle, (1859–1900). **£46** [DS 23/1280]

♦ A double boxed 2¾" wide triple iron m.p. by **Kinnear**, Perth (1820–45) to cut a reverse ogee, fillet and bead. The irons are echeloned and two mouths discharge to the right and left. **£180** [DP]

♦ A triple iron plane by **Manners**, Glasgow (1792–1822) to cut an astragal and hollow sash bar from the edge, the rear iron discharging through a top mouth. **£240** [DP]

♦ A triple iron m.p. by **Mathieson**, with single iron in front of the other two irons. G++ **£275** [AP]

♦ A rare four iron gothic sash plane by **Mathieson**. **£250** [DS 19/1329]

Cornice planes

♦ A 15" handled bench mouthed m.p. by **Burton** with round topped iron by **Marshall**, Manchester. **£187** [DS 23/1275]

♦ A twin iron 3¾" x 14¾" ogee cornice plane by **Welsh**, Dundee (1845–50), some worm. G **£154** [AP]

♦ A pair of handled cornice planes by **Rail**, Kelso. G **£555** [AP]

Bolection planes

♦ A twin iron, 3½" wide handled bolection moulding plane by **Barnes**, Worcester. **£95** [DP].

Drawing a bead

We all know what a bead is – a half-round moulding which is the most commonly used decoration in joinery. But there are many different meanings and contradictions in the terminology that has been used over the years. With a shoulder each side a bead becomes an astragal; if one of the shoulders is lower it may be a torus but the skirtings sold as torus today are equal each side. When more than one bead is present, the term "reeding" is usually applied – but by no means always! Let's look at bead planes.

A bead on the side

The majority of bead planes cut a bead at the side (edge) of the wood and are thus sometimes called side beads to differentiate them from centre beads which work in the centre of a surface.

Beads are the most practicable way of concealing a joint and are frequently used in other places – so they are sure to be there in any box of moulding planes. Centre beads are less common: at a guess, we see twenty or thirty side beads for every centre bead. Even less common are centre beads that cut a profile with bevelled or coved quirks.

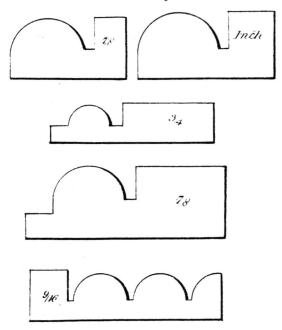

Two beads, an astragal, a torus and a reed [Varvill, 1879]. The sizing of beads and reeds is not always apparent and different manufacturers seem to have followed different practices. Here, in the top three, the size is measured overall, the width of the bead including the quirks and shoulders, but in the reed, it is the diameter of the bead only.

Price Guide

- A 9⅜" bead by **Okines** (1740–1834) (mark G+). G **£55** [DS 24/268]
- A 9¾" large bead by **Madox. £24** [DP]
- Two side beads by **Blizard** (1800–28). **£30** [DP]
- A 9¾" bead by **Moodey**, Worcester, (mark P) with intermittent boxing in lignum vitae, iron by **Edward Dingly**. G **£253** [DS 23/122]
- A rare 9⅛" bead fully slipped in lignum by **T. Child** (–1816–) (mark G–). G **£32** [DS 21/1235]
- A long set of eleven fully boxed or double boxed beads by **Routledge**, Birmingham. **£105** [DP]
- A set of eight boxed beads by **Moseley**, ⅛" to 1". **£72** [DS 24/757]
- A set of five bead/torus (removable slip) planes by **R. Nelson**, Edgware Road. **£34** [AP]
- A ½" centre bead by **Arthur**, Edinburgh, with eyed mouth. **£12** [DP]
- A very large (1") centre bead plane with coved quirks by **Currie**, Glasgow, with double shouldered stock, 10½" x 2¾". **£70** [DP]
- A ⅜" follow-on bead by **Griffiths** Norwich with this firms proprietary reversible fence to provide guidance for follow-on work. **£46** [AP]
- A triple boxed follow-on bead plane for ³⁄₁₆" beads by **W. Parkes** with removable fence to cover the guide bead. **£30** [DP]
- A rare 7½" twin iron double reed with double boxing by **McGlashan** (1827–49). **£42** [DS 24/716]
- A fully boxed ⅛" triple reed with integral fence by **S. King**, Hull. This form of fenced plane may have been used for shelf or table edges. **£30** [DP]
- A fully boxed triple reed by **Stothert & Walker**, Bath. F **£30** [DP]
- A 9½" x 2¾" fully boxed five reed plane by **Moseley. £57** [DS 24/754]
- A 9⅜" cluster reed by **Mutter ★ Moseley** (mark G+). **£66** [DS 24/267]
- A fully boxed small cluster reed by **Lund. £36** [DP]
- A cluster reed by **Varvill** at cabinet pitch with three ⅛" reeds in ½". **£60** [DS 24/701]
- A 9½" fully boxed cock bead plane by **John Hazey** (1765–85) (mark G–). G **£26** [DS 21/1238]
- A cock bead plane by **Gabriel. £36** [DP]
- A cock bead plane by **Varvill. £16** [DP]
- A 1" nosing plane by **Wm. Marples & Sons**. G++ **£18** [DP]
- A twin iron 1⅛" nosing plane by **T. Turner**, Sheffield. **£42** [AP]

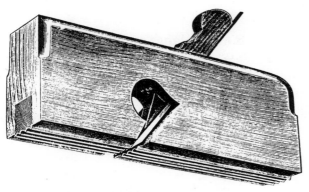

Top: a boxed and slipped bead and bottom: a fully boxed triple reed [Harding, 1903]. A slipped plane has a removable side so it can be worked even if there is a higher section in the moulding adjacent to the bead.

Carry on beading

At times beads became so fashionable that one or two just weren't enough. Planes to cut two, three or even four at once were the result. Multiple beads are more often called reeds.

The beads, or reeds, might be of any size from ⅛" to ½" in 1/16" increments and up to four or more in number and the planes either with or without a fence. So a comprehensive set would have been a large number of planes. Multiple reeding planes were also difficult to use as the irons were a problem to sharpen and keep in shape.

Far more flexible was the follow-on beading plane which will cut any number of reeds. These planes were developed around 1810, a period when reeds were particularly fashionable. There are several ingenious designs for these follow-on bead planes; some have a removable fence which allows the cut to be started at the edge, whilst others need to be started against a batten. The basic principle of all is that after the first bead is cut, a blind bead on the sole of the plane follows this whilst the next bead is cut. These planes are not always recognised – particularly if the removable fence has been lost. **Griffiths** of Norwich designed a particularly elegant arrangement, but all types are of interest.

Cock beads

Upstanding beads fitted around drawers, cupboard doors and the like are termed cock beads and are invariably made as a separate strip. Special planes to cut the full half-round needed had been developed by the last quarter of the 18th century. These planes, often 1" shorter than standard moulding planes, are nearly always fully boxed and are distinguishable from centre beads in that the bead is fully inset into the sole.

Nosing plane

The largest cock beads are ¼" diameter and most are smaller, so if you have a plane with a full half-round sole that is bigger than this, it will be a nosing plane. These were intended to round the edge of board for window boards and the like and may be sized from ¾" to 1½" diameter. The extremities of the irons rub horribly; I have never managed to use one successfully, which is no doubt the reason why most examples are in such good condition.

Collecting tips

➤ *If your nosing plane has a vertical or near vertical iron, it's a fishing rod or billiard cue plane. (See* Hollows, Rounds and Relations, *p.116)*

➤ *Most makers offered a bead and torus plane. This had a removable slip screwed onto the sole adjacent to the bead. With the slip in place, it produced a bead moulding; without it, a torus moulding.*

➤ *Boxing is not only attractive but is an indication of a quality plane. Superior quality boxing – full, keyed, dovetailed, etc. – is sought by collectors. (See also* Wooden Planes, *p. 107)*

Tongue and groove (match) planes with moving fence [Harding, 1903]. Supplied with three pairs of different width irons.

plane is a double fillister and was sometimes, pedantically, described as such. Another alternative older name, now seldom used, was match plane.

Most tongue and groove planes have integral fences which cannot therefore shift in use. These were planes to be worked hard, turning out floor boarding and the like and it was normal to have a pair for each thickness of board, but a small percentage do have moving fences.

An interesting variation is the combined T & G plane – in reality the two parts joined together side by side but reversed, therefore having irons and wedges facing both ways – always a fascination

Tongue and groove planes

The most commonly found grooving plane will be one half of a pair of tongue and groove planes. Of moulding plane format, it has a steel skate similar to a plough plane. The matching tonguing plane – which should always accompany it – has a flat sole with a deep groove. Strictly speaking, the tonguing

Drawer bottom planes and the like

At about 8½", shorter than moulding planes, these are grooving planes that cut a fixed size groove at a fixed distance from an edge. The tongues can be either boxwood or steel.

They are usually described as drawer bottom planes – for cutting a groove in the drawer side to

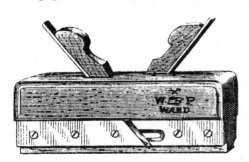

A combined T & G plane [Ward & Payne, 1911].

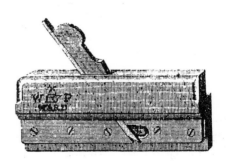

Drawer bottom plane [Ward & Payne, 1911].

Price Guide

Tongue and groove planes

Matched pairs of late 19th/early 20th C. date **£10–14**.
* A pair of T&G planes by **Mutter** (1766–99). F **£40** [DP]
* A pair of T&G planes by **Blizard** (1800–28) (marks G++), marked No. 4. **£33** [DS 24/702]
* A little used pair of ⅝" T&G planes by **G. Busby**, Leicester. G++ **£20** [DS 23/1312]
* A pair of adjustable T&G planes by **Howarth**. **£26** [DP]
* A pair of adjustable T&G planes by **Griffiths**, Norwich, complete with two pairs of alternate width irons. **£36** [DP]

* A pair of handled T&G planes by **King & Peach**, Hull, little used condition. **£65** [DP]
* A pair of handled T&G planes by **Eastwood**, York with adjustable fences. G++ **£55** [DP]
* A combination T&G plane by **Gleave**, Manchester. **£28** [DP]

Drawer bottom and grooving planes

* A rare form of grooving plane 7⅛" long with an iron skate, adjustable extra depth fence and thumb turn brass depth stop by **F. Willey**, Leeds. **£40** [DP]
* A 7½" drawer bottom plane with full boxed sole and facing to integral fence by **Buck**. G++ **£24** [DP]
* A rare 7½" boxed drawer bottom plane by **Gabriel**. G **£72** [DS 25/1240]

receive the drawer bottom. This type of plane must have often been used in making small boxes and other specialised work such as wooden cameras, as fully boxed versions, intended to cut really small grooves, appear with some regularity.

Dado planes

A comparative late-comer to the joiners' chest, the dado plane was used to cut cross-grain trenches. The oft-repeated suggestion that the name comes from its use to fit dado (low level panelling) into the floor would seem to be fallacious as this type of construction is almost unknown (see *Raglet planes* below).

It is quickly recognisable by its two wedges, the front one vertical. This secures a two-pronged nicker iron set forward of the main iron. The usual

sizes are from ¼" to ⅞" wide but sizes above and below these are occasionally found.

There are three types of depth stop: brass, operated by a thumb turn; screw fixed brass; and wood, morticed through the stock. The front wedge may be turned sideways and in the smallest sizes, the sole may be boxed.

Dado grooving plane [Harding, 1903].

Raglet planes

An interesting plane, capable of cross-grain trenching and, in that respect, having some resemblance to a dado plane. The body is however wide and the handle is offset and angled. The purpose was to cut a trench in the perimeter of a boarded floor to receive a rebated skirting board. The design keeps the knuckles away from the wall. This form of construction was only used in late 19th century Scottish houses of the best quality and these planes are always of this date and origin.

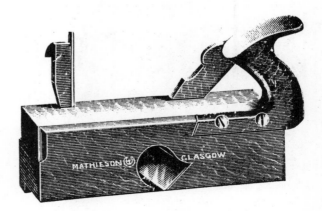

Flooring raglet plane by [Mathieson, c. 1900].

Price Guide

Dado planes
Single dado planes, mid 19th/early 20th C. with brass depth stops **£14–20**.
* An early 9⅞" long dado plane by **John Hazey** (1765–85). G **£77** [DS 23/139]
* An 18th century dado plane by **Wm. Moss** with full length depth stop. G **£75** [DS 21/1248]
* A wooden depth stop ⅝" dado plane by **J. Sym** (1753–1803). **£70** [AP]
* A ⅝" dado plane by **Gabriel** with wooden depth stop. **£40** [DP]
* A pair of little used dado planes, ⅝" and ¹³⁄₁₆" by **Kirk & Asling**. F **£38** [DS 22/319]
* A ⅛" dado plane by **Mathieson** with unusual boxed sole and side facing front wedge. **£32** [DP]
* An unusual ⅜" radius dado plane with adjustable curved fence, 2⅝" wide by **Griffiths**. G++ **£66** [DS 24/695]
* A set of seven dado planes from ³⁄₁₆" to 1⅜". Unmarked, the two largest with external moving fillister form of depth stop. **£187** [AP]

Raglet planes
* A 3" flooring raglet plane by **Storie**, Edinburgh, (1860–64), replaced wedge. **£40** [DS 20/405]
* A fine little used raglet plane with typical offset handle by **Mathieson**. The wood skate is steel shod. **£65** [DP]

Side rebate planes
* A pair of side rebate planes by **Shepley** (1771–1805). **£30** [DS 20/993]
* A side rebate plane by **Loveage** (1735–51). **£45** [DS 20/989]
* A pair of side rebate planes by **King & Co.**. G++ **£30** [DP]

Side rebate planes

If you plough a groove that is not quite wide enough you will need a side rebate plane. Wooden versions, which are of moulding plane format, are distinctive with their vertical irons and wedges and should be in handed pairs. The sideways pressure on the iron is considerable and they must have been tricky to use. (See also *Hollows, Rounds and Relations*, pp.114–116.)

Thumb size (3½" long) versions were also offered by several manufacturers. Like the full size versions, these should be in handed pairs.

Pair of side rebate thumb planes [Harding, 1903].

The difficulty of using the wooden type is, no doubt, why the metal versions seem to have quickly become popular after their introduction around 1900. They were made by the following: **Stanley** Nos. 98, right hand, and 99, left hand, and, from 1926, No. 79, combined right and left; **Preston** three types, one of which became the **Record** No. 2506. This latter model was also available with a depth stop (No. 2506S).

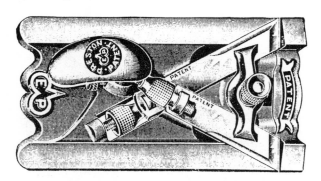

Patent adjustable side rebate plane No. 1369, combined right and left hand, nickel plated. The nose can be removed for working close up into corners [Preston, 1909].

Cill planes

Almost always craftsman-made, these planes often raise questions about their identification and use. The cutter is set to the side of the stock and is used to cut a semi-circular groove in the upstands of wooden cills and **not**, as is popularly believed, to cut a drip on the underside. The cutter and wedge are exposed on the side of the stock.

Collecting tips

➤ *Matched sets of more than three dado planes are rare.*

➤ *Check that T & G planes really are a match before buying. They are usually marked on the heel with the basic size but each size could be had with different size tongues and these sometimes become muddled.*

➤ *The earliest dado planes had wooden depth stops but, like many tools, as improvements were made, the earlier and cheaper type were still offered so a wooden depth stop is no guarantee of an early date.*

➤ *Some models of metal side rebate planes had depth stops. These are frequently missing.*

British Standard hags had only one tooth so it was perfectly logical that they should claim precedence in naming this tool. The formal name may be router plane but there are other names including old woman's tooth, granny's tooth, carver's router and the pedantic "depthing router".

Not strictly a plane as there is no restriction at the mouth, the purpose of the router, however named, is to clean up the bottom of recesses.

A few examples carved in the form of a face with the mouth acting as the escapement are known – perhaps these allude to the name though the appearance usually seems more cherub than granny.

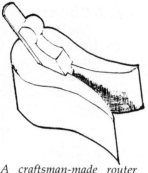

A craftsman-made router plane.

Router planes which can be easily made by the craftsman attracted much ingenuity so there are good artistic examples to be found.

With the coming of the iron age in manufactured tools (1880) **Stanley** and **Preston** produced many models.

Collecting tips

➤ *For craftsman-made examples, elegance and beauty are all!*
➤ *Commercially made wooden examples are not as common as one might expect.*

Price Guide

Craftsman-made router planes **£5–30** dependent on artistic merit and quality of finish. Boxwood and exotic woods will increase the price.

◆ A beech OWT router by **John Green**, York. Marked twice, once at each end. **£45** [DP]

◆ A beech OWT router by **Mathieson**. **£22** [DS 23/344]

◆ Four OWT routers by **Speight, Varvill, Marples** and **Gleave**. **£70** [DS 21/526]

◆ A hornbeam D-shaped router plane with rudimentary scroll decoration, the iron fixed with an early looking iron wing nut. **£38** [AP]

◆ A beech OWT router with decorative brass inlay around the iron and mouth. **£65** [DP]

◆ A craftsman-made mahogany and boxwood OWT router with inlayed decoration. **£56** [AP]

◆ A fine boxwood OWT router with profile moulding around the base. **£48** [DP]

◆ A miniature beech router and two other routers by **Varvill** and **Nurse**. **£35** [DS 22/575]

◆ A small (4" o/a) ebony router of swept open front form (see illustration left). **£50** [DP]

◆ A carvers' router, 4" long, by **Buck**. **£24** [AP]

◆ A small cast iron nickel plated **Preston** two handled router. Reg. Des. No. 534243. **£28** [DP]

◆ A **Preston** No. 1399P, cast iron, nickel plated two handled router. **£32** [AP]

◆ A **Preston** No. 2500P cast iron nickel plated two-handled grooving router with two fences. **£40** [AP]

◆ A **Preston** No. 22 (small type) router. **£20** [DP]

◆ A **Stanley** No. 71 router complete with three cutters and fence. **£18** [AP]

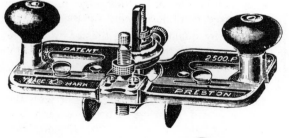

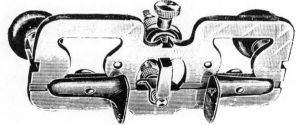

*Router planes: Left: two views of the **Preston** No. 2500P. Right: beechwood open front router plane [Harding, 1903].*

A rebate – often pronounced as rabbit – is a rectangular recess in the corner of a wooden section.

Fillister planes

Fillister is the old-fashioned name for a rebate and the name fillister has stuck to the plane which is normally used to cut rebates. The fillister plane is guided by a fence and has a depth stop. In these respects it differs from a rebate plane which has neither.

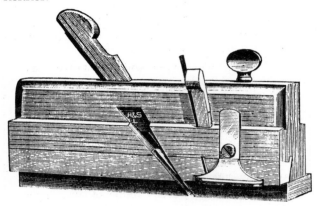

Above: Moving fillister [Harding, 1903]. This is a medium quality version. It has a thumb-turn depth stop but the boxing is simple.
Below: Improved moving fillister stop [Sheffield List, 1862].

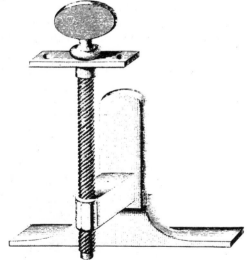

Standing fillister: The earliest form was standing, that is, cutting a rebate of fixed size. Although planes of this type are not common, they were made throughout the age of the wooden plane. They provide greater accuracy than the moving type and were often used in applications such as glazed display cases where the sections are small.

Moving fillister: This was one of the basic planes that any 19th century joiner would have had. Manufacturers made them in the widest variety of qualities. The best had thumb-turn brass depth stops and large boxing fixed into the stock with several dovetails, the cheaper planes had simpler brass stops and boxing whilst the bottom of the range had wooden stops and no boxing. There were also corresponding differences of quality in the fence.

The sash fillister

The sash fillister is, in essence, a moving fillister with the fence on arms. The clever part of the sash fillister is that the plane works on one side of the piece of wood being shaped and the fence rides on the other. The plane cannot therefore come off the work and the rebate formed and the moulding,

---

Price Guide

Moving fillisters
Late 19th/early 20th C. moving fillisters, of medium quality, plain fillet or shoulder boxing **£20–30**. Ditto with extra quality boxing **£25–35**.

• A 10¼" long moving fillister by **Owen**, no depth stop with possibly replaced sole fence. **£280** [DP]

• An early fillister with a brass depth stop by **Shepley** (1771–94). **£135** [DP]

• A moving fillister by **Frogatt** (1760 –90) (mark G++) with lignum vitae wear strips and side wood depth stop. A little worm. **£125** [DS 21/1288]

• A moving fillister by **John Green** (1768–1808), no boxing. F **£120** [DP]

• A moving fillister by **Routledge**, Birmingham, fillet boxed, condition as new. **£26** [DP]

• An extra quality moving fillister by **Moseley**, large boxing and three fence screws. G++ **£42** [AP]

• A rare handled moving fillister by **Mathieson** with double boxing to sole. **£145** [DS 24/1758]

• A rare **Mathieson** moving fillister with skew mouth and d/t boxing to both sole and fence, with long brass fence and two adjusters, minor chips. G **£130** [DS 24/1757]

Standing fillisters

• A standing fillister by **Mutter** (1766–99) with wood depth stop. G– **£26** [DS 21/1272]

• A ½" standing fillister by **Parry**, London. **£12** [DP]

Don't forget - fine condition will often radically increase the value.

Price Guide

Sash fillisters

Wedge stem sash fillisters, mid 19th C. or later **£20–30**. Ditto with extra quality boxing **£30–40**.

◆ An early sash fillister with d/t lignum wear strips by **Moody** (1755–76) (mark G). **£135** [DS 21/1294]

◆ A sash fillister by **Higgs** (1785–1820) with wood depth stop. **£45** [DP]

◆ A sash fillister by **John Green** (mark F) with box-wood scales to the stems and an unusual internal nicker iron. **£120** [DS 21/1300]

◆ A sash fillister by **Atkin & Sons** with set of eight irons in a roll. **£38** [DP]

◆ An extra quality sash fillister by **Buck** with box-wood sole and boxwood fence. **£46** [DP]

◆ A **Mathieson** No. 8 d/t boxed sash fillister with boxwood screw stems and nuts, one cracked. **£55** [DP]

◆ A beech screw stem sash fillister with boxwood stems and nuts by **J. Miller** Newcastle. G++ **£ 90** [DS 21/1257]

◆ A combined sash and moving fillister by **Mathieson**, No. 17, with two brass depth stops. **£105** [DP]

◆ A handled screw stemmed sash fillister by **Mathieson**, No. 11, with a boxwood fence. **£160** [DP]

◆ A rare handled bridle sash fillister with ebony stems by **Mackay, Burley & Heys**. **£290** [DS 19/1352]

◆ A handled sash fillister by **Kimberley** with patent metal screw adjustment, complete with key. **£260** [DP]

Left: A rare combined sash and moving fillister plane [Harding, 1903]. This type of plane is unusual in having an eyed mouth and depth stops on both sides. It can therefore be used to cut a rebate to either side of the stock. The normal sash fillisters will only cut a rebate to the fence side of the stock.

*Below: Kimberley pattern patent sash fillister [Harding, 1903]. **David Kimberley & Sons** were a Birmingham firm of planemakers who produced this metal-armed screw-adjustable form of fence. This fence was fitted to both sash fillisters and plough planes.*

which will subsequently be stuck on the other side, will be worked from the same (datum) side.

Like moving fillisters, sash fillisters were made in widely different qualities. Particularly desirable are handled versions and those with improved methods of fixing the stems. (See *Plough Planes* pp.134–136.)

You may see references to "on the bench" and "off the bench" sash fillisters. Both are used at the bench, the only difference being that the irons are skewed either to throw the shaving to the right, i.e. onto the bench, or to the left, onto the floor.

Collecting tips

➤ *Fillister–fillester–filletster–phillister. The same planes, different spellings!*

➤ *There is wide variation in the quality of both moving and sash fillisters. Best quality planes can quickly be identified from the multiple dovetailed boxing.*

➤ *Fielded panels have a flat perimeter; raised and fielded panels have a sloping one. The names used for the planes should reflect this difference.*

➤ *Small panel fielding planes of moulding plane format often pass unrecognised.*

Price Guide

Rebate planes

Beech rebate planes, commercially-made, 1850 or later **£4–8**. Ditto, 2" or more wide **£8–12**.

◆ A rare 11½" x 1" rebate plane by **Owen** (mark G–), two of three lignum wear strips remain. **£210** [DS 22/640]

◆ A 10" x 1¼" skew rebate plane by **Gabriel** (mark G–). G **£57** [DS 23/115]

◆ A solid boxwood rebate plane 9½" x 1⅜" with good golden patina. F **£45** [DP]

◆ A set of four skew rebate planes from ½" to 1½" by **Stewart**, Edinburgh. F **£24** [DP]

◆ A coachmakers' T-rebate plane by **Buck**. **£18** [DP]

◆ A set of three coachmakers' T-rebate plane straight, compassed and curved side, by **Berry**, London. **£45** [DP]

Badger planes

Badger planes, late 19th/early 20th C., no boxing or simple fillet boxing **£10–18**. Extra quality boxing **£16–25**.

◆ A badger plane with ebony slip and offset handle by **Thos. Bradburn & Son**. Some stains. G **£24** [DS 22/353]

◆ A solid boxwood skew mouth badger plane, 16" x 3" with mahogany handle, scrolled wedge and brass mouth, spur damage to handle. **£55** [DS 21/1431]

Rebate plane

A wooden plane of moulding plane format with an eyed mouth (a hole right through the stock) and an iron, often skewed, which extends to both sides of the plane. To cut a rebate from scratch with this plane, it is necessary to give guidance with the fingers or a batten; this was not the plane's usual purpose, which was to clean up or enlarge rebates that had already been started. (See also *Metal Planes* pp.144–148)

Badger plane

The largest of the rebating planes, this is similar to a jack plane but the iron is skewed to bring the leading corner to the edge of the plane.

The better quality examples are boxed, the best, dovetail boxed. The handles can give a clue to quality: the cheaper are open, the better closed. Badger smoothers were also made but are surprisingly rare; ideally they should be in pairs.

A badger plane with fences can be used to raise panels; whilst planes of this type occasionally surface, they are not as common as might be expected.

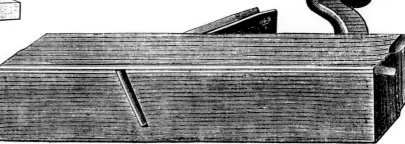

Above: coachmakers' T-rebate plane [Harding 1903]. Also available compassed. Right: badger plane [Harding, 1903]. This plane has a boxed slip at the wearing but most do not.

Panel fielding/panel raising planes

These planes were used to field and raise panels for all manner of joinery. A rebate in one edge of the stock exposes the side of the iron. Most examples include a depth stop and a fence, which may be fixed or moving.

There are three formats for this plane: a) Size similar to a jack plane and incorporating an open or closed handle. b) The most commonly found type – a parallel-sided plane, about 10" long. c) A plane of moulding plane form to cut smaller panels for cupboard doors and the like.

Where the panel was raised as well as fielded, the body of the plane was normally canted over at an angle ("working with spring") but sometimes the sole of the plane is just formed at an angle.

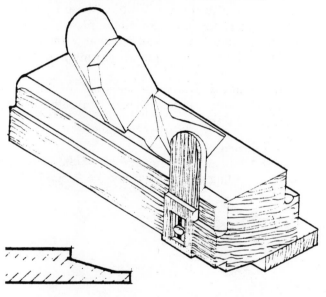

Section of panel.

A panel raising plane of typical late 18th century form. The stock is 8½" long with a skewed mouth and wooden depth stop. This iron is single. This plane cuts a raised and fielded panel as shown in the detail.

Cock bead fillister

Often overlooked, the plane, at first sight, is similar to a dado plane, with a second nicker iron and wedge set forward of the main iron. Unlike the dado plane, however, it is only a single bladed nicker. The function of this plane is to cut a deep but narrow rebate (fillister) to accommodate the cock bead on a drawer. At the ends the work is cross grain, hence the nicker.

Price Guide

Badger planes cont.

♦ A rare handled beech badger plane by **Gleave** with boxwood mouth trimmer and scrolled wedge. **£75** [DS 22/556]

♦ A handed pair of 5½" long beech steel soled coachmakers' badger planes by **Wm. Cowell**, Newcastle, of smoothing plane form. **£110** [DP]

♦ An 8" long beech badger smoothing plane by **Armour. £32** [DP]

Panel fielding/raising planes

♦ A beech panel raising plane 14" x 3¼" with off-set handle and vertical wood depth stop by **Jennion** (mark G–, first two letters hammered) (1738–78). G **£390** [DS 24/303]

♦ An 11⅝" panel fielding plane by **T. Darbey** (1767–85) with full length fence, vertical wood depth stop lignum d/t wear strips and strongly offset handle. **£105** [DS 24/124]

♦ A panel raising plane 8½" long by **Higgs**, London (1785–1827) with round topped iron by **John White** and round topped vertical wood depth stop. **£105** [DP]

♦ An unusual panel raising plane with round edge (to produce a coved raised panel) by **Shillinglaw** (1793–1814) with full length side and sole fences. **£75** [DS 22/351]

♦ A panel raising plane by **Varvill** with full length fence and full length depth stop. **£50** [DP]

♦ An 8½" panel fielding plane by **Currie**, Glasgow, with vertical type wooden depth stop. **£65** [DP]

♦ A large handled (badger style) panel raising plane, 15½" x 3¾" by **Moseley** with full length fence and full length depth stop. Replaced wedge. **£95** [DS 24/1112]

♦ A little used handled panel raising plane by **Moseley** with full length fence and stop, with the price 17/6 still on the sole. F **£120** [DS 22/309]

♦ A panel raising plane of moulding plane form by **Marples** with nicker iron with boxwood wedge. F **£36** [DP]

Cock bead fillisters

♦ A cock bead fillister by **Varvill. £55** [DP]

♦ A cock bead fillister by **Gabriel** (mark G). G++ **£88** [DS 23/107]

♦ A cock bead fillister with unusual form of inset brass depth stop secured by two screws by **Holbrook**, Bristol. **£60** [DP]

Remember - condition is G+ unless otherwise shown.

"They are thumb planes," I said. The customer looked interested but there was a long pause and I thought I detected the wheels going round – slowly. "You mean for removing corns and the like?" he asked.

What exactly many of the small planes were used for is a question that is not easy to answer. Many, I think, were made more for pleasure than work. Miniatures in many fields seem to have a special appeal and small planes are no exception; they are keenly collected today.

Thumb planes

Many tradesmen faced with a small problem of smoothing or shaping made a small plane. The choice of wood was usually boxwood but ebony and anything else suitable and to hand might be used. We well remember a beautifully made group of small planes for which the maker had chosen Tufnol (a tough but unprepossessing brown, fabric-reinforced plastic) – excellent for quiet running gears but not for collectable planes. If only he had used boxwood!

Craftsman-made planes are sometimes very small, maybe only 1¼" long, whereas commercially made thumb planes were standardised at 3½" long. The types listed by **Preston** were smoothing, rebate, compassed rebate, side rebate, side rounds,

Collecting tips

➤ *Recently made examples of small brass planes abound. Mostly poorly finished they should not mislead the careful.*
➤ *Similarly many recently made boxwood thumb planes are being offered. Many of these are well made but don't be fooled – they are not old.*

side rebate and hollows and rounds – all made in boxwood. But other makers also offered beech thumb planes.

Trades that used the small planes included patternmakers and that rather specialised joiner, the handrailer. Planes frequently turn up with the heel drawn out to form a handle. These look like miniature versions of coachmakers' planes but it is doubtful if they were used by that trade – the handles are merely an aid to gripping a small plane.

Carvers' routers: Miniature versions of the old woman's tooth router plane, typically with rectangular bodies.

Violin planes

These did have a very specific use – the shaping and thicknessing of the front and back boards for

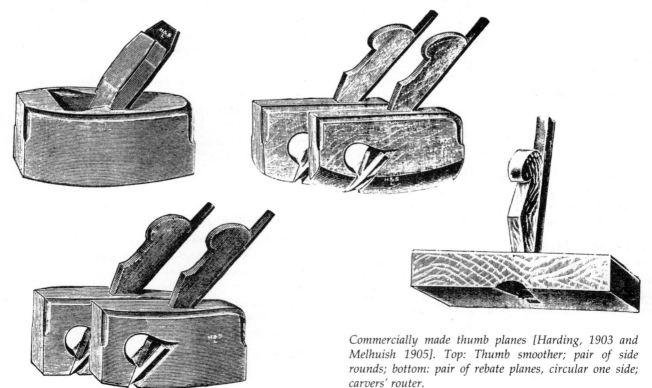

Commercially made thumb planes [Harding, 1903 and Melhuish 1905]. Top: Thumb smoother; pair of side rounds; bottom: pair of rebate planes, circular one side; carvers' router.

violins, guitars, etc. The commercially made examples are of gunmetal with screw caps to secure the iron, whilst unnamed examples are mostly wedge fixed. The principal English maker was **Ed. Preston**, who introduced them in 1912. This range contained five sizes from 1" to 2⅛" long. Each size came flat or round soled and had both a plain and a toothed iron.

Violin planes are still available today, but, like most things, the style and quality seems to have declined. Modern planes are likely to have a hole in the back to fix an extended handle.

Gunmetal violin makers' plane [Preston, 1912].

Price Guide

Smoothing plane type

• A rosewood smoothing plane, 3¾" x 1½", craftsman-made. **£36** [AP]

• A 3½" beech smoothing plane by **Currie**, Glasgow. **£36** [DP]

• A 3½" boxwood smoothing plane by **Mitchell**, London. **£55** [DP]

• A miniature beech smoothing plane, 2⅜" x ¹⁵⁄₁₆", patination indicates some age. **£45** [DP]

Hollows, rounds and beads

• A set of five patternmakers' miniature boxwood round planes, 3¼" x 1⅛" to 2¼" x ⅝"; top mouthed, hair crack to rear of one. G++ **£198** [DS 25/958]

• A pair of miniature beech H&R planes, 1⅞" x ½". **£65** [DP]

• A 4" x ⅝" boxwood hollow. **£30** [DP]

• A pair of boxwood miniature tailed compassed H&R planes with ebony wedges, 2½" x ⅜". **£75** [AP]

• A pair of No. 16 beech H&R planes by **Moseley**, 3½" long, with boxwood wedges. **£85** [DP]

• A 4½" beech left and right side bead plane with brass front to cut a ⅛" bead. **£40** [DP]

• A boxwood miniature tailed compassed hollow plane, 4" o/a. **£55** [AP]

Rebate planes, etc.

• A set of three skew mouth boxwood rebate planes, flat, compassed and radiused, 4" x ⅝". **£85** [DP]

• A set of miniature beech planes, 3½" x ⅝" comprising a pair of left and right radiused rebate, a pair of concave and convex, flat, and round both ways planes. G++ **£201** [DS 24/1127]

• A 3½" boxwood compass rebate plane by **Gleave**, Manchester. **£50** [DP]

• A pair of 3½" boxwood compassed side round planes by **Preston**. F **£90** [AP]

• A pair of miniature tailed compassed and radiused grooving planes with offset handles and steel skates. **£145** [DS 22/325]

• A miniature boxwood rebate plane, 1⅞" x ⅜", with ebony wedge. **£40** [DP]

• A miniature rosewood OWT router plane with ivory wedge and thin ivory sole. **£62** [AP]

Violin makers' planes

• A **Preston** violin maker' gunmetal plane, 1¾" x 1", round sole. G++ **£140** [DS 24/871]

• A **Preston** violin maker' gunmetal plane, 1⅛" x ⅝", round sole. G++ **£165** [DS 24/872]

• A **Preston** violin maker' gunmetal plane, 1" x ½", round sole. G++ **£140** [DS 24/874]

Remember - condition is G+ unless otherwise shown.

Two craftsman-made tailed miniature planes.

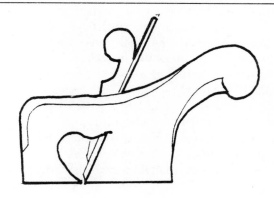

If a plane to cut a groove has a fence on arms, it will be a *plough* plane; if the guidance is integral with the stock it is a *grooving* plane. (See *Planes for Grooving*, pp.124–126.)

Development

The earliest plough planes had a wooden skate formed as part of the stock and cut a groove of one size only. A plane of this type has been recovered from the *Mary Rose* shipwreck of 1545. From the description given by Moxon, writing in 1680, it is clear that, by then, the plane had a metal skate but the inference is that the joiner still had a separate plane for each size of groove. His description of the fence and arms, which were secured with small wedges if needs be, seem to be extremely similar to the known examples of early 18th century ploughs. It would thus seem that the basic design of the English plough plane had already been fixed by the end of the 17th century. But this is also true for most wooden planes.

Fixing the fence

There is, in Moxon's description, a hint of the problem that was ever present in the plough – the design of an efficient and durable system of securing the fence arms.

The simplest method of fixing an adjustable fence is to make good tight mortices, backed up with captive wedges, and this is a pretty effective system but not easy to adjust. Once brass fittings became freely available, around 1770, makers started to add caps to the ends of the stems which had been prone to damage during "adjustment" with the mallet.

Improvements

Also around 1770, the first English screw stem ploughs were made and, from then on, this was always an alternative for the better quality planes. But wooden screw stems have a serious drawback: they chip and the nuts break.

Manufacturers seem to have been remarkably resistant to adopting the obvious solution which was to replace the stems with metal – hardly surprising as they were *wooden* planemakers. The only

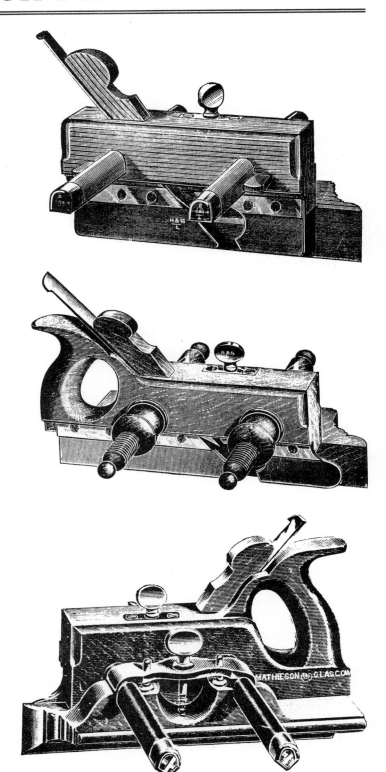

From top: A wedged stem plough (this is the standard form and most ploughs to be found are of this type); a screw stem plough with handle and extended plate [Harding, 1903] and a handled bridle plough with ebony stems [Mathieson, c. 1900].

metal stem design that was made in quantity was the **David Kimberley & Sons'** patent screw adjustable system. This was applied to both ploughs and sash fillisters and incorporated an additional feature. Whilst the fence is stabilised by two thin metal arms, sliding freely through the stock, it is adjusted by the central screwed arm.

Amongst the very few metal arm ploughs is the **Mathieson** No. 10 brass stem plough, actually a moving fence arrangement as the fence moves along the stems and is clamped to them.

The principle of moving the fence along the arms had been used for many years in the so-called bridle ploughs. These have wooden arms, fixed permanently into the stock with a movable fence clamped to them. These planes are comparatively rare and are therefore sought by collectors.

Improved plough stop [Ward & Payne, 1911]. The side screw projects through the side of the plane stock and bears against a small brass arch topped plate. Ploughs with this feature are sometimes known as church window ploughs.

Collecting tips

➤ *Plough planes were a fertile area for innovation, particularly those used by coachbuilders, so unusual types are to be found. These include a plough with long metal screws within the stems (Gabriel); a design with a deep (full height) fence (John Green and other York makers); a curved plough (John Sym) and the two patent plough planes (Kimberley).*

➤ *The screws and nuts of screw stem ploughs, usually made of boxwood, are liable to chipping and almost impossible to repair – look carefully.*

➤ *There is a range of designs of steel bridles which secure the fence of some superior quality ploughs.*

Price Guide

Plough planes

Beech plough planes, wedge stems, 1850 or later **£20–30**. Ditto with set of irons **£30–40**.

◆ An early beech plough by **Cogdell** (1730–52) (mark G–) with wooden depth stop. Replaced wedge. G **£275** [DS 24/132]

◆ A plough plane by **Mutter** (1766–99) with wooden depth stop. The stems are uncapped and are friction fit with no wedges. **£220** [DP]

◆ A beech plough by **Gabriel** (mark G). G– **£100** [DS 21/1253]

◆ A rare beech plough by **Holtzapffel** with brass fittings and unusual original brass strengthener behind wedge. Replaced wedge. **£65** [DS 23/1420]

◆ A beech plough with full length boxwood scales to stems by **Holbrook**, good replacement wedge. G++ **£40** [DS 23/1429]

◆ A beech plough by **Preston** with a set of eight **Preston** irons. **£45** [DP]

◆ A little used beech plough by **Greenslade** in rare original box, lacking lid, with set of eight **Hearnshaw** irons. F **£55** [DS 21/1256]

◆ An unusually small (6") beech plough for ¹⁄₁₆" grooves with proportionally small stems and fences by **Lund**. **£110** [DP]

◆ A craftsman-made plough in rosewood compete with brass church window depth stop and brass capped stems with boxwood wedges. This could be taken for a commercially made plane except for minor eccentric details. **£160** [DP]

◆ An ebony plough by **Varvill & Son**, York. The handled stock with steel skate, normal brass fittings, ivory key wedges, rules in stems and wedging to stem ends. Some minor defects. **£1500** [AP]

Screw stemmed ploughs

◆ A "York pattern" early screw stem beech plough by **John Green**. The boxwood stems are fixed through the full depth fence and terminate in large diameter and decorative turned ends. The securing nuts are boxwood. **£350** [DP]

◆ A little used handled screw stem beech plough by **Malloch**, Perth, with boxwood stems and nuts and projecting skate front with a set of irons by **Colquhoun & Cadman**. F **£120** [AP]

◆ A screw stem plough by **Eastwood** with boxwood stems and nuts. **£50** [DP]

◆ A handled screw stemmed beech plough by **Varvill** with pronounced curled front skate and set of eight irons. **£125** [DP]

◆ A handled **Kimberley** patent plough with skate front and key. **£140** [DS 24/1962]

Mr Falconers' Plough

➤ *Mr Falconer was the inventor, in 1846, of a special form of plough plane which incorporated a flexible steel fence. Coachmakers' plough planes, with either one or two arms and capable of circular work, had been made many years before but the adjustable fence enabled the plane to work circular grooves as well as compassed grooves. Confusingly, planes designed to do both sorts of curve are called circle on circle planes. Falconer's design was published but planes to the design are very rare. Rosewood or ebony was the preferred material.*

➤ *Prices for similar examples sold recently:*

• *A fine quality mahogany coachbuilders' plough with brass plates to top and underside and with squirrel tail handle, the single sliding stem holding the mahogany fence, both stock and fence with decorative mouldings. F £1,100 [DS 24/1697]*

• *A fine and rare tailed ebony and brass coachbuilders' plough with brass fence, adjustable along single slotted arm, 6½" overall. G++ £2,300 [DS 22/1502]*

• *A fine Falconer type coachbuilders' plough in rosewood with squirrel tail and single stem, the flexible fence, mounted on rosewood, is adjusted by screws at either end, similar to the Sym plough. F £1,750 [DS 19/1361]*

Mathieson No. 10 patent brass slide stems from the c. 1900 catalogue.

Price Guide

Bridle plough planes, etc.

• A beech bridle plough by **Greenslade**, Bristol, with steel bridle secured with brass thumbscrew with wide and well-shaped fence. **£75** [DP]

• An unusual beech bridle plough by **Bewley**, Leeds. The fence slides in grooves in the underside of the stem and is secured with an elaborate bridle. **£140** [DP]

• A handled ebony stemmed bridle plough with skate front by **Malloch**, Perth. G++ **£220** [AP]

• A beech bridle plough with brass tipped slotted stems d/t into body and attractive bridle with boxed fence by **Bewley**, Leeds. **£80** [DS 24/1694]

• A beech bridle plough by **Griffiths**, Norwich. The stems are uncapped, the stock is fitted with a turned handle at the side (original) and the bridle is beech. **£120** [DP]

• A beech plough by **Moseley** (early mark). The fence operates with long screws concealed within the stems turned by brass thumb turns at the end of the stems which also include ivory scales. **£450** [DS 24/1689]

• An unusual **Barnes**, Worcester, plough plane. The fence moves along the slotted arms and is secured with wing nuts through the arms. **£95** [DP]

• A rare adjustable beech plough by **Wallace**, Dundee. The central screw through the stock is attached to both stems above the fence. Some pitting to cross bar. G **£150** [DS 23/1413]

• A rare **Mathieson** No. 10 handled beech plough with adjustable hollow brass arms and fence supports with skate front. Some bruising to wedge. **£400** [DS 22/1378]

Coachbuilders' plough planes

• A rare coachbuilders' tailed plough by **Gabriel**, the fence moving along a single thick stem. Screw operated depth stop. G **£1600** [DS 19/1377]

• A beech plough by **Cockbain**, Carlisle, with interchangeable short skate to adapt plane for curved work. **£220** [DP]

• A beech tailed coachmakers' plough, unmarked, 7" o/a. The single thick stem is heavily brass plated top and bottom and carries a small fence. **£650** [DP]

• A beech coachmakers' plough by **Moseley**. The stock 5¾" long with short skate and internal brass depth stop. The two stems are brass tipped and secured by wedges. F **£350** [DP]

Don't forget - fine condition will often radically increase the value.

Salaman, in his *Dictionary*, lists no less than sixteen different names for this tool. Pleasing are fork shaft rounder, moot, stail engine and witchet. Avoid rounding plane, the name given to a plane with a hollow sole which, in fact, produces the same result as a rounder but by planing along the grain. The purpose of the rounder was to produce poles which were used in a variety of products – tool handles, ladder rungs and the like.

Form

Rounders are of two basic forms. The throats of both taper so that the tool can be gradually worked onto the wood.

Solid stock: Commercially made versions are of two styles – a spiral body with handles of normal proportion or a square body with stubbier handles. Country made examples are often appealingly primitive.

Adjustable stock: More often called stail engines, these consist of two blocks, one with a mouth for the iron, the two halves secured together with handled wooden screws. These are for larger work than the solid type and have the advantage that tapering handles can be made.

Collecting tips

➤ *Inspect the screws of stail engines care-fully, as they are often damaged.*

➤ *Commercially-made rounders have double slotted irons – generally true.*

Fishing rod planes

This plane is constructed of two blocks, one of which is fitted with an iron. In this respect is similar to the adjustable form of rounder but the two blocks are hinged together at one end with handles at the other. The plane can therefore be squeezed together during use to cut a tapering rod. (see also *Hollow, Rounds & Relations*, pp.114–116.)

Price Guide

Solid stock rounders

◆ A graduated set of three matching solid stock craftsman-made rounders in ash. **£44** [AP]

◆ A pair of square bodied solid stock beech rounders with plain turned handles. **£32** [DP]

◆ A small solid stock rounder in boxwood, 6½" o/a, the cylindrical stock and squat handles both turned. F **£48** [DP]

◆ A spiral form solid stock handled rounder by **King & Co.**, Hull. **£30** [DP]

◆ A small spiral form solid stock beech rounder by **Fenton & Marsden**, Sheffield, incorporating two boxwood screws through stock providing adjustment of size. **£40** [DP]

◆ A beech solid stock rounder by **Marples**. **£20** [AP]

◆ A square bodied, top mouthed ladder makers' beech rounder, 1½". **£44** [DS 24/1094]

Adjustable stock rounders

◆ A beech adjustable stock rounder by **T. Turner**, the mouth with heavy brass plating. **£55** [DP]

◆ A beech adjustable stock rounder by **Marples** with boxwood screws. **£35** [DP]

Fishing rod planes

◆ A fishing rod makers' trap plane and six matching graduated solid stock rounders. **£198** [DS 23/1290]

◆ A fishing rod trap plane by **Mathieson**, the iron secured by a wedge into a hinged stock with two turned integral handles. **£260** [AP]

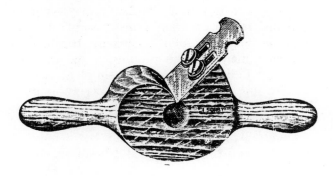

Left: Wheelers' beech solid stock rounder. Above: Beech fork shaft (adjustable stock) rounder with beech screws. [Turner Naylor, 1928].

Specialised, sometimes *very* specialised, planes were needed to speed work in an age of hand production. There is space here to review only a few but comprehensive details can be found John M. Whelan's *The Wooden Plane* (see p. 19).

Airtight case planes

Planes to cut the beaded airtight joints in the opening doors of museum cases. The beads may be single or double. The minimum set of planes is a pair – a hollow and a round. Often the bead was formed by inserting a tongue which was then rounded, so there may also be a plane to plough a groove. As the detail at the top and bottom of the door was slightly different from the sides, there was another plane for this. Making the joints fully airtight required great accuracy so the fences are either integral or firmly screwed to the side of the stock – adjustable fences are a rarity.

Hook joint plane: Typically 10" long, full boxed, with two irons and a full length brass depth stop, this plane cuts an interlocking joint – the hook joint – used on the meeting stiles of airtight cases.

Price Guide

Airtight case planes

H&R airtight case planes, not in pairs, have little appeal £15–25, but other types may be singles.

• A part set of six airtight case planes by **Buck**, one pair of double reeds and four others without matching plane. G++ **£159** [DS 24/1130]

• A rare set of eight airtight case planes by **T. Turner**, Sheffield, and a standing fillister made from the same tree as the set as the annual rings precisely match one of the planes in the set. The fillister would have been used to make the rebates for the glass. All irons marked **T. Turner** with their trade mark. F **£1250** [DS 21/1414]

• Two pairs of fully boxed airtight case planes, single H&R and double H&R, by **Greenslade**, removable fences. **£240** [DP]

• A double round airtight case plane by **Nurse**. **£18** [DP]

• A pair of double airtight case planes, unboxed but with integral fences, by **Preston**. **£90** [DP]

• A fully boxed (double d/t) double iron hook joint plane by **Moseley**. Full length brass fence. **£100** [DP]

• A boxed double trenching plane by **Gleave**. **£45** [DP]

*Remember - condition is **G+** unless otherwise shown.*

Collecting tips

➤ *Hook joint planes are best if with the associated airtight case planes.*

➤ *The best quality airtight case planes are boxed. Surprisingly, for a plane intended for the most accurate of work, some are not.*

➤ *All airtight and hook joint planes date from 1850 or later.*

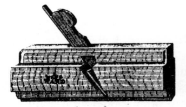 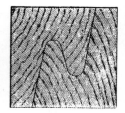

 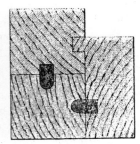

An airtight case plane and details of the cuts made by the planes. From left: a hook joint plane, a double airtight joint and a single airtight joint with the jamb and head detail alongside [Ward & Payne, 1911].

Handrail plane

The name is usually applied to a bench mouthed plane around 7" long which cuts a large ogee. The fence may be fixed or moving.

Around 1790 the design of handrails changed radically. Newel posts were done away with and, at changes in direction, the handrail was curved smoothly round, in trade terms "swept". For this type of handrail, solid mahogany became the preferred material. The profiles of the rail also changed radically, becoming simpler, with a curved top and ogee sides, a design popular throughout the 19th century. The handrail planes cut the sides of this type of handrail.

Handrail planes should ideally be in matched pairs, left and right hand, and, best of all, should also have a matching shave to use on the sections too tightly curved for a plane. Compassed handrail planes were made but are very rare.

Price Guide

Handrail planes and shaves

♦ A handrail plane by **J. Miller** to cut a simple ogee, with full length adjustable fence. **£60** [DP]

♦ A small fenced compass and radiused handrail plane, 5" long, by **Moseley**. **£330** [DS 22/340]

♦ A pair of left and right handrail planes by **Buck**, not adjustable, with round top irons. **£135** [DP]

♦ An adjustable handrail plane by **Higgs**, to cut a deep ogee and fillet. **£50** [DS 21/482]

♦ A pair of compassed radiused handrail planes by **Griffiths**, Norwich, with round topped iron by **John Green**. G++ **£200** [DS 19/672]

♦ A rare set of four handrailing tools by **Moseley** consisting of right and left hand planes and right and left hand beech shaves. **£220** [DP]

♦ A boxwood handrail shave by **Gleave**, cuts a deep ogee both ways with triangular section iron and adjustable fence. **£60** [DP]

♦ A pair of left and right beech handrail shaves with adjustable brass fences by **Lunt**, Liverpool. **£140** [DP]

Spill plane

There are just a few planes where the shaving is the product and not just the waste – in this case, the shaving comes from the plane tightly rolled up to form a spill useful for lighting a candle or your pipe from the fire. To the uninitiated spill planes are not easy to spot.

Some are of moulding plane format and might at first sight be mistaken for a fixed fillister but the key is the sharply skewed iron – this rolls and discharges the spills to the side. Many are owner-made and many are not of moulding plane format – some versions are used sole side up, fixed into a vice with the wood moved across the tool.

Spill plane [Mathieson, c. 1900].

Collecting tips

➤ *Another type of plane where the shaving is the product, not the waste, is the spelk plane, used in basket making.*

Price Guide

Spill planes

♦ A spill plane by **Griffiths**, Norwich, of moulding plane form with skew iron and shaped mouth. **£28** [DS 24/700

♦ Two craftsman-made bench standing spill planes in beech. One has a skewed iron to cut spiral spills, the other a straight mouth to cut spelks. **£39** [DP]

♦ A well conceived craftsman-made spill plane of jack plane form with heavily skewed iron with scroll wedge and open handle. **£70** [DS 22/567]

♦ A spill plane of moulding plane form by **Mathieson**. **£30** [DP]

♦ A beech spill plane by **T. Turner** . G **£49** [AP]

♦ A craftsman made beech spill plane with an iron by **Buck**. **£35** [AP]

♦ A **Preston** patent bench standing cast iron spill plane. **£26** [DP]

♦ A mahogany bench standing spill plane, 11½" long with well shaped ogee feet. **£35** [DP]

♦ A craftsman-made spill plane 16" long. Open handled stock with grooved bottom – spill discharge through side aperture. **£45** [DP]

Table joint planes

Planes to cut the rule joint on drop leaf tables must be in pairs. Again, they are not always easily recognisable, and for this reason often separated, they are in essence a hollow and a round cutting a segment of 90°.

Later examples incorporate fences and stops and are even less recognisable. A clue to identity is a circle or part circle inscribed on the toe.

Price Guide

Table joint planes

♦ A pair of table joint planes by **Stewart**. G++ **£85** [AP]

♦ A pair of table joint planes by **Mathieson**. **£185** [AP]

♦ A pair of large table joint planes, unnamed. **£65** [DP]

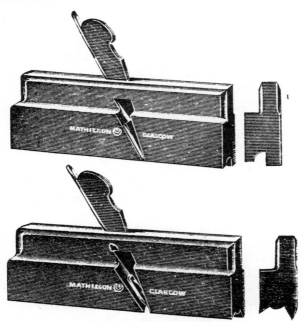

A pair of table planes [Mathieson, c. 1900].

Chamfer planes

Much ingenuity has been applied to the design of these planes which are intended to cut a 45° chamfer of fixed size on the corner of a section. The most usual design consists of a sole that can be moved up and down within the V-shaped body of the plane, although there are many variations on this theme.

An alternative approach is a plane with a movable fence and a brass face. The internal sides of the fence are at 45° and can be moved to expose the iron for the width of chamfer required. This type has some advantage as it can be used close to the end of an inset chamfer.

There are also a number of other designs based on a moulding plane format.

*Below left: An alternative form of chamfer plane [Melhuish, 1905]. Below right: A box-type chamfer plane by **C. Nurse** showing his claim as inventor, circa 1907. There is, however, no real evidence to support this claim.*

Price Guide

Chamfer planes

• A rare handled **Hart's** Patent, London, adjustable chamfer plane, 10" x 1¾" with internal brass depth stop and brass strengthening bar. (One of only three known). G++ **£262** [DS 25/842]

• A beech chamfer plane by **Nurse** with internal box-type sole secured with a fixing screw from the side of the plane, marked C. Nurse, Sole Inventor. G++ **£55** [DP]

• A **Preston** No. 360 movable fence type beech chamfer plane with scrolled wedge. **£40** [DP]

• A movable fence type beech chamfer plane by **Melhuish**, the iron set at a low angle. **£42** [DP]

• An adjustable box-type chamfer plane, 7" long. **£38** [DS 24/1142]

• A **Preston** No. 362 chamfer plane, with a cast iron box forming an adjustable sole positioned in front of the iron and secured to it with a screw. A rare type with few examples known. **£120** [DP]

• An unusual sloping front craftsman made chamfer plane with gunmetal fittings. **£55** [AP]

• A ⅜" side chamfer plane of moulding plane format, fully boxed (to cut a fixed size chamfer only) by **Malloch**, Perth. **£15** [DP]

• A "Moseley" type steel box chamfer plane by **John Moseley & Son**. The double cut iron is secured with two screws directly to the wooden stock. **£65** [AP]

• A boxwood craftsman-made adjustable box type chamfer plane, 5¼" long with a decoratively shaped brass plate for the securing screw. **£38** [AP]

• An attractive craftsman-made mahogany chamfer shave, 13½" o/a, with turned integral handles and two adjustable fences. **£32** [DP]

*Remember - condition is **G+** unless otherwise shown.*

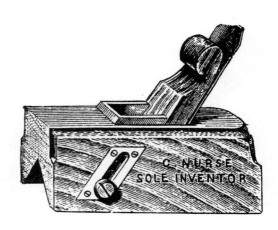

The idea of a combination plane – a plane where a number of uses are combined in one tool – is appealing but not easy to achieve.

There had been a few attempts to make such planes before the **Stanley Rule & Level Co.** marketed the Miller's patent adjustable metallic ploughs in 1871 (the No. 41+ 42). These planes were capable of ploughing, fillistering and tonguing but seem to have sold in small numbers as today, they are rare. They were withdrawn from sale in 1895.

Stanley 45 & 50

The first really successful combination planes were the Stanley No. 45 and the smaller No. 50, both introduced in 1884. These were a significant advance on the No. 41 as they had a double skate bottom, giving greater stability to the cutter and better resistance to digging in. The Achilles heel of all plough planes is that there is no sole in front of much of the cutter and the wider the cutter, the worse the problem. For hollow and round cuts, it is necessary to use special full length bottoms; these were marketed for the 45 in 1885.

The Stanley 45 was a very popular tool continuing in production until 1962. Over time, the decoration on the castings changed, gradually becoming plainer; to the initiated, this will indicate the approximate date of manufacture.

Record 405

Introduced in 1932/33 the Record 405 was a close copy of the Stanley 45. Production continued until around 1980. After a short break it was resumed by **Clico** of Sheffield.

Stanley 55

"The universal combination plane" introduced 1897 and withdrawn in 1962. Larger and more complex than the No. 45, this had the added advantage of a third sole and a vertical adjustment on this and one main sole, allowing the use of asymmetrical cutters. Anyone who has used a No. 55 will know that it is a large, heavy plane to handle and keep level – in truth, not very practical compared with its smaller brethren.

Stanley No. 55 set for making an asymmetrical moulding.

Collecting tips

➤ *The nickel plate finish on all Stanley planes, and to a lesser extent Record, was not durable. Earlier examples that retain most of the plating are the exception and command premium prices from collectors.*

➤ *If you want a combination plane for use – and they are useful in any workshop – don't worry too much about the finish.*

Price Guide

* A **Stanley** No. 45, complete in original box. **£121** [DS 23/442]

* A **Stanley** No. 45, complete but not in box. G **£65** [DP]

* A **Stanley** No. 50, complete but not in box. **£30** [DP]

* A **Record** 405, complete with instructions in original wooden box. **£75** [DP]

* A **Record** 405, complete with instructions in original wooden box. F **£84** [DS 24/1246]

* A **Sargent** No. 1080 combination plane in original box. **£90** [DS 22/985]

* A **Stanley** No. 55 combination plane with 55 cutters and instructions in original cardboard box, some plating coming away. G **£209** [DS 24/462]

* A **Stanley** No. 55 combination plane with 51 cutters in original box (lid split), some plating worn. **£190** [DS 21/684]

* A **Stanley** No. 55, complete in owner made box, usable not collectable standard. **£120** [DP]

* A **Howkins** Model B combination plane complete with fences, cutters and instructions in craftsman-made box. **£165** [DP]

* A little used **Stanley** No. 46, skew cutter combination plane with full set of eleven cutters. **£65** [DP]

Production of Record planes started in 1930. The makers were **C. & J. Hampton Ltd.**, an established Sheffield manufacturer of vices, cramps, etc. Most were copies of Stanley Bailey planes. Unlike other Stanley imitators, of whom there have been many, Record never compromised on quality and the planes rapidly gained acceptance. By 1934 the **Ed. Preston** planemaking activities had been incorporated into the firm.

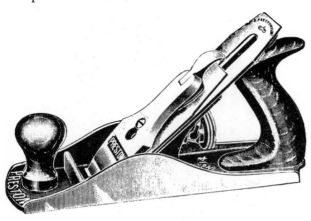

Preston No. 14 smooth plane [Churchill, 1935]. This was the last type of plane made by Preston. The Churchill 1935 catalogue lists Nos. 13 to 18.

Collecting tips
➤ *This is a field with increasing interest. Buy only the best – and now.*
➤ *Packaging is important. Items still in the original dark blue cardboard/cloth boxes with gold labels are especially sought. Later packing is white cardboard with light blue printing.*

Record planes are today still a market leader. The planes, like those of their principal competitor, the Stanley Works, have over the years been updated in small ways. Collectors' interest is in examples made before 1960.

Stay-set
Stay-set cap irons – an unfixed and quickly removable section of the cap iron allowing the iron to be touched up (sharpened) without loosening the screw – were an option introduced in 1933 and withdrawn in 1966. The plane stocks are exactly the same with the exception that the lever cap is marked SS. These letters were also used as suffix to the model numbers.

Price guide
All prices are for planes made before 1960 and in good collectable condition (G+ or better) with original little used irons. The more common items will have the original packaging.

No.		Type	Price
Bench planes			
02	7¼"	Smooth	**£150–200**
03	9¼"	Smooth	**£ 20–25**
04	9¾"	Smooth	**£ 15–20**
04½	10¼"	Smooth	**£ 15–20**
05	14"	Jack	**£ 18–25**
05½	15"	Jack	**£ 20–25**
06	18"	Fore	**£ 30–35**
07	22"	Jointer	**£ 35–45**
08	24"	Jointer	**£ 45–55**
T5	13"	Technical Jack	**£ 35–45**
Corrugated base bench planes			
03C	9¼"	Smooth	**£ 25–30**
04C	9¾"	Smooth	**£ 25–30**
04½C	10¼"	Smooth	**£ 30–35**
05C	14"	Jack	**£ 30–35**
05½C	15"	Jack	**£ 30–35**
06C	18"	Fore	**£ 40–50**
07C	22"	Jointer	**£ 50–60**
08C	24"	Jointer	**£ 60–70**
Rebate bench planes			
010½	9"	Smooth rabbet	**£ 35–40**
010	13"	Jack rabbet	**£ 40–50**

Planes with stay-set iron **+£10–30** depending on rarity of plane.

Record jack plane, No. 05½ C (corrugated bottom) [Churchill, 1939].

RECORD PLANES

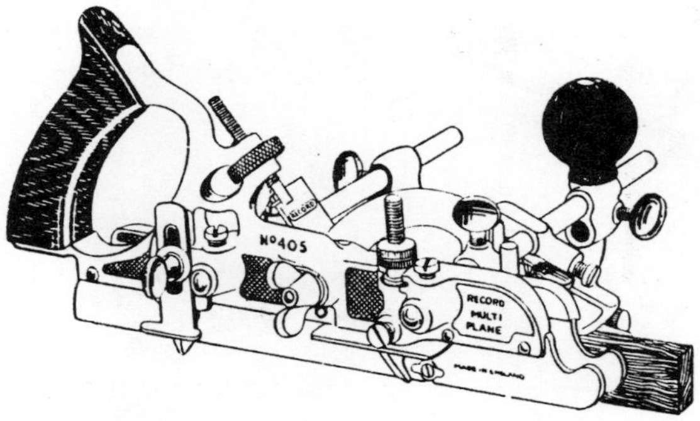

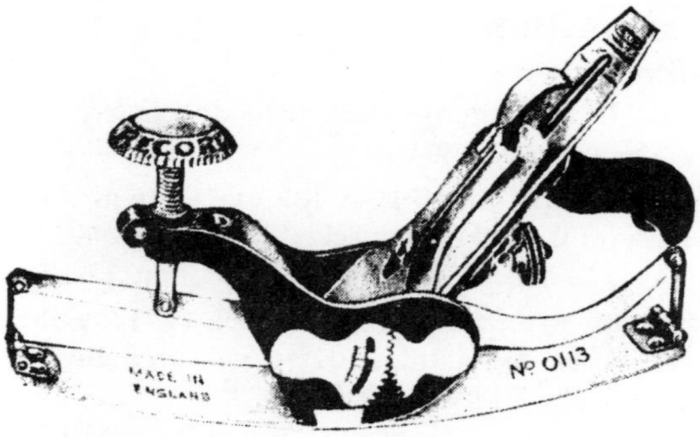

Top: No. 405 combination plane. Bottom: No. 0113 compass plane.

No.	Type		Price
Block planes – adjustable mouth			
09½	6"	Adjustable block pl.	**£ 18–22**
016	6"	Adjustable block pl.	**£ 18–22**
015	7"	Adjustable block pl.	**£ 18–22**
017	7"	Adjustable block pl.	**£ 18–22**
018	6"	Knuckle cap adjust.	**£ 25–35**
019	7"	Knuckle cap adjust.	**£ 25–35**

016 and 017 have nickel-plated lever caps.

Block planes – simple		
0110	Non adjustable	**£ 8–10**
0120	Adjustable	**£ 10–14**
0220	Adjustable	**£ 12–15**
0130	Double end	**£ 8–12**
0102	Non adjustable	**£ 6–10**
0230	Knuckle joint	**£ 20–25**

No. 713 skew rabbet plane.

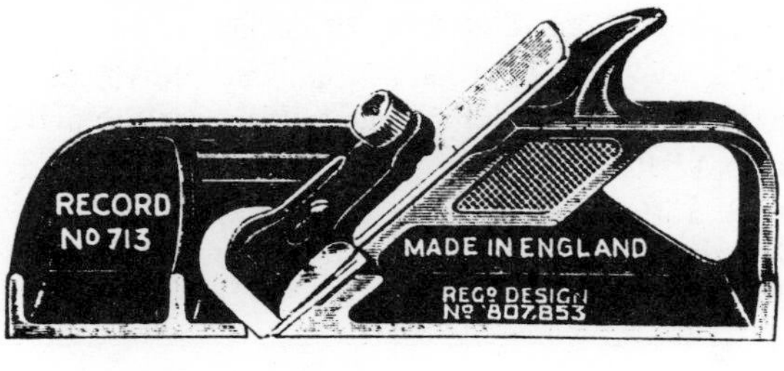

Plough and combination planes		
040	Plough	**£ 12–15**
043	Plough	**£ 12–15**
044	Plough	**£ 20–25**
050	Combination plane	**£ 30–40**
405	Multi-plane	**£ 60–80**
Pairs of bases for 405 with irons		**£ 25–35**
735	Soft & hardboard pl.	**£ 20–25**
730	Softboard plane	**£ 16–20**
Router planes		
722	Small	**£ 6–10**
071	Open throat	**£ 20–25**
071½	Closed throat	**£ 18–22**
Compass planes		
020	Centre screw	**£ 60–80**
0113	Forward screw	**£ 100–120**

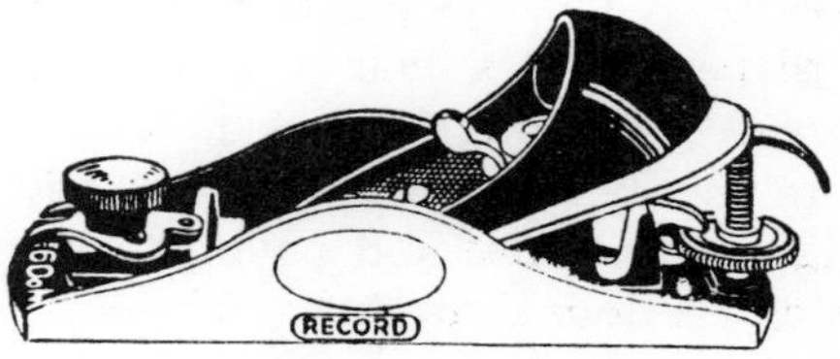

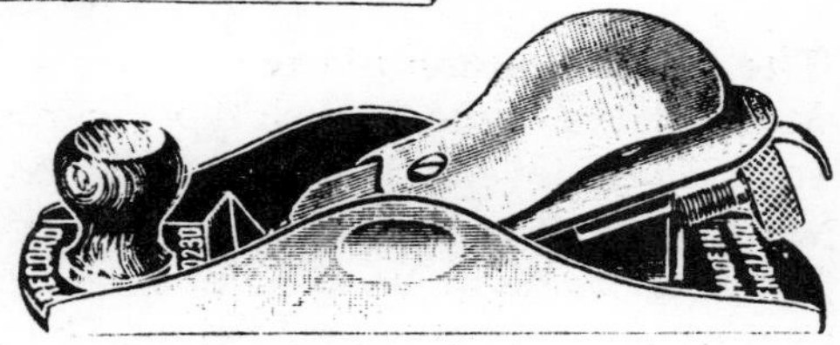

Top: No. 09½ (6") and No. 015 (7"). Bottom: No. 0230 Knuckle joint.

Rabbeting planes		
1366	Small bull-nose	**£ 35–45**
075	1" width bull-nose	**£ 6–10**
076	1⅛" width bull-nose	**£ 15–20**
077A	1⅛" width bull-nose	**£ 20–26**
311	3 in 1 plane	**£ 35–45**
2506S	Side rabbet	**£ 20–28**
041	5" shoulder	**£ 30–40**
042	8" shoulder	**£ 30–40**
072	8⅛", 1" width shoulder	**£ 50–60**
073	8⅛", 1¼" width shoulder	**£ 60–80**
074	8⅛", 1½" width shoulder	**£ 60–80**
712	8⅝", 1¼" width skew rabbet	**£ 40–50**
713	8⅝", $1\frac{1}{24}$" width skew rabbet	**£ 40–50**
714	8⅝", 1¾" width skew rabbet	**£ 40–50**
778	Rabbet	**£20–25**
078	Fillester	**£18–25**

Dovetailed steel mitre plane [Mathieson, c. 1900].

We have all become so used to the cast iron plane with knob and lever adjustment of the iron that it comes as a shock to many to be told that this now universal (Stanley) type of plane only gained general acceptance in Britain at the time of the First World War. Tradesmen are a conservative lot and it has been suggested that the slow changeover to metal planes was a result of this attitude. The reason is perhaps less obvious.

Until the middle of the 19th century metal planes were a rarity – as also was woodworking by machine. Metal planes, besides their cost, had one particular drawback – although easier to adjust they require more energy to push. When virtually all woodworking was by hand and payment was by results, this was a serious drawback.

The English metal plane

Metal planes of the English type – wrought iron or steel plates dovetail jointed together to form a box which was then "stuffed" with a wooden infill – had been made in small numbers during the 18th century and probably long before. The early planes are of mitre plane shape, i.e. a parallel sided, round heeled box, occasionally in early planes with the front plate extended to form a tote.

Around 1840, **Stewart Spiers** of Ayr started to make metal planes in quantity, producing not only planes of mitre type but also smoothing, panel and rebate types. Other makers soon followed and by the end of the century, the traditional form of metal plane was to be increasingly found in the tool kit.

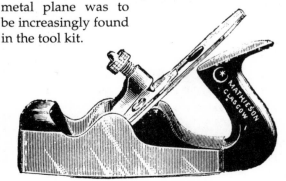

Handled steel smoothing plane [Mathieson, c. 1900]

Price Guide

Mitre planes

• A 10" brass mitre plane with dovetailed steel sole by **Gabriel**. £550 [DP]

• A cast iron mitre plane, 10½" x 2⅛", with rosewood infill and **Ward** snecked iron. £110 [DP]

• A brass mitre plane, 10¼" x 2¼", with d/t steel sole and protruding toe and heel, by **W. Dibb**, with rosewood infill and wedge (probably replaced). G £594 [DS 23/1322]

• A small steel mitre plane, 8" x 2", by **Rt. Towell**, London, with fine mouth, rosewood infill and wedge behind cupid's bow bar. £440 [DS 25/656]

• A d/t steel mitre plane, 9¾" x 2½" by **Buck**, Tottenham Ct. Rd., with rosewood infill. Well shaped cupid's bow steel bridge, fine mouth and snecked iron. £320 [AP]

• A d/t steel mitre plane by **Holtzapffel & Co.**, 10" x 2⅜" with rosewood infill and scrolled wedge, cupid's bow bar and snecked iron by **Sorby**. Some discolouration to sole. £1815 [DS 25/1694]

• A cast iron mitre plane by **Spiers**, Ayr, marked on the rosewood infill, 10½" x 2¾" with brass lever cap and snecked iron by **Ward**. £400 [DP]

Panel planes

• An iron panel plane by **H. Slater**, Maker, London, 18" x 3", with figured rosewood infill and handle and brass lever. Minor pitting to one side, otherwise G++ £242 [DS 25/635]

• A d/t steel panel plane by **Mathieson**, 13½", with rosewood infill and handle and original **Mathieson** iron. G++ £396 [DS 25/1501]

• A late model 17½" **Norris** A1 panel plane with figured rosewood infill and handle, 90% of **Norris** iron remains. G++ £396 [DS 25/1673]

• A 15½" d/t steel panel plane by **Norris** with rosewood infill and original little used **Norris** iron. £360 [DP]

• A 15½" adjustable cast iron panel plane by **Norris**, late model with beech stuffing. £240 [DP]

Dovetail steel jointer plane [Mathieson, c. 1900]. **Mathieson** *describes all planes above 20½" as jointer planes.*

Steel/cast iron

The dovetailed steel plane will not shatter if dropped or accidentally hit with a hammer when setting the iron. Drop a cast iron plane and you are likely to have a large chunk break out. By 1900 some makers were describing their planes as "malleable iron" but this is just heat treated cast iron – a little less brittle but certainly not unbreakable.

Manufacture of a dovetailed steel plane was a lengthy and skilled process. Unlike wooden dovetails, which can be drawn apart, metal dovetails are two directional – the metal is spread with a hammer during assembly and cannot come apart – we have never seen one that has.

Surprisingly in 1900 the difference in price between a dovetailed steel 14½" panel plane (30 shillings) and a cast iron plane (28 shillings) was little. The cost of a Stanley No. 5½ taken from the same catalogue was 10 shillings.

Collecting tips

➤ *Inspect cast iron planes with the greatest of care. Cracks are very difficult to see if the plane is dirty – and will show clearly after cleaning.*

➤ *Is it cast iron or dovetailed? The dovetails will always be visible in a careful inspection.*

➤ *Unless gunmetal planes have become corroded or green, do not give them more than a light rub with a rag. If you try to clean them further you will be left with areas that you have cleaned and areas you cannot reach. The result looks patchy.*

Price Guide

Panel planes cont.

• A d/t steel panel plane by **J. Buck**, 13½" x 2¾", with rosewood infill and handle. **£137** [DS 23/1337]

• A 25½" d/t jointer plane by **Norris** with rosewood infill and 80% of original early **Norris** iron. Minor pitting spots. **£1320** [DS 24/1712]

• A 14" cast iron craftsman-made panel plane with well finished mahogany stuffing and attractive brass bridge. **£80** [DP]

Smoothing planes

• A cast iron adjustable coffin smoothing plane by **Norris**. Late closed handle model, with beech infill shedding varnish. With original almost unused **Norris** iron. **£120** [DP]

• A d/t steel coffin smoothing plane by **Tyzack** unhandled, with rosewood infill, brass lever cap and screwed sides. **£104** [DS 25/633]

• A d/t steel smoothing plane by **Preston** unhandled, with rosewood infill and brass lever cap. Minor pitting to sole. **£104** [DS 25/1000]

• A late model **Norris** A5 smoothing plane in original box (poor), 90% of original **Norris** iron remains. G++ **£242** [DS 25/1679]

• A d/t steel coffin smoothing plane by **Mathieson** with rosewood infill and handle, light pitting spots. **£99** [DS 25/1497]

• A cast iron parallel smoothing plane by **Marples** with rosewood infill. G **£66** [DS 23/1330]

• A gunmetal parallel smoothing plane with steel sole and ebony stuffing by **Miller**. **£275** [DP]

Shoulder planes

• A craftsman-stuffed cast iron shoulder plane, 8¼" x 1⅜", rosewood stuffing and wedge. **£40** [DP]

• A cast iron shoulder plane by **Slater**, 7¾" x 1⅜" with rosewood stuffing. Little used. **£65** [DP]

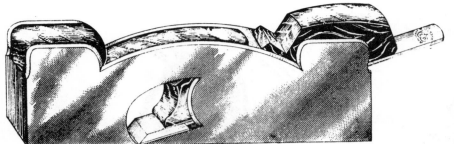

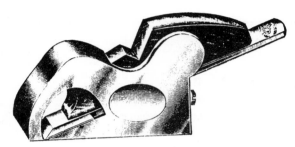

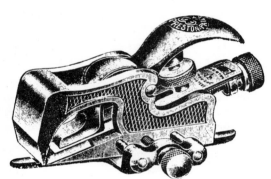

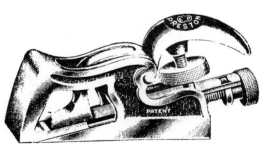

*From the top: 1) Typical turreted design commercially made shoulder plane. These were available in both cast iron and gunmetal. Craftsman-made examples are frequently more elaborate and may incorporate rhino-horn totes. 2) Bull-nose rabbet plane of round nose form. Again available in cast iron or gunmetal, there were three different widths. 3) **Preston's** patent adjustable bull-nose side rabbet and chamfer plane. Introduced in 1914, this inventive design allowed the plane to be used for chamfering when two additional fences were fitted. 4) One of the many **Preston** designs for adjustable bull-nose planes. [Preston, 1914]*

Gunmetal

Gunmetal and brass do not lend themselves to dovetailing so well as steel so most planes made of these metals will have been cast. The advantage of gunmetal is that it is not brittle but it is really too soft to make a long wearing sole and therefor the best quality planes will have a steel sole. Gunmetal planes are certainly fine looking and today will command higher prices than the equivalent cast iron plane.

Mitre planes

The earliest form of metal plane, with 18th and early 19th century examples to be found. Later examples (post 1890) may be of the so-called "improved type". Here the sides are cut away and the impression is less box-like. The mouths of mitre planes should be very fine. If they have been refinished, this will be apparent in the mouth. The earlier mitre planes had wedges to secure the iron but after the development of the lever cap in the 1840s these were fitted to mitre planes, although wedged examples were also made until late in the century.

Shoulder/rebate planes

The exact definition of a shoulder/rebate plane is a little difficult as they are in most respects similar and many makers seem to have been imprecise in their terminology.

Shoulder plane: Originally used for trimming up tenons, these have a low angle iron to give a good cut on end/cross grain. In general, the plane will have a turreted or otherwise shaped top (this allows it to be held when working on the side), will be made from a casting and will be 1¼" wide.

Rebate plane: Characteristically of dovetail steel (although later proprietary examples are cast), these are of simple rectangular shape. The angle of the iron may be a little higher than that of shoulder planes but it should be remembered that the angle of cut in all these bevel up planes is dependent on the angle of the bevel much more than the angle of the iron.

Bull-nose plane: These are the short nosed brethren of the rebate plane family allowing the plane to be worked close to an obstruction. Almost without exception the planes are castings. The traditional forms (square nose and sloping nose) were made by most of the metal planemakers. In the 20th century **Ed. Preston** produced a surprising number of different varieties of these planes, both with and without screw adjustment.

Chariot planes

Small planes around 3" long with the mouth well to the front. Often made in gunmetal, of pleasing proportion, these were the forerunner of the smaller one-hand types of block plane and were rendered obsolete in the 20th century by the cheapness of the cast iron mass produced product.

Chariot plane by **Norris** *[Buck & Ryan, 1932]. This plane cost 24s. The equivalent cast iron block plane cost 3s.*

Metal planemakers

Stewart Spiers, Ayr. 1840-1938: Until at least 1919 it would seem that all the bench planes were of dovetailed steel construction. Thereafter cast planes were added to the range. Some of the types made only in the last days of the firm are very rare.

Thomas Norris, London, 1872-1943: Products of this firm are more highly esteemed by both users and collectors than those of any other metal plane manufacturers. The reason for this is that Norris was the only manufacturer of traditional type English planes to develop an adjusting mechanism – and very smooth acting it is. That they were the last manufacturer of traditional style planes and were working within the memory of many craftsmen still alive also seems to have earned this firm a special place in the esteem of tool buffs.

John Holland, London, 1861-1892: A name to be found mostly on gunmetal shoulder and bull-nose planes.

Henry Slater, London, 1868-1909: Manufacturer of cast iron and gunmetal planes; the shoulder planes are numerous.

George Miller, London, 1890-1914: Gunmetal planes seem to have been something of a speciality.

George Buck, London, 1838-1980: BUCK/TOTTENHAM COURT ROAD is to be found on many metal planes, but this firm, like other principal tool retailers of

Price Guide

Shoulder planes cont.

◆ A gunmetal shoulder plane, 8" x 1½", by **Badger**, London, with ebony infill and wedge. **£104** [DS 23/1323]

◆ A d/t steel **Norris** shoulder plane, 7¾" x 1½", rosewood stuffing and original iron. **£140** [DP]

◆ A steel soled gunmetal **Norris** shoulder plane with ebony wedge and infill. 30% **Ibbotson** iron remains. **£297** [DS 25/1683]

Rebate planes

◆ A d/t steel rebate plane by **Buck**, 9" x 1", with rosewood stuffing. **£65** [DP]

◆ A d/t steel **Norris** rebate plane, 9" x ¾" with rosewood infill and wedge, fine mouth. 50% **Ward** iron remains. G++ **£104** [DS 25/1682]

◆ A **Preston** No. 1367, nickel plated cast iron rebate plane, 5" x ⅝". **£55** [AP]

◆ A d/t steel rebate plane, 9" x ¾" by **Spiers** with rosewood infill. **£60** [AP]

◆ An unusually small iron rebate plane by **H. Slater**, Clerkenwell, London, 4¾" x ½", with ebony infill and wedge. **£132** [DS 25/622]

◆ A small and elegantly shaped gunmetal rebate plane, 6¼" x ⁹⁄₁₆", with decorative shaped top and projecting ebony stuffing. **£110** [DP]

Chariot planes

◆ An cast iron chariot plane by **Slater**, 3¾" x 1½" with rosewood wedge and brass bar. **£75** [DP]

◆ A fine steel sole gunmetal chariot plane, 4" x 1⅞", by **Miller**, Grays Inn Rd. with ebony infill and wedge. F **£429** [DS 24/1074]

◆ A gunmetal chariot plane, 4½" x 1¾", with ebony wedge but with no steel sole. **£55** [DP]

◆ A cast iron chariot plane by **Preston**, 3¼" x 1⅜", with rosewood wedge and brass bar and original **Preston** iron. **£77** [AP]

Bull-nose planes

◆ A **Norris** A27 adjustable steel sole gunmetal bull-nose plane with rosewood wedge. 10% original **Norris** iron. **£1045** [DS 25/1688]

◆ A gunmetal bull-nose plane, 3¾" x 1⅛" by **Slater**, square front type with rosewood wedge. **£85** [DP]

◆ An iron bull-nose plane, 3¾" x 1⅛", of sloping nose type, marked MALLEABLE IRON but without maker. **£36** [DP]

◆ A cast iron **Preston** adjustable bull-nose plane, 4" x 1⅛", nickel plated. F **£40** [DP]

*Remember - condition is **G+** unless otherwise shown.*

the late 19th/early 20th century whose names are to be found on metal planes, did not actually make them.

Considerable numbers of metal planes were also made by some general tool manufacturers – **Alex. Mathieson** and **Ed. Preston** come immediately to mind – and there were also a number of less prolific firms. As much as is known about all these firms is detailed in *BPMIII*.

Stanley Rule & Level Co.

In America there had been fitful, then accelerating efforts to produce effective metal planes for some decades before 1869. In that year, the **Stanley Rule & Level Co.** purchased **Bailey, Chany & Co.**, leaders in this quest, who were already manufacturing planes in small numbers. Thereafter the numbers of metal planes made by this company grew rapidly though it was not until the 1880s that significant numbers started to come into Britain.

The rest, as they say, is history. Within 30 years both the wooden planemakers and the makers of English type metal planes were under severe competition.

Prompted by the success of **Record**, in 1936 **Stanley** started to manufacture the better selling block and bench planes in Sheffield.

Stanley planes are now much collected on both sides of the Atlantic, with the earlier and rarer items fetching some of the highest prices of any collectable tools.

The failure, before 1930, of any British manufacturer to produce bench planes to compete with **Stanley** is surprising – particularly as **Ed. Preston** had by 1900 produced a fine range of shoulder planes. But this firm seems to have had something of a mental blockage when it came to bench planes, whilst the other makers of traditional metal planes were mostly just too small to pay for the patterns, tooling and machinery needed – and in any case couldn't see the point of making a plane that was not as good as the traditional planes they were already making.

Collecting tips

➤ *Many firms sold cast iron castings unfinished or partly machined for finishing by the craftsman. This is the source of a considerable number of metal planes. For this reason, they are often very similar to the commercially made cast iron planes, particularly as brass lever caps were also supplied.*

➤ *What wood has been used for the stuffing? Mahogany stuffing was virtually never used by commercial makers but often by the craftsman. Commercially made planes are stuffed with rosewood, ebony or beech, varnished over.*

➤ *Metal planes of jack plane size are termed panel planes and those of try plane size are described as jointer planes.*

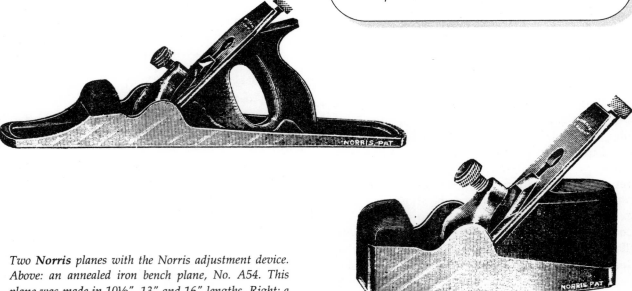

*Two **Norris** planes with the Norris adjustment device. Above: an annealed iron bench plane, No. A54. This plane was made in 10½", 13" and 16" lengths. Right: a dovetailed steel smoothing plane, No. A4.*

Measuring

MEASURES

In 1963 the British government decided that the country would change over to the metric system of measurement. This seemed quite simple; all that was being done was changing to a well-established system, one already in use in science and medicine.

Over thirty years later, the road signs still indicate miles, petrol prices are shown in gallons – whilst the pumps discharge in litres – and draught beer is dispensed in pints but canned beer is sold in millilitres. The populace of Britain is not alone in resisting change; the Paris garment trade were still making in French inch sizes well into the 20th century, eighty years after the adoption of the metric system.

Perhaps the desire to hold onto what was known was one of the influences that defeated successive English monarchs who had tried to establish countrywide and accurate systems.

Eventually order came. In 1491, Henry VII and his parliament decreed that new standards of length, weight, and capacity should be made and in 1495 copies of these standards were distributed to principal towns. However something seems to

have gone wrong as only one year later the standards were recalled as they were defective. In 1497 new standards were distributed to forty-three English towns for which they again had to pay.

Some weights and the yard standard of this series have survived and are now in the Science Museum in London. The yard is an octagonal brass rod, the measure being from end to end. It is divided into 3 feet, one of which is further divided into 12 inches.

Elizabethan advances

By the time Elizabeth I came to the throne sixty years later it was apparent that differences had again developed in the standards used in various parts of the country. This resulted, by 1588, in the issue of new sets of standards to around sixty towns in England and Wales. The standard weights were used from 1588 until 1824 for the purpose for which they were intended, i.e. checking the weights used in trade throughout the kingdom, and quite a number are now in the Science Museum. They were of remarkable accuracy; when weighed in 1873 most were accurate to around one part in 3,500.

MEASURES

At the same time, in 1588, a new yard – a bronze rod ½" square giving the measure end to end – and a cloth ell of 45" were also constructed. A bronze bed to contain these was also made and this was used to check the length of the subsidiary yards.

The degree to which the size of the yard was maintained from 1497 in the reign of Henry VII until 1824, when the advancement of science made new standards desirable, is remarkable. The 1497 yard was only 0.037 inch and the 1558 yard 0.01 inch short of the standard set in 1824.

Today, volumes are derived from linear measurement but, since the earliest of times, volumetric measures had been needed and in 1601, new standards for capacity were also set as part of the Elizabethan redefinition of standards. For dry measure (principally corn), the Winchester bushel and gallon and, for liquid measure, the ale gallon (268 cu. in.), quart and pint were set. The wine trade had traditionally used a smaller gallon of 231 cu. in. and this caused confusion, so the wine gallon was recognised in an act of 1707. This was the gallon (20% smaller than the 277 cu. in. imperial gallon of 1824) adopted in the United States.

1824 standards

The 1824 standards were set on a much more scientific basis, weight being defined by one two-pound troy weight and length by reference to a yard measured between inscribed lines. Other weights and lengths were then derived from these standards. From this point on science took over, and as collectors, we need no longer be concerned with the standards, only with the application.

Glorious confusion

If everything in daily life had been measured on the basis of the primary standards all would have been straightforward, but the whole history of

Set of grain weights: 10 grains and its sub-division into 1,000 parts [Harris, 1896].

weights and measures seems to be a long story of governments with the best of intentions setting standards which, for reasons ranging from prejudice to criminality, were ignored or subverted by the people. The happenings of the past 30 years are but one more turn of the wheel.

Trades have always used what is easiest and most effective for them with little regard for authority and in so doing have produced measurements which may be unintelligible to those outside the coterie – but then, this can be of use in keeping outsiders at bay.

Hats were measured using a notional diameter even though most were oval; the purity of silver was expressed in pennyweights; the thickness of glass was described in ounces; paper sizes included glorious names like half-imperial and double-elephant and slates had somehow become related to queens and duchesses. For the collector such complexities are a world to explore.

British Imperial Liquid and Dry Capacity Measures – after the Act of 1824.								
Cu. inches	Gills	Pints	Quarts	Gallons	Pecks	Bushels	Quarters	Chalders
8.669	1							
34.677	4	1						
69.355	8	2	1					
277.420	32	8	4	1				
554.840	64	16	8	2	1			
2,219.360	256	64	32	8	4	1		
17,754.880	2,048	512	256	64	32	8	1	
79,896.960	9,216	2,304	1,152	288	144	36	4.5	1

Before 1824, in England, the ale gallon contained 282 cu. in. and the wine gallon contained 231 cu. in. After the Imperial Weights and Measures Act of 1824, the gallon used for all dry and liquid measure contained 277.42 cu. in.

Wine measures and cask volumes before 1824

Pints	Quarts	Gallons	Rundlets	Barrels	Tierces	Hogsheads	Puncheons	Butts	Tuns
1									
2	1								
8	4	1							
144	72	18	1						
252	126	31.5	1.75	1					
336	168	42	2.33	1.33	1				
504	252	63	3.5	2	1.5	1			
672	336	84	4.667	2.667	2	1.33	1		
1,008	504	126	7	4	3	2	1.5	1	
2,016	1,008	252	14	8	6	4	3	2	1

Ale measures before 1824.

Ale was measured only in pints, quarts, gallons and barrels and hogsheads, which were of different volume to the wine casks.

Pints	Quarts	Gallons	Barrels	Hogsheads
1				
2	1			
8	4	1		
272	136	34	1	
408	204	51	1.5	1

Customs & Excise

Since 1643, successive governments have levied excise duties on the production of a whole range of products and even before this date, customs duties were levied on the import and export of some goods. For centuries duties were the backbone of government revenue – duties on various goods came and went but beer, wines, spirits and tobacco remained favourite targets.

The need to accurately measure quantities and assess strengths of alcoholic drinks was basic to the application of duties and this need spawned a whole range of rules, slide rules and equipment. Indeed a principal application of the slide rule in the 18th century was calculation for excise purposes

By as early as 1700, the revenue system in England was an extremely well-organised and effective system. Officers regularly visited all manufacturers within their areas assessing what was due. The assessors were forbidden to collect any monies, and separate collectors received them. These were well-paid jobs in an agency which, unlike most at that time, did not operate on a system of institutionalised graft. During the 18th and 19th centuries, numerous manuals were published to instruct the officers in the performance of their duties, in both gauging and administration.

It is not surprising therefore that equipment has survived in some quantity. The earliest dates from around 1700 or even before, the latest being only 20 or 30 years old.

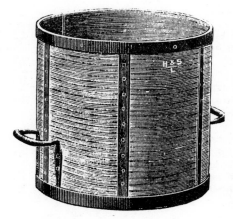

Corn measure (Harding, 1906). This measure came in bushel and ½ bushel sizes and was government stamped.

British Imperial Linear Units of Measurement – after the Act of 1824

Inches	Gunter's links	Feet	Yards	Fathoms	Rods	Gunter's chains	Cable lengths	Furlongs	Miles	Nautical miles	Leagues
1											
7.92	1										
12	1.515	1									
36	4.545	3	1								
72	9.091	6	2	1							
198	25	16.5	5.5	2.75	1						
792	100	66	22	11	4	1					
7,200	909.091	600	200	100	36.36	9.091	1				
7,920	1,000	660	220	110	40	10	1.1	1			
63,360	8,000	5,280	1,760	880	320	80	8.8	8	1		
72,960	9,212.160	6,080	2,026.680	1,013.340	368.486	92.122	10.133	9.212	1.1515	1	
190,080	24,000	15,840	5,280	2,640	960	240	26.4	24	3	2.605	1

European inches

Almost every major trading centre in Europe had its own local foot and inch – indeed there were over one hundred in use in the early years of the 19th century. The inches shown in the table below are commonly found on rules, often two or three on one rule.

Rules bearing a very large inch, the Russian archine, 1¾" long, are relatively common and frequently perplex buyers who cannot imagine an inch this large and think that this must be an oversize scale. Rules marked with other inches used in the Baltic timber exporting ports are also to be found. It is interesting to note that most of these rules are of English manufacture.

Comparison of the English inch with some of the European inches in use in 1821.

	Ins.	Cm.
English Inch	1	2.54
Revel (Estonia)	0.88	2.23
Riga (Latvia)	0.90	2.28
Madrid	0.93	2.36
Amsterdam	0.93	2.36
Stettin (Poland)	0.93	2.36
Antwerp	0.94	2.38
Hamburg	0.94	2.39
Danzig (Poland)	0.94	2.39
Groningen (Netherlands)	0.95	2.43
Swedish	0.97	2.47
Konisberg (Prussia)	1.01	2.56
Copenhagen	1.03	2.61
Rheinland (Germany)	1.03	2.61
Vienna	1.04	2.63
Paris	1.07	2.71
Lisbon	1.08	2.74
Archine (Russian)	1.75	4.45

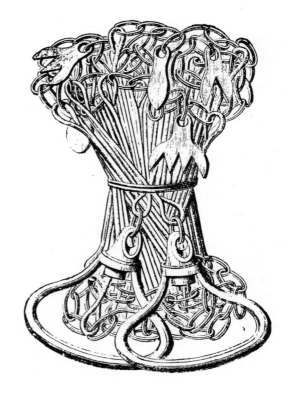

Land chain, 4 poles long, divided into links and tallied every ten links [Chesterman, 1880].

Two of the earliest British rules known to us are not made of boxwood A 24" rule from the 1545 wreck of the *Mary Rose* is made of oak and another rule, dated 1552, is made of ebony. Both are laths, that is without any joint. The *Mary Rose* rule is marked on both sides with various sections divided into inches, half-inches and quarter inches. The most surprising feature is that it also bears a line of timber measure. (See also *The Carpenters' Rule*, pp.160–162.)

Boxwood

By the end of the 17th century and probably long before, boxwood – hard-wearing and capable of holding precise clear markings – had become the preferred material for rules of all types. The exceptions are few; some of the largest gauging rods and other instruments were occasionally made in satinwood or lancewood but these are very similar to boxwood.

By the middle of the 19th century the South American Maracaibo boxwood had become the usual material for making rules. The supplies of European boxwood could never have been sufficient for the huge numbers made thereafter.

The process

Whether in ivory or boxwood, the first step was to properly season the material. The boxwood logs were seasoned under cover and then sawn into the small laths needed for rulemaking. The wood was seasoned for a further period. The sticks were then prepared – cut to length, slotted and rebated to receive the various brass fittings.

The small brass plates, disks, tubes etc. that are needed for the joints were originally cast, filed, sawn, etc. by hand but by 1800, and possibly even earlier, the Birmingham makers had started to mechanise this process, using punches and presses. This type of small-scale metal working technology was already well established in 18th century Birmingham in trades such as button and silver-plate making. **John Rabone & Son** had gradually been introducing this production technology from the 1820s onwards. By 1871, when they moved to their new Hockley Abbey works, most of the process had been mechanised.

The rule was assembled by inserting brass pins into holes drilled through the brass fittings and the boxwood. If the ends had brass plates, the pins were hammered to form small rivets. The rule was then filed to bring the brass fittings and the boxwood level and make the edges square. This process was also mechanised by the use of sanding machines, though some filing of areas that could not be reached by machine, and on small run production, continued into the 20th century.

Marking

The finished but blank rules were polished with a shellac-based polish/sealer. The markings were then applied with a marking knife, guided by a square, working from a master rule placed alongside. Whilst this all sounds very slow, it is recorded that a good worker could mark the 540 lines of a Gunter's scale in 10 minutes – about one line per second – so a straightforward rule might have only taken a couple of minutes. The numerals and other lettering were then stamped on with sharp-edged punches.

When the marking was complete, a mixture of carbon black in oil was rubbed into the markings – the first coat of polish protected the general surface – and the result was a very durable and clearly marked rule. Finally a second coat of polish was applied.

Hand marking of small run orders continued as long as quality boxwood rules were made – and this was until about 1960. For major lines, mechanical marking, using dividing engines or pantograph-type machines capable of marking a number of rules at one time (some sources suggest up to fifty), were in use by the middle of the 19th century.

Collecting tips

➤ *A **rule** is an instrument for measuring and/or calculating. A **ruler** is for drawing straight lines, but may also be marked with some measures.*

➤ *The quality of rules is important so check carefully. Like most tools, the quality of rules started to decline with the First World War and never stopped.*

➤ *An example of the care which rulemakers took can be seen in the knuckle joints. Look carefully: on the best quality rules you will see that the plates in the two legs are made of dissimilar metals. The belief that dissimilar metals should be used in bearings applied, even if they only moved as little as a joint in a rule.*

Decline

Printing of the numerals onto the surface of the rule, as opposed to the much more durable inscribing, seems to have started around 1910 with the so-called blind man's rules.

Thereafter, this spread to the cheaper lines and by the 1930s it had become common practice also to print the markings. Rules made by this process have neither the durability nor preciseness of the better quality rules – the collector will have little difficulty in identifying them. The wood surface is not flat, having been deformed by the press, and the much poorer quality of the lacquer finish will also betray late rules – they should be avoided.

RULE MAKERS.

Aston Thomas	Harris T.	Rabone Elizabeth
Bakewell Richard	Ludlow Thomas	Rabone John
Betts James	Morris W.	Richards John
Betts John	Onions Benjamin	Salt Abraham & Son
Cox George	Parke Samuel	Shaw William
Farrol John	Powell Robert	
Harriman Thomas	Powell ———	

List of rule makers from Wrightson's Triennial Directory of Birmingham *for 1823.*

Birmingham

In the 19th century, Birmingham became the principal centre of rule making. The London makers who survived were forced to specialise in the low volume part of the market. In Birmingham, the largest rule maker was undoubtedly **John Rabone & Sons.** Pick up a handful of rules and the chances are that nine out of ten will have been made by them.

The Rabone firm

The firm was founded in 1784 by Michael Rabone and his wife, Elizabeth, in Bull St., Birmingham; by 1803 they had moved to Snow Hill. Michael died shortly after and for some years, the business was conducted by Elizabeth. As soon as their son, John, was old enough he joined the business and by the 1820s he had assumed control and had moved it to St Pauls Square.

The firm was already growing. By 1834 a steam engine had been installed to drive machinery and indications are that from then on there was a steady increase in mechanisation with the machines designed and made in house.

In 1845 John Rabone, Junior joined his father, and the name was changed to John Rabone & Son. In 1877 when John Junior's sons became partners the name changed to "& Sons".

Market leader

In 1871, the firm moved from central Birmingham to a new factory at Hockley Abbey to the north of the city (hence the trade name). The principal partner of the time, John Rabone Jnr., appreciated the importance of introducing as much machine-working as possible, and this he proceeded to do – but not without considerable opposition to the extent that he was stabbed by a disgruntled worker.

The degree of mechanisation by the 1870s is clear from a lengthy description of the Hockley

*A **John Rabone & Sons** facsimile label from the 1892 catalogue. These were reproduced in this catalogue because "A certain unscrupulous Merchant having issued spurious labels, somewhat similar and in imitation of ours, upon inferior goods, the subjoined fac-similes are given for the protection of buyers and the public."*

Price Guide

Boxwood arch joint rules, late 19th/20th century: 2ft. 4-fold general purpose rules with engraved (not printed) markings **£6–£10**. Ditto in fine or mint condition **£10–£20**. 3ft. 4-fold rules **£10–18**. Ditto in fine or mint condition **£18–25**. The original quality of rules is important. Round joint and printed rules are of lesser value.

Boxwood 12" 4-fold rules

• A 12" 4-fold boxwood and brass rule, round joint, marked **Hockley Abbey**. **£16** [DP]

• A 12" 4-fold boxwood and brass rule, arch joint, by **Rabone** in original leather case. F **£28** [DP]

Boxwood 2ft. 4-fold rules

• A 2ft. 4-fold boxwood rule by **Rabone** with arch mid-joint. **£25** [DP]

• A 2ft. 4-fold boxwood and brass rule of 18th C. appearance, thick section and large round joint. **£65** [DP]

• A 2ft. 4-fold fully brass bound boxwood rule by **Rabone** No. 1101. G++ **£38** [DS 23/1190]

• A 2ft. 4-fold boxwood rule by **Buck,** inside and outside edges brass bound and marked. **£30** [DP]

• A 2ft. 4-fold boxwood and brass rule by **Preston** No. 3444 with brass depth slide. G++ **£38** [DS 24/857]

• A 2ft. 4-fold architects' rule with single bevels by **Preston** with edge plate joint. **£24** [DP]

• A 2ft. 4-fold boxwood "EesEseE" rule by **Preston. £28** [DP]

• A 2ft 4-fold boxwood Braille rule by **Rabone** – lettered and marked with small brass pins. **£28** [DP]

• A 2ft. 4-fold boxwood rule by **Preston** with advertisement for Jennings (Bristol) Ltd., Joinery Manufacturers, **£18** [DP]

• A 2ft. 4-fold square section boxwood rule, marked with metric measure and London, Paris and Rheinland inches. **£45** [AP]

• A 2ft. 4-fold boxwood rule by **Smallwood,** Birmingham, twenty-five Gamla inches long also marked in metric. **£30** [DP]

Boxwood 3ft. 4-fold rules

• A recent 3ft. 4-fold combination boxwood and brass rule and level with protractor hinge by **Rabone** No. 1190. G++ **£22** [DS 24/1307]

• A 3ft. 4-fold boxwood rule with spirit level and disk joint by **Rabone. £24** [DP]

• A 3ft. 4-fold boxwood architects' rule by **Rabone. £32** [DP]

Abbey machinery written for a local newspaper. The following is a brief extract.

…One is an adjustable steam punch which stamps with astonishing quickness pieces of thin brass for the large hinges… The pieces stamped out have to be turned or otherwise dressed by machinery, so as to bring the edges to the utmost nicety… Elsewhere in the room, 3 beautifully constructed automata are busily at work unattended, turning out small brass washers of perfect shape and size for the double joints. At another machine another part of the joint is being prepared; and at others the various parts are being placed together on the central wires… After being placed together roughly, the joints and woods are submitted to a machine which fixes [holds] them firmly and in an exact relationship, after which the rivet holes are struck through. One girl puts the pins in; another administers a few taps with a hammer which effectually closes the rivets and the joint is completed... the rule passes to a machine which rubs off all ugly excrescences; then to others which brings all the edges square; and finally to an arrangement of wheels, rubbers, whereby it is dressed up to a beautifully finished smoothness…

The article then goes on to describe the marking of the divisions and the numerals which were also performed by machine. It also comments upon the extreme subdivision of labour whereby each person carried out only one of the up to one hundred separate operations needed in the manufacture of a rule. Even with help of all the machines it must have been soulless toil, for the wages were mostly piecework payments – and were dependent on results.

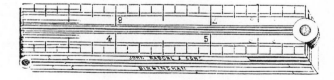

*Top: 12" 4-fold **round joint** ivory rule with brass mounts. Bottom: 12" 4-fold **arch joint** ivory rule with German silver mounts.*

In 1945 **John Rabone & Sons Ltd.** were producing 27,000 boxwood rules per week. The immediate post-war period was a busy one but even so, this figure is an indication of the quantities made by this firm.

For the collector these numbers are of significance – boxwood rules are common. Were it not that boxwood rules used on the building site deteriorate quickly, with death from a fractured femur a frequent occurrence, the 4-fold rule would be so common as to create no interest amongst collectors. And this is not the only hazard; oils will darken rules to the point of illegibility; damp will quickly rust the iron hinges, causing black and immovable stains, and chips and cracks propagate readily.

Considering the numbers made, boxwood rules that have survived in fine condition are surprisingly few.

The 2ft. 4-fold

The 2ft. 4-fold is the commonest form; the 3ft. 4-fold less so. The 4ft. 4-fold, which was described as a carriage builders' rule, is rare and most of the examples seen are warped, always a problem with boxwood and a common fault even with the 2 and 3 ft. rules.

All 2 ft. 4-fold rules are not the same so look carefully. Like most products before the rationalisers got to work, makers sold the widest variety of types and quality.

Getting it straight!

> The terms 2-fold and 4-fold can often cause confusion: 2-fold rules have only one joint and therefore two sections; 4-fold rules have three joints and four sections.

> 3-fold and 6-fold rules were also made, though these are rare; they are usually ironmongers' rules.

> Rules where the sections are face to face and joined by pivots are known as zig-zag rules.

The 1ft. 4-fold

Charming small versions of the 4-fold rule, intended for the gentleman's pocket or the lady's handbag, are quite common. They are often in leather cases.

Quality

The cheapest rules had "round" joints (no reinforcing plates) and were narrow in width. The next quality had plated joints, the cheaper being straight-ended, the better with an arch.

*Similar but different. By **Ed. Preston**, from the top: A ¾" wide round joint rule, this was the cheapest. An arch joint rule with "extra strong" small joint – note the number of leaves. A 1⅛" wide rule with three arch joints. A 1½" wide rule with brass bound edges. All except the first were also available as 3ft. rules [Preston, 1909].*

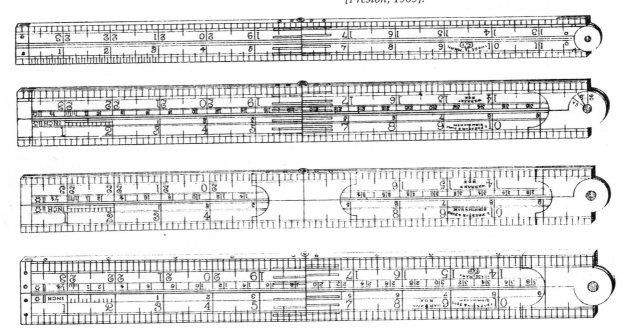

The next step upward in quality concerned the mid-joint. The basic quality had two metal leaves set into the boxwood; a better quality, three, and the best also had edge plates. An alternative improvement was the application of a brass arch at the mid-joint, though this is rare.

In general, the wider the rule, the better the quality. The cheapest were only ¾" wide; the best 1⅝". A further option was brass edge-binding. Whilst this provided additional strength it frequently seems to cause looping between the pins – rarely possible to straighten.

Feature rules

A common added feature is the small level built into one leg usually with a large diameter disk joint graduated in degrees. Another, though rarer, feature was the addition of a slide. These may be marked in inches for use as a depth gauge or they may terminate with a calliper.

Architects' rules

Characterised by single or double bevelled edges on the inside faces which are marked with drawing scales. Contrary to the name, some are marked with surveyors' scales and judging by the numbers to be found they must have been bought by master builders and foremen as well as the professionals. (See also *Scales*, pp.176–177)

Ironmongers' rules

These can be 2ft. or 3ft. 4-fold or the unusual 12" 3-fold. They always include a calliper slide and are marked with a variety of tables including pence per hundredweight and the weight per foot for different shapes (square, flat, etc.) and sizes of iron bars. The larger versions also have tables giving the volume and weight of saucepans and coppers.

Ivory

Most types and sizes of rules were available in ivory as an alternative to boxwood. The 2ft. 2-fold plain types were not in general made in ivory but the engineers' rules were.

The price of ivory rules was about six times that of boxwood so they were mostly bought by professionals – thus architects' rules in ivory are almost as common as plain rules.

Availability dwindled in the 1920s and 30s but some types were still listed until 1939.

The coiled serpent strikes

The 4-fold rule used by carpenters was a kind of trade magic wand – or so at least it seemed to me

Collecting tips
➤ *Rules with heavily rubbed corners are evidence of long term service to owners. Touching maybe but such rules are of no interest to the collector.*

as a child when watching carpenters clad in dungarees that even had a special long pocket on the bib! Whipping out the 4-fold, marking off 6ft. or 8ft. lengths by stepping along the wood with a pencil at the rule's end and then swiftly refolding it and returning it to the pocket was done with panache. It was never put down; the experienced knew that this was how rules became broken.

But today, this is just a memory – the triumph of the coiled steel spring measure is complete.

Price Guide

Boxwood rules

• A 3ft. 4-fold boxwood ironmongers' rule by **Jas. Rowe**, Birmingham, with calliper and tables for saucepans, coppers and iron. **£85** [AP]

• A 1 metre 4-fold boxwood rule by **W.F. Stanley**, London, with nickel silver fittings, also marked in inches. **£30** [DP]

Ivory rules

• A 6" 2-fold ivory rule (from a drawing case). **£20** [DP]

• A 12" 4-fold thin section ivory rule from an etui. **£26** [DP]

• A 12" 4-fold ivory rule by **Rabone** with brass round joint. G++ **£55** [AP]

• A 12" 4-fold architects' ivory rule with German silver fittings. F **£90** [DP]

• A rare 12" 4-fold ivory rule with brass fittings advertising Blundell Spence & Co., Paints, Colours, Oils etc. F **£85** [DS 24/1511]

• An unusual ivory 12" 3-fold rule with brass fittings. G++ **£94** [DS 24/1493]

• A 2ft. 4-fold ivory rule with German silver fittings by **Bradburn & Sons**, **£120** [DP]

• A 2ft. 4-fold architects' ivory rule with German silver fittings by **Rabone**. **£143** [DS 24/1504]

• A 2ft. 4-fold ivory rule with nickel silver fittings and calliper slide. **£175** [DP]

• An early 3ft 4-fold ivory rule with brass fittings by **T. Ludlow**, maker, Birm. G **£115** [DS 24/1523]

• A 3ft. 4-fold ivory rule by **J. Buck**, London, narrow section, marked in inches and metric. **£121** [DS 24/1464]

Steel

The originator of the steel measuring tape was almost certainly James Chesterman. Having moved to Sheffield from London, in 1821 he started in business and was soon making a variety of products that used springs, including such things as clockwork roasting spits. Amongst these was a linen tape measure that included a spring to return it to the case.

Exactly when the firm came to concentrate on the manufacture of steel tape measures is less easy to determine; it has been suggested that the spur was the demise of the fashion for crinoline skirts which were supported on "crinoline wire", a speciality of the firm. This was around 1860. By 1880 the firm was making almost nothing but measuring equipment and the product range had expanded to include fabric tapes, land chains, and steel rules.

Price Guide

• A 100ft. tape in a leather case by **Rabone** No. 2601 in original box. G++ **£31** [DS 24/31]

• A 50ft. steel dipping tape by **Chesterman** in a handled brass frame with brass weight. **£18** [DP]

• A 78ft. lawn tennis tape measure by **Dean**. G++ **£44** [DS 24/79]

• A **Rabone** No. 260 66ft. linen tape in leather bound case in original box. G+ **£18** [DS 24/376]

• An **Rabone** No.4381 33ft. steel tape in steel case in original box. G++ **£22** [AP]

• A **Rabone** "Flexlet minor" steel spring rule. F **£8** [DP]

• A **Rabone** architects' steel scale rule (marked ft., ins., and feet and chain scales) in nickel silver case. F **£16** [DP]

• A military map scale steel tape by **Chesterman** in a nickel silver case also stamped J.A. Steward. **£22** [DP]

• A **Rabone** 12ft. ¼" steel tape in a nickel silver case with recessed handle. **£40** [DP]

Fabric tapes

Fabric tape measures have been in use for tailoring and dressmaking for hundreds of years. There are records that as early as 1747, tapes were being woven for this purpose at Tean near Stoke-on-Trent, and by 1818, Edward Dean was producing painted and printed tapes in Whitechapel, London, the founder of a trade name that is still in use today. The problem is that tape measures move – considerably – so, until metallic threads were introduced into the weave, by **James Chesterman & Co.** in 1867, users distrusted long tapes, and surveyors wanted accuracy, and the like, who used chains.

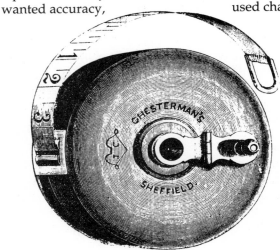

A ⅝" wide, 66ft. linen tape in a leather case with folding handle [Chesterman, 1880]. A nice aspect of the long tapes was the well-finished hard leather cases,

Special purpose tapes

From the catalogues it is clear that tapes were made for many different purposes – timber girthing (see *In the Timber Yard*, pp.204–206), cattle gauging, shoe making, horse measuring, undertaking (7ft. long!) laying out tennis courts and cricket pitches. But the numbers made must have been few, as today all of these are rare.

Top: Steel tape with scale for conversion of pounds to kilos and vice versa. Bottom: steel tape with circumference and diameter scales.

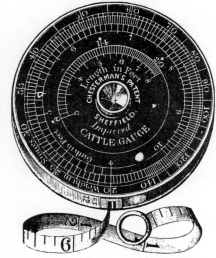

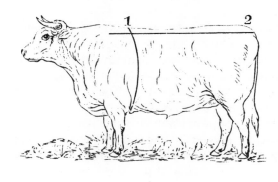

CHESTERMAN'S
PATENT SPRING MEASURING TAPE,
12 FEET LONG,

Marked English feet and inches on one side of the tape and hands on the other, with

CATTLE GAUGE

Attached, for ascertaining the weight of Cattle by Measurement.

To the Farmer and Grazier this is an almost indispensable article, as by its use he may tell the dead weight of the four quarters, either in the London stone of 8lbs., the country stone of 14lbs., or the score of 20lbs.

INSTRUCTIONS FOR USE.

Take the length from the foremost upper corner of the shoulder-blade bone in a straight line to the hindermost point of the rump, by the tail, then take the girth close behind the fore legs.—Suppose the length of the beast to be 5 feet 6 inches, and the girth 6 feet 3 inches.—Turn the slide till the figure $\frac{8}{8}$ is over 5 feet 6 inches on the inner circle, then look at 6 feet 3 inches on the slide, and immediately over that (on the outer circle) will be found the weight, viz., $89\frac{1}{2}$ stones of 8 lbs. to the stone.—If the weight be required in stones of 14 lbs., turn the slide till $\dot{1}4$ is over 5 feet 6 inches, and referring as before, the weight will be found 51 stones; or by placing $\dot{2}0$ over the length, the weight will be found $35\frac{3}{4}$ scores.

NOTE.—*Very fat* cattle will weigh a little more than shown by the scale, and *very lean* ones a little less.

No. 31.	Brass Case, with Linen Tape			7s. 0d. each.
„ 31SS.	„	„	„ and Spring Stop	7s. 6d. „
„ 277.	„	„ Steel Tape		9s. 6d. „
„ 277SS.	„	„	„ and Spring Stop	10s. 0d. „
	With Nickel-Plated Cases, 1s. each extra.			

*Illustration of the patent cattle gauge tape from the 1880 catalogue of **James Chesterman & Co.**, Bow Works, Ecclesall Road, Sheffield. Note the reference to the "London stone" of 8lb, the "country" stone of 14lb. and the "score" of 20lb. Until recently, butchery, like many other trades, used distinct measuring standards, which were often quite local.*

THE CARPENTERS' RULE

Origins

Rules adapted to assist the carpenter have a long history – the rule recovered from the wreck of the *Mary Rose* (1545), marked with a line of timber measure, is a solid 2ft. lath. It is likely that it was well into the 18th century before folding joints became common.

The carpenters' rule described by Edmund Stone[1], writing in 1758, is a folding rule which has not only a line of timber measure but also a line of board measure and also a Gunter's line (log scale). (See *The Slide Rule*, pp.169–171.) The Gunter's line is inscribed but there is no slider.

By the middle of the 18th century, specialist rule makers had emerged in London and makers such as **Ed. Roberts**, **W. & J. Rix** and **Thomas Cooke** (all principally known for duty slide rules) were producing in quantity.

Developments

Also to be found on carpenters' rules is the girt line, which enabled the volume of timber to be calculated using Coggeshall's method (invented around 1677). This uses ¼ girths (in inches) and the length of timber (in feet) and gives around 20% under-measurement which was intended to allow for the loss during squaring. This feature was so commonly provided that the carpenters' rules is often known as a Coggeshall rule.

The advantage of the 2ft. 2-fold rule is that it has plenty of space for the provision of a moving slide with a Gunter's line, lines such as 8-square (E&M scales) and timber, and also additional information such as timber load tables. For this reason, the 2ft. 2-fold rule remained the preferred form for the carpenters' rule right into the 20th century, even though it is clear from surviving examples that the more convenient 2ft. 4-fold type was available by 1750.

E & M scales

Literally, edge and middle scales – sometimes also known as the 8-square lines or mastmakers' scales. These give the dimensions needed to set out an octagon onto a square timber.

Without trigonometry, and few 18th century carpenters would have been able to tackle such calculations, the setting out of an octagon was a matter of trial and error – unless you had these

1. Bion, *Construction and Principal Uses of Mathematical Instruments*, translated and supplemented by Edmund Stone, London 1758, reprinted by Astragal Press, 1995. As far as the carpenters' rule is concerned this is all Stone, not Bion.

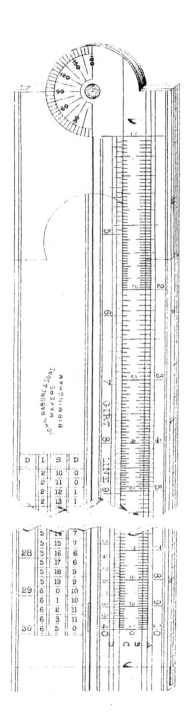

A 2ft. 2-fold boxwood carpenter's rule with a brass slide. Note the girt line and the price per load table giving the cost in £.s.d. of a load of 50 cu.ft. at up to 30 pence per cu.ft. A book of instructions was sold with this rule [Rabone, 1892].

Line from 16" on E scale crosses inch scale at 4¹¹⁄₁₆".

Line from 16" on M scale crosses inch scale at 3⁵⁄₁₆".

*A 2ft. 2-fold rule by **Ed. Preston**, No.5535, showing an 8-square lines (E& M scale) [Preston, 1914]. In this catalogue illustration, the rule has not been drawn very accurately. The 16" mark is about ⅛" out of position.*

scales on your rule. They are direct reading, i.e. by looking across the rule to the point on the main measure opposite the number given on the E & M scales (which is the size of the square timber needing to be octagonized), the distance in from the edge of the square or the distance out from the mid-line can be read off.

Price Guide

2ft. 2-fold rules

Boxwood arch joint 2ft. 2-fold rules, late 19th/early 20th C., plain or with 8-square (E&M) lines **£12–18**. Ditto with brass measuring or log slide **£15–20**.

♦ An early (c.1820) 2ft. 2-fold boxwood rule with straight brass joint, marked with 8-square (E&M) lines. **£80** [AP]

♦ A 2ft. 2-fold boxwood rule by **T. Bradburn** with nickel silver protractor hinge including patent bolt. G++ **£48** [AP]

♦ A 2ft. 2-fold boxwood rule with drawing scales. Quality arch jointed, marked in degrees and brass edge binding. **£42** [DP]

♦ A 2ft. 2-fold boxwood rule by **T. Webb**, Bristol, with timber line and table. **£110** [DP]

♦ A 2ft. 2-fold boxwood rule marked E.P. Registered by **E. Preston & Sons**. Brass slide, log and girt lines and 8-square lines. **£30** [DP]

♦ A 2ft. 2-fold boxwood rule marked **Wellington Fecit Crown Court Soho**. Brass log slide and girt lines. **£94** [AP]

♦ A 2ft. 2-fold boxwood rule by **Rabone**. Vulcan Registered Design (which relates to the joint) with brass slide and girt line. Also includes 8-square lines (E&M scales) and drawing scales from ¼" to 1" to the foot. **£55** [DP]

♦ A 2ft. 2-fold boxwood rule by **Sampson Aston**, Birmingham (1833–70), with ivory slide, log and girt lines and drawing scales from ¼" to 2" to the foot. **£140** [DP]

♦ A 2ft. 2-fold boxwood rule by **T. Aston**. Brass slide, log and girt lines and timber load tables at 50 cu.ft. per load. **£125** [DP]

Measurement from M scale 3⁵⁄₁₆"

Measurement from E scale 4¹¹⁄₁₆"

16" square timber

Diagram to show how an octagon is set out using the 8-square lines (E and M scale).

Board rules

Some timber was priced by the "board foot" which is the amount of timber in one square foot of one inch thick board .

The simplest form gives the length of board of a given width that measures one square foot. The length needed is read off the rule opposite the appropriate width on the line of board measure. For small widths, when the length exceeded the length of the rule, the length could be read from a table positioned at the top end of the line. Later

Collecting tips

➤ *Early carpenters rules' (made before about 1820) are likely to have the following features:*

• *The body is thick – about ¼".*

• *The joint is straight, not arched.*

• *The circular part of the joint is small in diameter – ½" rather than the 1" of later rules.*

• *The slide may be boxwood.*

versions had a number of scales to allow for boards of varying thickness.

Timber line/table

A line of measure found on early carpenters' rules, it is also normally accompanied by a small table. Both give a direct reading of the length of different sizes of square timber that contain one cubic foot. Obviously once this length is known the volume can be speedily and easily counted by stepping along with dividers.

Table of timber measure giving the lengths of timber which contain one cubic foot.

1	2	3	4	5	6	7	8	*(size of square)*
144	36	16	9	5	4	2	2	*(feet)*
0	0	0	0	9	0	11	3	*(inches)*

The table gives the lengths for the small sizes (from 1" x 1" to 8" x 8") as these exceed the length of the rule, whilst the larger sizes are read off the rule. For example, a 10" x 10" timber contains 100 cu. in. per inch run, so for 1 cubic foot (1728 cu. in.) a length of 17.28" (1728 ÷ 100) is required and the 10" mark on the timber line will be found opposite 17¼" on the rule.

Price Guide

 • A 2ft. 3-fold boxwood rule by **J. Tree**. Folds to 9" with a 6" fold out in one leg, the other incorporates an ivory slide rule. **£280** [AP]

 • A 2ft. 3-fold boxwood rule, folds to 9" with a 6" extension. Includes a timber load table, a slide rule with a boxwood slide and drawing scales of ¼" and ½" to the foot. **£325** [DP]

 • A 3ft. 4-fold boxwood rule by **Preston**. Brass slide, timber table and log and girt lines. **£95** [DP]

 • A 3ft. board measure lath with projecting rectangular brass end stop marked **Bentley Taylor**, London (maker?). **£65** [AP]

Ivory rules

 • A 2ft. 2-fold ivory rule with log and girt lines and drawing scales, with brass fittings. **£400** [DP]

 • A 2ft. 2-fold ivory rule with nickel silver fittings and nickel silver log slide. Also marked with a range of drawing slides which occupy the whole of one leg. **£400** [DP]

 • A 2ft 4-fold ivory rule marked **Wharton & Evans** with nickel silver mounts and log slide (in one leg only) and hundredweight table and drawing scales. **£300** [DP]

Price per load tables

The extensive tables to be found on many carpenters' rules are price per load tables. A "load" of timber was not always the same. A "load" originated as the weight that could be carried on a wagon, which had different volumes according to the density of the timber. Mostly, it seems to have been 50 cu.ft. but 40 cu.ft. was also used and there were other standards. The first column in the table is the price per cubic foot in pence; the next three give the cost of the load in £.s.d.

Hundredweight tables

Not to be confused with price per load tables, these give a conversion of the price per pound in pence (expressed in farthing increments) to the price per hundredweight (1 cwt = 112 lbs.) in £.s.d. The tables can of course be used to convert in either direction. Strictly speaking, not a carpenters' table, but remember that the carpenter was often the leader of the building team and would have needed this for pricing many items.

If you think these tables weren't useful, try calculating the cost of a hundredweight of material at 5¾d per lb.

D	L	S	D
	0	2	4
	0	4	8
	0	7	0
1	0	9	4
	0	11	8
	0	14	0
	0	16	4
2	0	18	8
	1	1	0
	1	3	4
	1	5	8
3	1	8	0
	1	10	4
	1	12	8
	1	15	0
4	1	17	4
	1	19	8
	2	2	0
	2	4	4
5	2	6	8
	2	9	0

A section of hundredweight table: From this you can see that the cost at £2 13s 8d per cwt is 5¾ d per lb. Not all tables cover the same range of values – a smaller value may indicate an early rule.

Almost always 2ft. 2-fold, these are marked with factors and other information for engineering calculations which were undertaken using the log slide which is always part of the rule. All types are somewhat similar and draw inspiration from Routledge's rule. In the late 19th century there was a change to a more formula-based approach to calculation and they then became obsolete. Production had stopped by the 1920s.

Like all specialist rules, they are largely unintelligible without instructions – original booklets are very rare but, for Routledge's rule, reprints are now available.

Routledge's rule: The oldest and most common of this type of rule, designed around 1811 by Joshua Routledge of Bolton. It was intended for practical engineers and mechanics. Amongst the many gauge points are those for calculating the weight of volumes of various materials and the duty of steam-driven water pumps.

This is the only engineers' rule which was also made in both the normal 2ft. 2-fold format and, rarely, as a 2ft. 4-fold.

Slater's rule: Intended for cotton spinners, this rule was designed by another Bolton man – William Slater.

Hawthorne's rule: Invented by Robert Hawthorne, a civil engineer in Newcastle upon Tyne, the gauge points and other information are generally directed towards structural calculations.

Wilkinson's Improved rule: Substantially similar to Routledge's rule, but this rule includes a direct reading log/number scale. The layout varies.

Carrett's rule: The rarest of the named engineers' rules. The information given is mostly weights, densities and measures, which are intelligible without the help of an instruction book, and not the factors of the earlier engineers' rules.

Price Guide

♦ 2ft. 2-fold boxwood and brass Routledge rule by **Rabone** with protractor hinge. **£60** [DP]

♦ A 2ft. 4-fold boxwood Routledge rule with German silver fittings, slide and protractor hinge by **W.H. Brown**, Birm. Engraved on hinge W.H. Prime Rochdale. G **£66** [DS 24/1365]

♦ 2ft 2-fold ivory Routledge rule marked **Norris**, Optician, Birmingham, with nickel silver fittings. **£360** [AP]

♦ 2ft. 2-fold boxwood and brass Slater's rule by **T. Ashton** "The Original Maker". **£110** [DS 23/1267]

♦ A 2ft. 2-fold ivory and nickel silver Slater's rule by **John Rabone & Son**. G++ **£1200** [DP]

♦ A 2ft 2-fold boxwood and brass rule "Improved and Arranged by Rt. Hawthorne, Civil Engineer, Newcastle Tyne". **£125** [AP]

♦ A 2ft 2-fold ivory rule with German silver fittings, "Improved and arranged by Rt. Hawthorne, Civil Engineer, Newcastle Tyne". Maker **W.H. Brown**. G++ **£900** [DS 19/391]

♦ A fully brass bound Wilkinson's Improved boxwood rule by **Higgison**, Birmingham, with steel tips and brass slide. **£85** [AP]

♦ A 2ft 2-fold boxwood and brass rule by **Ed. Preston** marked "Improved Rule W.E. Carrett, Engineer, Leeds". **£120** [DP]

♦ A 2ft. 2-fold boxwood and brass slide rule marked "Improved, Arranged and Manufactured by James Noble" **£110** [DP]

Noble's rule: There are some similarities with Hawthorne's rule – both include tables related to steam pumps – but these rules are unusual in being marked "improved, arranged and *manufactured* by James Noble". Slide rule versions (not made as a folding rule) are also known – both types are rare.

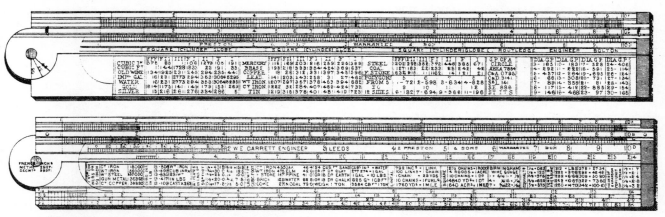

Top: Routledge's rule. Bottom: Carrett's rule [Preston, 1909].

All barrels are barrel shaped – but, whilst the cooper no doubt just shaped the staves the way he always had done to make a good tight job, the eagle eye of the gaugers could see that the shapes were different and knew that barrels with the same three basic dimensions (length; head diameter; bung diameter) could contain differing quantities.

Varieties

For this reason gaugers classed barrels into varieties according to the shape. **Charles Leadbetter,** who, in 1739, wrote one of the most acclaimed 18th century manuals, *The Royal Gauger,* describes six varieties, although most instructors suggest that, for practical purposes, three were sufficient. These are the 1st, 2nd, and 3rd varieties that are marked on many gaugers' slide rules. Each variety was considered to approximate with a geometric solid.

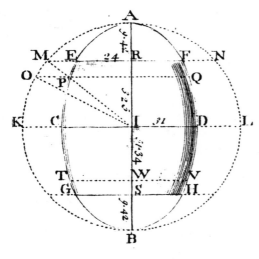

The geometry of a spheroid variety of barrel: From C. Leadbetter, The Royal Gauger. *5th Ed. 1760 .*

Gauging rods

Inserted through the bung hole of a barrel, the rod measures the diagonal and this measurement alone will give a direct and acceptably accurate measure of the volume of barrels that are within normal proportions – neither excessively long and thin nor short and fat.

Gauging rods can be readily identified by their square section and the taper end, designed to fit into the corner. The usual length is four or five feet, though shorter versions for the smaller sizes of barrels were also made. They may be one piece, take apart into sections or fold into four, six or even occasionally eight.

Round section take-apart versions of gauging rods will also occasionally be found but most round measures are dipping rules.

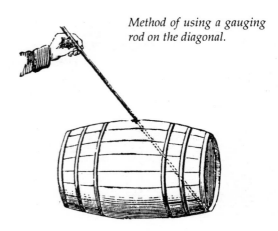

Method of using a gauging rod on the diagonal.

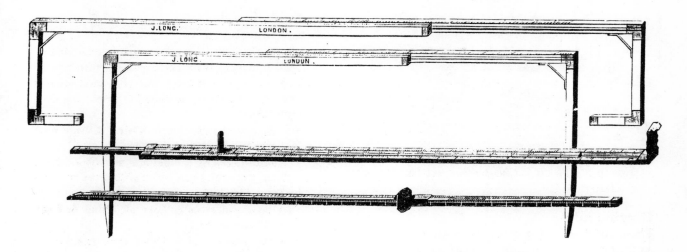

A set of gauging instruments, as supplied to the Customs & Excise [Long, 1922]. All are made from boxwood with brass fittings.
From the top: A long calliper, a cross calliper, a head rod and a bung rod and slide.

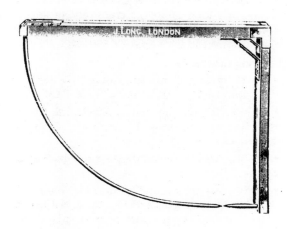

Left: The rare endometer, or head calliper.

Barrel callipers, etc.

For customs and excise purposes, a more accurate assessment of barrel contents was required than could be obtained using the gauging rod, so the excise men used four large callipers. The first three types are quite rare.

The long calliper: For taking the length of the barrel. This instrument included returns to reach over the chimes and had a built-in allowance for the thickness of the heads. Callipers of this type are shown in *The Royal Gauger* but the three following instruments are a 19th century development.

The cross calliper: For taking the diameter – earlier practice used a large square.

The endometer: These delightfully shaped callipers could be inserted through the bung hole and the thickness of the barrel heads measured.

The head rod, content and ullage rule: More commonly found is this device. It is a handsome thing, 3ft. 8in. long, incorporating a moving slide with upstanding brass measuring points which are used to take the diameter of the head. It also doubles as a slide rule for the usual excise calculations of content, ullage, etc.

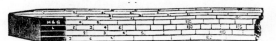

A gauging rod with dipping scales [Harding, 1909].

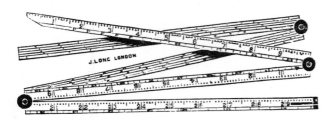

A 5ft. 6-fold dip rod also marked for every size of beer cask [Long, 1922].

Dipping rules

These rules are often confused with gauging rods as they give direct readings in gallons but they are inserted vertically through the bung hole of a cask. They were used to dip part-full barrels to establish the content, known as "wet inches", and are marked with measures for standing or lying barrels of the standard English sizes – butts (126 gal.), puncheons (84 gal.), hogsheads (63 gal.), etc.

They were made in much the same forms as gauging rods – solid rods, folding rods, and in round, sectional, form. These dipping scales were very often also added to gauging rods.

Very thin versions of these rules – about ³⁄₁₆" diameter – are spile rods, small enough to be inserted through the spile hole in the bung.

A set of spile rods [Hicks, c. 1890].

Occasionally dipping rules will have a sliding brass collar. This was set to the diameter of the barrel before dipping so that an allowance could be made for barrels that had become misshapen.

Ullaging rules

These are short (up to 15" long) with some form of shoulder designed to measure down from the bung hole and indicate the "ullage" – the amount that is missing from the barrel. Some versions, for the whisky trade, are marked to show the annual allowance for evaporation through the staves.

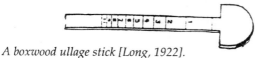

A boxwood ullage stick [Long, 1922].

Bottle rules

Strictly speaking this isn't the right place to mention these as they are for bottles and not barrels but their purpose is really the same as barrel gauging rules – to measure the diagonal and give a direct reading of the content.

The rules are 2-fold or 4-fold and are scaled to 6 gallons. The bottles were stoneware, either cylinder or the intriguingly named bulge (or bouge) bottles. There are usually three scales for each – presumably, like barrels, there must have been varieties.

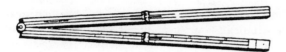

A bottle rule for bouge or cylinder stone bottles or jars [Long, 1922].

Makers

The trade in gaugers' rules was dominated by a handful of firms who were specialist suppliers to the brewery, wine, and spirit trades. Principal amongst these were **Dring & Fage** but the names **Buss**, **Long** and **Loftus** will also frequently be found. Whether any of these firms, other than **Dring & Fage**, actually made the rules they sold is debatable; it is clear, however, that the big rule manufacturers in Birmingham did not dominate this sector of the market in the same way they did for the general purpose rules.

Collecting tips

➤ *Confusion exists about what is a **gauging** rod and what is a **dipping** rod or rule – not helped by the terminology in some suppliers' catalogues.*

➤ *Measures with a **triangular** end, used to measure a barrel diagonally to establish the capacity, are **gauging** rods. But they are often also marked for standing or lying barrels so they can be used as **dipping** rods.*

➤ *Measures that have **square** ends are intended to establish the amount in a standing or lying barrel – these are **dipping** rods.*

➤ *Shorter folding versions of **dipping** rods were usually known as draymen's rules.*

Dring & Fage

Founded in 1790, the firm traded until the 1970s. For most of this period, it was situated in Tooley Street, London, just south of the river. It was a principal supplier to the Customs & Excise.

The firm's name is to be found on every type of rule, hydrometer and other instrument needed for gauging and for the wine and spirit trades. They were also suppliers to the Port of London Authority – their name is found on many cargo callipers, timber rules and dock rent rules.

A footnote – the last managing director has said that the marking and stamping of the firm's rules was still carried out by hand until it closed.

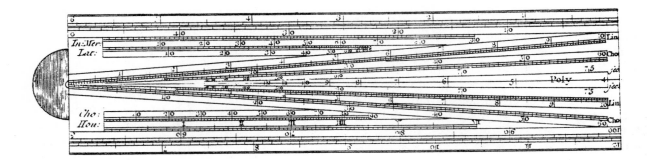

The sector

Virtually every box of mathematical instruments (drawing instruments) made before about 1870 contained a sector. Pocket cases had small versions, 4½" long. For more serious calculating, 12" examples were made but far and away the most common size was 6" opening to 12". Few collectors, however, seem to have much idea as to how they were used.

Sectors are made like a folding rule. They are inscribed with a variety of mathematical lines (scales) which are of two types; plain or geometric lines (line of equal parts, chords, rhumbs, hours etc.) and log scales (tan, V-sin, sin, numbers). Sectors had been in use long before Napier invented logarithms in the early years of the 17th century but thereafter the log scales were added.

Calculations were made using dividers. However, from the pristine state of virtually all sectors, it would seem unlikely that many were ever used. In 18th century manuals, instructions on how to use the sector occupy many pages, so this is not the place to try. I will only make the point that the sector could be used closed, fully open or folded, in which mode the scales which originate at the hinge point can be used to very easily resolve problems involving ratios.

Collecting tips

➤ *Sectors changed little over 120 years and are therefor not easy to date unless signed – most are not.*

➤ *10% or less have arch joints.*

➤ *Sizes other than 6" sell at a premium.*

➤ *Examples in both ivory and boxwood are found in approximately equal numbers.*

A typical sector: This illustration is from John Robertson, Treatise of such Mathematical Instruments as are usually put into a Portable Case, 1775. This book devotes about 100 pages to explaining the use of the sector.

This is a 6" sector with a round joint. Some sectors were made with arch joints similar to those used on rules.

Price Guide

♦ A 6" boxwood round joint sector. **£18** [DP]

♦ A 6" boxwood round joint sector by **C.W. Dixey**, Math'l Inst Maker to Her Majesty. F **£30** [DS 21/314]

♦ An early 12" boxwood sector with brass arch joint by **Underhill** St Mary's Gate Manchester. G **£253** [DS 23/189]

♦ A boxwood sector, 6" when folded but with additional folding sections to each leg giving an overall length of 15", marked **Gutteridge** Patent London. **£120** [DP]

♦ A 4½" round joint ivory sector by **Carpenter & Westley**, London. **£52** [DS 23/808]

♦ A 4½" ivory sector with a brass straight rule-type joint. **£40** [DP]

♦ A 6" arch-joint ivory sector by **Archbutt**, Westminster Bridge, London. **£53** [AP]

♦ A 6" round joint ivory sector by **W.H. Harling**. **£45** [DP]

♦ A 6" round joint sector in ivory and German silver by **F. Robson**, Newcastle. **£50** [DS 24/1542]

♦ A 6" ivory sector with German silver arch joint. F **£72** [DS 24/1539]

♦ A 6¼" brass sector by **Thomas Wright**, (1686–1748) signed *T. Wright fecit*. **£315** [AP]

Don't forget - fine condition will often radically increase the value.

Disputes about who was the inventor of the logarithmic slide had already broken out in the 17th century and the competing claims seem to have caused a degree of confusion ever since.

John Napier published his invention of logarithms in 1614. By 1623 Edmund Gunter had made a "line of numbers" divided as a logarithmic scale.

In a log scale, the distance from 1 to the point scribed and numbered for any other number is a linear representation of the logarithm of that number. To multiply, the lengths representing the numbers are simply added together and the result read from the scale. Even with a single log scale, multiplication and subtraction can be undertaken speedily with a pair of dividers – or even the edge of a piece of paper.

Logarithms

➤ *The addition of the logarithms of two numbers has the effect of multiplying the numbers and subtraction will divide. There are two methods by which this process can be effected.*

➤ *1. The logarithms are looked up in a book of log tables – added or subtracted and then reconverted via the tables.*

➤ *2. By using a slide rule. The log scale on a slide rule is a linear representation of logarithms. Thus by adding or subtracting the lengths representing the numbers, multiplication or division can be undertaken and the answer read off directly. Thus no log tables are needed to convert the numbers into or out of logs.*

The slide rule

If two similar log scales rules are placed alongside one another, the dividers can be dispensed with and you have a slide rule.

By 1633 William Oughtred had published a description of a slide rule, intended for gauging, constructed from two rules held together by loops of brass. By the last decades of the 17th century rules with a slide or slides moving in a solid body were more usual.

Opposite: Everard's Sliding Rule illustrated in C. Leadbetter, The Royal Gauger, 1760.

Price Guide

• A rare **Burts Patent** timber slide rule No. 4359 by **Dring & Fage** consisting of a 2ft boxwood single slide rule with brass fittings, attached to which is a matching boxwood and brass calliper rule for measuring the diameter of timber, marked "For Uneven Sided Timber, Wood & Stone of every Description" and "Solid Cubic Ft of English and Foreign Timber etc. by Hoppus". With many more instructions in a copy of M.B. Cotsworth: *Railway and Timber Trades Measures and Calculator.* F **£550** [DS 24/101]

• A 19" boxwood slide rule by **Robinson** for calculating the weight of steel. **£65** [DP]

• A 13⅜" boxwood double slide rule by **Stanley** London stamped length, breadth, thickness, cubic content, also Frederick A. Sheppard Patentee No. 155, slight chip on one slide. **£60** [DS 24/507]

• An 8½" boxwood slide rule marked "Ewart's Cattle Gauge adapted for any market" by **Tree**, London, with ivory slide and nickel silver handled cursor. **£160** [AP]

• A rare 6" boxwood double slide machine gunners' calculating rule dated 1937 by **A.G.T. Ltd.** G++ **£77** [DS 24/1357]

• A 26" slide rule with two slides for paper calculations by **J. Tomlinson**, Partick, Glasgow. **£60** [DS 22/235]

• A 34" Harrow Mark Reducer boxwood slide rule with brass tips by **ESA** to help teachers at Harrow School find averages when marking papers. **£65** [DP]

• An unusual 8¾" boxwood slide rule by **John Taylor Sons & Santo Crimp** Westminster. Reg. No. 409323. Marked depth, velocity discharge. G++ **£65** [DS 21/1541]

• A 16" boxwood dock rent rule with brass slide marked **Dring & Fage**, marked with a two year calendar. Indicates the number of weeks between any two dates. **£135** [DP]

• A 9" boxwood slide rule with two boxwood slides inset into face marked "Arranged by Chas. Hoare" also marked **J. Fenn**, 105 Newgate Street, London. **£75** [DP]

• A 12" narrow section (only ¾" wide) boxwood and brass slide rule by **E. Preston & Sons**. Intended for the textile trade the rule is marked "Compiled by James Holmes, M.S.A., Burnley" with tables and information on back. **£130** [DP]

• A 14" boxwood slide rule by **J. Tree**, Maker, London, with single boxwood slide inset into face. For calculations relating to the speed and power of ship engines. **£175** [DP]

Special purpose rules

Early inventors of slide rules directed their efforts towards producing rules for specific purposes, navigation, the gauging of barrels and other vessels or the measurement of timber being common.

Prominent amongst these men was Henry Coggeshall who, in 1682, published his system for measuring timber using "A Two-Foot Rule, which Slides to a Foot". His system was republished many times during the 18th century and the term "Coggeshall rule" came to mean a two foot rule with a slide in one leg intended for use by carpenters. (See *The Carpenters' Rule* pp.160–162.) Also prominent was Thomas Everard, an excise officer, who, in 1683, wrote a work on gauging. This included a design for a slide rule set out for this purpose, which formed the basis for many future excise rules.

18th Cent. Calculations

➤ *18th/19th century gauging relied heavily on factors (for multiplying) and divisors (for dividing) to undertake calculations and, in so doing, reduced the number of calculations that needed to be made. Gaugers' manuals contain hundreds of such divisors/factors usually with both the divisor and its reciprocal allowing the same calculation to be done by dividing or multiplying. The most frequently used were marked directly on the rules with small brass pins or were figured on the back of the slides.*

➤ *Without the help of a gaugers' guide, and there were many published from 1700 onwards, the interpretation of these factors is impossible.*

➤ *For example, one often-marked factor is for G[reen] Starch. This is the number of cubic inches of green, that is only partly dried, starch that will yield 1lb. avoirdupois of dry and finished starch (incidentally this was 34.8). So, as a hard-pressed excise man, you could determine the duty which was to be levied on each pound of dry starch even if the manufacture was not finished or, more likely, determine if any had disappeared between the two stages!*

Price Guide

♦ An **Otis King** Model K calculator complete with instructions in original cardboard box. F **£35** [DP].

♦ A **Fowler's** double sided long scale calculator "Vest Pocket Model" with instructions in original tin case. **£82** [DS 24/191]

♦ A **Fowler's** Jubilee New Model Extra Long Scale calculator in original leather case. G++ **£125** [DS 23/570]

♦ A **Fowler's** "Magnum" Textile calculator complete with instructions in original leather case. **£70** [DP]

♦ A **Halden** "Calculex" circular slide rule in original aluminium case with instruction booklet. **£75** [DP]

♦ A **Dyson's** Textile calculator of pocket watch form. **£165** [DP]

♦ A Professor **Fuller's** patent spiral slide rule by **Stanley**, London with stand in original box. G++ **£160** [AP]

♦ A **Lord's** patent circular pocket calculator by **W. Wilson**. **£253** [DS 24/1004]

♦ An unusual and complete Fuller's Computing Telegraph made by **John Fuller** in 1852. On the reverse of "Palmer's Computing Scale Improved by Fuller" is "Fuller's Time Telegraph" which reckons the number of days from one day to any other day in the same year. It is rare to find a double-sided example complete with a 22 pp. instruction book (binding loose). **£550** [DS 23/95]

*Remember - condition is **G+** unless otherwise shown.*

FULLER'S SPIRAL SLIDE RULE

Equal to a Slide Rule 83 feet long,

Giving logs, multiplication, division, proportion, &c., in case, **£3**.

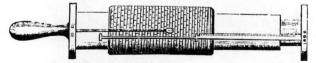

Now used by all practical Calculators, Government Officials, C.E.'s, Electricians, &c., &c.

1903 advertisement for Fuller's Rule.

General purpose slide rules

From earliest times virtually all slide rules were directed towards the needs of a particular group. Rules made for specific purposes, sometimes very specific, continued to be made until the arrival of the electronic pocket calculator in about 1970 caused the speedy demise of the slide rule.

The general purpose slide rule, the terror of schoolboys but the engineer's workhorse, was not widely available before the middle of the 19th century and it was the end of the century before these were made in large numbers for school and college use. By 1900 the normal method of construction had become celluloid strips fixed to a wooden core, and all-plastic rules followed in the 1940s.

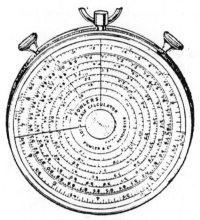

Front dial of Fowler's Long Scale pocket calculator [West, 1930].

Circular slide rules

The principle of both the circular slide rule – useful, as the calculation can't fall off the end of the scale – and the spiral slide rule – which has a scale much longer than can be put onto a straight rule, thus giving greater accuracy – had both been described as early as the 17th century. However, neither type was made in any numbers until the late 19th century.

Amongst circular calculators – it seems wrong to call them slide rules but many do – the **Halden Calculex** and the various **Fowler** models are quite common whilst amongst the spiral patterns the large size **Fuller's** calculator is also to be frequently encountered. This device was made in large numbers by **W.F. Stanley** of London (not to be confused with the American planemaker). It came in a good mahogany box, but the later examples had plastic ends.

Excise rules

The most commonly encountered slide rules (other than 20th century plastic items) are ones used by the Excise and by the wine and spirit trades. As already described, the gaugers' rule of the 18th century was a solid boxwood, very occasionally ivory, stock about 1" by ¾" with the slides running in slots.

The most frequently used gauge points (the number of cubic inches in a malt bushel, 2150, and an ale gallon, 282) are indicated by small brass pins inserted into the scales.

These rules were intended for the calculation of the contents of a wide variety of containers from malt floors to barrels both full and ullaged. By 1840 the standard form of duty rule had changed to the more modern flat, double sided 2-slide form.

Types and uses

Although usually lumped together as "Excise" rules, there are several distinct types with different uses. In general, excise rules are common.

Early type, solid stock gaugers' rules: These may have 2, 3, 4 or occasionally even more slides. The most common length is 10" but 6", 12", 15", 18" and even 24" were made.

Later type, flat construction gaugers' rules: These have two slides visible both sides. There are many variations on these rules but, in general, they were used for gauging, ullaging, valuing and reducing. Most often 12" long but versions from 6" to 24" were made. Ivory examples are common.

Ullaging rules: Two slide rules quite similar to type B above but these only perform the ullaging function. Should not to be confused with ullaging dip rods.

Comparative or reducing rules: Single slide rules to perform only the reducing function of the

Tho. Everard. *The Stereometry or The Art of Gauging Made easy, by the Help of a Sliding-Rule*, 7th Ed. 1712. In the preface, it is explained that the principles applied to the use of the normal sliding rule are equally applicable when using this instrument. Whether the slide rule was circular or linear is not clear but this must surely be the first pocket calculator.

complete gaugers' rule – this is, determining the amount of water that should be added to spirit to bring it to the chosen strength and the duty payable on spirits above or below proof strength. The vast majority are 9" long and were originally in hydrometer sets, but longer versions were also available.

Temperature rules: For use with the Sykes' hydrometer (which from 1818 was the only type acceptable in law). It provides the temperature adjustment needed when using the hydrometer on liquids not at 60° and thus avoids the need for a book of tables.

Saccharometer rules: The saccharometer is similar to a hydrometer and can be mistaken for such. It was used to measure the specific gravity of worts (boiled-up beer before fermenting). As this was often hot, adjustments had to be made for

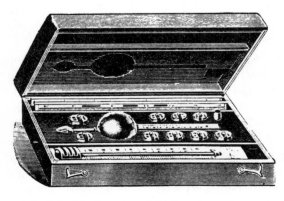
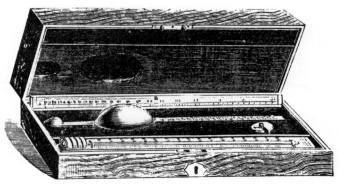

Left: Sykes' hydrometer; right: Long's saccharometer [Long, 1922].

temperature using a slide rule. At one time it was the practice to express this in pounds per barrel (54 gals.).

Wine & spirits trade rules (Farmar's rule): These perform the same tasks as the combined gaugers' rule but were angled to the merchant and were supposedly easier to use. All are 20th century and there are a number of variants and different makers or suppliers. The sliding cursor is often missing.

Price Guide

Early type

• A 12" 2-slide boxwood gaugers' rule of early appearance, unsigned. **£140** [DP]

• A 12" 3-slide boxwood gaugers' rule by **Rix** with scales and gauge points on underside of each slide. **£175** [DP]

• A fine 12" 4-slide boxwood gaugers' rule by **Cook**, Maker to the Hon. Board of Excise, Late Wellington, Crown Ct. Soho, No. 1590 and dated in ink 1841 and 1856. **£110** [DS 24/1355]

• A 4-slide boxwood gaugers' rule by **Edw. Roberts,** Maker, Dove Court, Old Jewry, London. **£185** [DP]

• A rare 6" 4-slide boxwood gaugers' rule with scales and gauge points on the underside of the slides by **Edw. Roberts**, Dove Ct., Old Jewry, London. G++ **£220** [DS 22/955]

Later type

• A 9½" boxwood and brass double slide gaugers' rule by **Dring & Fage**. **£42** [AP]

• A 12½" boxwood and brass double slide gaugers' rule by **Loftus**. **£65** [DP]

• A large size, 18", boxwood double slide gaugers' rule by **I. Aston,** Maker Soho, of transitional type (flat but slides inset into face of rule). **£93** [DS 23/506]

• A 9½" ivory double slide gaugers' rule with German silver fittings. **£130** [DP]

• A 12½" ivory double slide proof rule with German silver fittings by **Loftus**. F **£286** [DS 24/1494]

Don't forget - fine condition will often radically increase the value.

Price Guide

Comparative or reducing rules

• A 9" boxwood reducing rule (from a hydrometer set). **£14** [DP]

• A matched pair of 9" boxwood reducing and temperature rules by **Dring & Fage**. G++ **£35** [DP]

• A very unusual 4" long ivory reducing rule by **J. Long**, London. The stock is solid with the slide inset. **£75** [DP]

• An unusually short 5" ivory proof slide rule with German silver fittings by **Farrow & Jackson**, London. G++ **£65** [DS 23/374]

• A 7⅜" long ivory reducing rule marked "Dicas Patentee Liverpool" with two slides, one each side, joined together through the rule by two brass screws. **£85** [DP]

Temperature rules

• A 9" boxwood temperature rule by **Loftus**. **£16** [DP]

• A 9" ivory temperature rule. **£40** [DP]

Saccharometer rules

• A 13¼" boxwood saccharometer rule with brass fittings and ivory slide shorter than the rule, exposed both sides by **Langley**. F **£110** [DS 23/567]

• An 11" boxwood saccharometer rule with brass fittings by **Loftus**, marked with various scales expressed as lbs. per barrel. **£55** [DP]

Wine & spirits trade rules

• A 23" Farmar's boxwood and brass wine and spirit merchants slide rule with Royal Letters Patent for Gauging and Stocktaking, Overproof, Underproof, etc., minor chip to end of slide. G++ **£88** [DS 24/57]

• A 20" Farmar's wine and spirit merchants slide rule in ivorine (plastic). **£65** [DP]

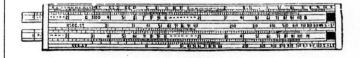

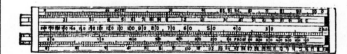

Ullage rule [Long, 1922].

Gauging, ullaging, valuing and reducing rules [Long, 1922].

I started to write a definition of a gauge but discovered pretty quickly that it was long, boring and, in any case, wasn't accurate. You all know what a gauge is – any slot or hole or recess into which the subject can be placed or any object of known size that can be inserted into the space or even an object of unknown size that can later be measured or … let's talk gauges.

Hole – slot – taper

Hole gauges: Less used than one might suppose – almost all are to measure drills, but as these come in fractional, letter, number and metric sizes there is a little variety.

Slot gauges: The most common form of gauge made to measure everything from lead sheet to woodscrews.

Taper outside: Quick and easy to use, this type seems to have been the choice for many watchmakers and jewellers.

Taper inside: Of an acute triangular shape, they must be used with considerable care as the area of contact is small. Not very common.

Calliper gauges: Small size sliding callipers made of metal or boxwood, sometimes with vernier scale (see p.231).

The joy of gauging

Gauges are not collected as much as they should be considering they are small, there is much variety and they generally do not cost a lot. If they have a drawback it is that they aren't much to display – fine for a secret collection in the bedside drawer.

*Two **L.S. Starrett Co.** gauges: Top: No. 280 music wire gauge – Washburn & Moen standard. Bottom: No. 31 boiler plate gauge. Contrary to appearance, this is a locking sliding gauge, not a micrometer, intended to enter small holes to check for loss of metal .*

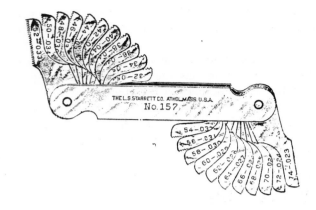

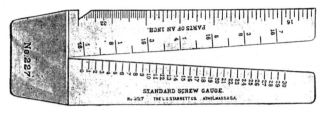

*Two **L.S. Starrett Co.** gauges: Top: No.157 screw pitch gauge – this is for the fine threads used on bicycles. Bottom: No. 227 screw and wire gauge.*

A gauge for every purpose

Wire: Imperial standard wire gauge usually abbreviated to SWG; Birmingham wire – originally called **Stubbs** *iron wire*; **Stubbs** *steel wire* (letter sizes); American wire – originally **Brown & Sharpe**; **Washburn & Moen** music wire; **S. & W. Co.** New American music wire; English music wire; copper electrical conductor wires, etc.

Wood screw: Although made from wire, the screw manufacturers (this means **Nettlefolds**) adopted their own series of gauges.

Drill/tapping: For selecting the correct size of drill for clearance or tapping.

Thread: Multi-bladed, serrated gauges for measuring the pitch of threads.

Metal sheets: Birmingham sheet and hoop; U. S. standard gauge for sheet and plate, iron and steel, lead (lbs. per sq. ft.), zinc, etc.

Glass: Ounces per sq. ft. and plate thickness.

Collecting tips
➤ *Look for the products of **The L.S. Starrett Co.** and **Brown & Sharpe** for fine quality.*
➤ *The more obscure the purpose the better.*

Shoemakers: "Substance gauges" to measure the thickness of leather soles. These are calibrated in "irons"– 48 to the inch – and may be of slot or taper type but, confusingly, some gauges have a built-in allowance for compression of the leather and the gauge is therefore smaller than the marked size!

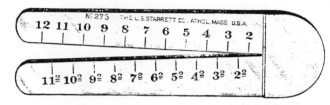

L.S. Starrett Co. No. 273 sole gauge.

Billiard ball: When billiard balls were made of ivory, which is not a stable material, they would shrink and warp, so callipers to check the balls were considered essential. These are not as rare as might be supposed.

Stave gauge: Used by gaugers to determine the thickness of barrel staves so that accurate measurement of contents could be made. These are short rules with a hook on the end and a large diameter ring handle to prevent them dropping into the barrel.

Stave gauge [Long, 1922].

Rope: Try measuring out 120 fathoms of 2½" (circumference, for this is the way ropes are sized) hempen rope and you will quickly appreciate the need for a rope gauge. The quick way of checking the length is to weigh it; the rope gauge calliper will give a direct reading of the circumference and the tables the weight per fathom of various ropes and chains. Some also give breaking strains. A rare variant is a combined gauge and slide rule.

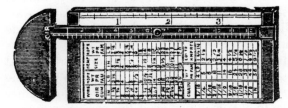

Rope gauge [Buck & Ryan, 1930].

Cork gauges

Listed by **Buck & Hickman** 1923, these are small boxwood sliding calliper gauges. The clue to their identity is the marking which is in English and Spanish inches divided to ¹⁄₁₆ths.

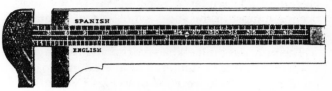

Cork calliper gauge [Buck & Hickman, 1935].

(See also *Measures for the Trades.* pp.178–183)

Price Guide

♦ Ten steel gauges for a variety of materials, zinc, lead, wire, cables, iron, etc. by **Wynn & Timmins, Nurse, Moore & Wright** and others. **£52** [AP]

♦ A substantial steel wood screw slot gauge, Nos. 1 to 24. **£12** [DP]

♦ Multi-leaf thread gauges by **Moore & Wright** or other makers. **£3 – £5** [DP]

♦ A three-leaf plumbers' gauge for zinc, lead and glass, penknife form in brass case. **£24** [DP]

♦ A glaziers' single leaf brass cased glass gauge marked "Pilkington Bros. Ltd.". **£18** [DP]

♦ A "St Crispin" iron shoemakers' substance gauge marked in irons (¹⁄₄₈") and mm. **£18** [DP]

♦ A billiard ball gauge in leather covered case by **Chesterman** also marked "Burroughes & Wattes, London". **£40** [DP]

♦ A 4" boxwood and brass calliper rope gauge by **Rabone** made for Donaghy and Sons, Melbourne & Geelong. G **£33** [DS 24/66]

♦ A 3" boxwood calliper gauge for corks. Marked English, Spanish and metric. G **£20** [DP]

♦ A **Browne & Sharpe** taper slot jewellers' gauge measuring ¼" to ¹⁄₁₀₀₀". F **£52** [AP]

♦ A boxwood watch glass measure, 5" long marked with diagonal scales. **£16** [DP]

♦ A Starrett No. 31 boiler plate gauge. F **£35** [DP]

♦ A weavers' gauge – two points, 1" apart in a beech handle with cover. **£15** [DP]

♦ A 4" boxwood and brass calliper gauge marked "Olney Anderson & Sons, London, Buyers". Marked with ten different systems of gauging tapes and ribbons. **£36** [DP]

♦ A boxwood and brass calliper gauge marked in metric and imperial with tables of weights per foot of sizes of round and square steel. **£36** [DP]

Ninety years ago a building contractor tendered for the construction of a four-storied pub with an elaborate front and bar fittings on the basis of a single double-elephant sized sheet of drawings and a specification of six pages. The plans, sections and elevations, drawn at a scale of ⅛" to the foot, occupied most of the sheet, whilst a few details at ½" to the foot filled the right hand side.

The pub was built – I had a drink in it when, as a young architect, I undertook alterations to the building and these were the drawings I was given. Today, if this building were to be constructed, the contractor would need fifty sheets of drawings in a variety of scales and a 150-page bill of quantities.

What this tale illustrates is that, as traditional ways of doing things died, the need to instruct the contractor in more and more detail has mushroomed. But this was a change that was already taking place by the 19th century – the 18th century documents for the same job would have consisted of a small sketch on the corner of a one page description.

Working to scale

18th century practice was to use dividers to transfer dimensions to the drawing, either from a rule marked with scales or, more likely, from a scale drawn onto the sheet when the drawing was started. Architectural drawings were normally drawn in scales of 10, 20, 30, etc. feet to the inch which are the scales usually given on the combined protractors and scales to be found in drawing sets.

The provision of scale rules marked at the edge gave a considerable improvement in speed and accuracy but, as most architects would have used eight or more different scales needing four or five rules, this was a costly solution.

Top: A universal scale; this layout was largely superseded by the Armstrong layout. Below: A triangular scale; never very popular, this design solved the problems of space and getting the scale tight to the paper but was awkward to reach over [West, 1930].

Each to his own

Scale rules may, at first sight, seem unintelligible. However, the use broadly falls into four groups.

Architects' scales: All architectural scales were related to the foot.

Scale	Ratio	Metric Equiv.
⅟₁₆" to 1'	1:192	1:200
⅛"	1:96	1:100
¼"	1:48	1:50
⅜"	1:32	
½"	1:24	1:20
¾"	1:16	
1"	1:12	1:10
1½"	1:8	1:5
3" to 1' (Qtr. full size)	1:4	
Half full size	1:2	1:2
Full size	1:1	1:1

With the change to metric in the building industry in the late 1960s, conversion scales, by which metric dimensions could be taken directly from imperial drawings and vice versa, were produced.

Surveyors' scales: On the ground, surveyors used chains to take dimensions. Although chains of 100 ft. were available, the traditional surveyors' chain, sometimes referred to as Gunter's chains, was 22 yards divided into 100 links of 7.92 inches.

The surveys were normally plotted to the scale shown below. However the scale rules were marked as 10, 20, 30 etc. suggesting that they could also have been used at ten times these scales, i.e. 10 chains to 1" which equals 1:7920 etc., but surveying manuals show that the ratios listed were those normally used.

Scale	Ratio
1 chain to 1"	1:792
2 chains to 1"	1: 1584
3 chains to 1"	1: 2376
4 chains to 1"	1: 3168
5 chains to 1"	1: 3960
6 chains to 1"	1: 4752

(and less frequently 7, 8, 9, and 10 chains to the inch).

Whilst some surveyors' scale rules are marked only with the chain scales, it is also common to have scales marked with both this and a scale at the same ratio but reading in feet. This latter scale would have been used if the survey had been made using a 100 ft chain.

Engineers' scales: When engaged on work related to buildings, engineers normally drew to the same series of scales as architects. However on other works they sometimes used the surveyors' scales but translated these to a scale of 10 ft. to the inch (giving a ratio of 1: 120), 20 ft. to the inch (1: 240), and so on.

Mapping scales: The large scale Ordnance Survey maps, widely used in the civil engineering and construction industries, were, and still are, produced to the following ratios: 1:1250, 1:2500, 1:10560, which are very close to 50", 25" and (exactly) 6" to the mile.

Dimensions can be taken from these ratio-expressed scales in whatever unit of measurement is preferred (as long as you have the appropriate scale!). Scales reading in links, chains, feet, yards and metres for all of these ratios are to be found. These are marked thus: 1:2500 Links, Feet, etc.

Unfortunately other map makers did not limit themselves to these ratios so scale rules in a wide variety of other ratios are to be found.

Offsets

Small scale rules 2" long often attract attention – and fanciful explanations. These are offsets. They were placed at right angles to the side of a normal 12" scale to measure off distances from the edge, thus mimicking the survey method where offsets from a base line are taken.

Collecting scales

After 1840 when direct use scales came into general use the bevel edge type was quickly followed by the oval (double sided) and triangular forms.

Boxwood was the preferred material. It is stronger and more stable than ivory – leave an ivory scale on a drawing for an hour and the ends are likely to be ¼" off the paper – turn it over and the same thing will happen!

The words *engine divided*, showing that the rule had been marked with a dividing engine giving greater accuracy than hand work, is an indication of a best quality scale.

The number of scales needed often resulted in insufficient space being available in the drawing instrument box so special mahogany boxes holding up to 12 rules and offsets were supplied.

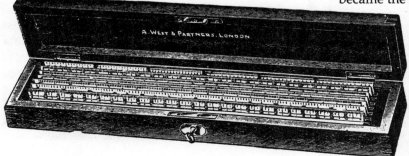

In 1892 **John Rabone & Sons** illustrated 12" *steel* scales – engine divided. Durable and accurate these may have been but steel rusts and would leave marks on the drawings. They didn't catch on – I have never seen one.

Armstrong scale

The desirability of putting all the scales commonly used by architects onto a single rule resulted in the "universal" scales marketed in the late 19th century, but these were not ideal, as the available length of the larger scales was short. Perhaps the ever-increasing pressure on architects to produce more large-scale drawings of details was the reason why, from the 1930s, the Armstrong scale, with the full length of the rule for each scale, became the most widely used type.

Sets of scales were sold in mahogany boxes accommodating up to 10 scales and offsets [West, 1930].

Glaziers' rules

Sliding rules for glaziers, designed to take internal dimensions, are mentioned in 17th and 18th century literature. Occasionally sliding rules do appear, though these are probably later and may be intended for the internal measurement of brewers' vessels and not glazing. The 19th century catalogues show only two types of rule for glaziers: simple boxwood laths up to 6ft. long, distinguishable from general straight laths by being only ⅛" thick, and boxwood T-squares. The latter are readily identifiable, being marked with measurements and having ramps for the glass cutter.

Horse measures

Horses are measured in hands (4 inches) and inches. The measurement is made to the withers which is the point at which the back meets the neck.

Horse measures fall into two groups: substantial rods with a sliding arm for use at equine establishments and portable types which are either concealed in walking sticks or are in the form of a weighted tape measure, with an arm to reach the withers. The better quality measures include a spirit level in the arm.

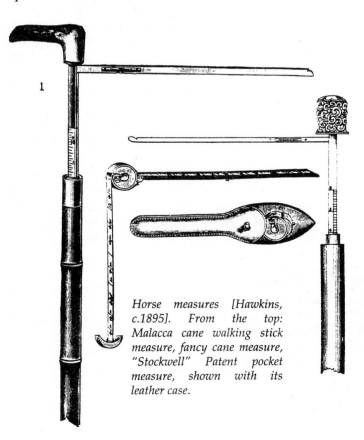

Horse measures [Hawkins, c.1895]. From the top: Malacca cane walking stick measure, fancy cane measure, "Stockwell" Patent pocket measure, shown with its leather case.

Contraction rules

Used by pattern makers, contraction (shrinkage) rules are, contrary to the description, marked with a measure **larger** than standard. Patterns, in essence wooden models of the castings, need to be made slightly larger than the required casting as the metal is at a high temperature when cast and shrinks as it cools. The exact amount of extra size needed is mainly dependent on the type of metal but the form of the casting is also relevant.

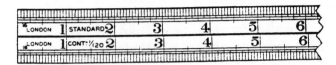

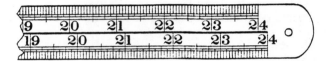

A 24" steel contraction rule. ¹⁄₁₂₀ was the smallest contraction [Rabone, c. 1925].

Some contraction rules are simply marked "contraction for iron" (usually ⅛" to 1ft. or 1:96) or "double contraction for..." Others are more specific, with the ratio marked.

Rules with much larger ratios of contraction (1:10 or 1:20) are known. These were for clay which shrinks considerably when fired.

Glovers' rules

Easily identified by their tapering square section form, 14" long, these rules are often marked with Paris and English inches. Reputedly the thin end was for insertion into the fingers of the glove.

Often mistaken for shoe measures are hand gauges with folding arms. These are very similar to shoe sticks but shorter. Measuring at the knuckles, they indicated the size of glove needed.

Shoemakers' measures

Bespoke shoes are made on a wooden last patterned to the customer's foot. To make a last the shoemaker takes measurements around the foot with a tape or even strips of paper. In the 18th century if you couldn't afford bespoke shoes what you got would have been ill fitting – some sources say that the right and left shoes were often the same.

Shoemakers' size stick with wooden "foot" - the foot is the part of the gauge that folds up when in use [Rabone, 1892]. These could alternatively be had with brass feet.

Shoesticks had been used since the 17th century and probably long before but with 19th century standardisation of sizes (to ⅓ in.) they became essential for both makers and sellers of shoes.

Shoesticks fall into two groups: compact folding designs intended to go into the pocket and larger versions – counter sticks – for use in shoe shops. Eagerly sought are examples with the stops carved in the form of shoes or feet.

The 1892 **John Rabone & Sons** catalogue lists shoemakers' tapes – only 24" long, marked with inches one side and shoe sizes on the other.

Rarely found, and even then often overlooked, are factory sticks – square section rules with a small brass stop at one end, marked with inches and shoe sizes, measured along the sole of the *shoe*, not the foot. Similar are cloggers' size sticks. Clog sizes were, at one time, two sizes larger than shoes.

Opticians

The Orthops rule by **Raphaels** (a London wholesaler of opticians' supplies) is the most commonly encountered.

When lenses were smaller, it was important to position them centrally in front of the eye, so fitting of frames was important. With this rule, the optician could take many measurements including the distance between the eyes, the bridge width and the length of sides. The rules developed considerably over the years – there are many versions, including some in ivory.

Smaller rules intended for fitting spectacle sides and "near point" measures are also to be found.

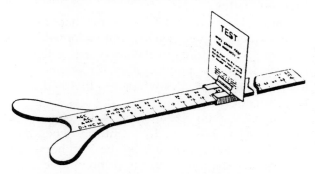

Near point measure [Allen & Hanbury, 1930].

Price Guide

Glaziers' rules

♦ A 36" boxwood glaziers' T-square. **£18** [DP]

♦ A 60" boxwood glaziers' T-square by **Preston**. **£36** [DP]

Horse measures

♦ A patent horse measure being a weighted tape in a brass case with a folding boxwood arm with a level in the original leather case. **£95** [AP]

♦ A horse measure consisting of a boxwood faced mahogany lath with a boxwood arm, figured to 18 hands to one side and to 6ft. 6in. on the other. **£110** [DP]

♦ A boxwood horse measure by **Smallwood** scaled to 18 hands and 72" with beech and brass bar with level. **£55** [DS 24/1371]

♦ A horse measure contained in a Malacca walking stick with silver mounts. **£180** [DP]

Contraction rules

24" boxwood and brass contraction rules by good makers **£6–10**. Ditto in steel **£3–5**.

♦ A 26" 2-fold boxwood and brass contraction rule marked "Contraction for Clay" by **Rabone**. **£40** [DP]

Glovers' rules

♦ A glovers' hand gauge in boxwood with folding stops. Scaled for boys and girls, ladies and men. Marked Patent No. 29036. **£42** [DP]

♦ A tapering square section glovers' rule, 14" long, by **Rabone**. **£20** [DP]

Shoemakers' measures

♦ A boxwood and brass shoe stick by **Ullathorne** stamped No. 3. G++ **£18** [DS 24/498]

♦ A boxwood and brass shoemakers' stick, (folding stop) with ivory tip by **Preston**. **£28** [DP]

♦ A boxwood counter stick with shaped rosewood stops. **£14** [DP]

Shoemakers' measures cont. on p. 181

Bristle rules

Apparently pig bristle doesn't come longer than 7" so these rules are always that long. They are just simple solid rules in boxwood, nicely made with brass capped ends and almost always bear the name of a bristle merchant or paintbrush maker.

Button gauges

The English measure was in "lines" of ¹⁄₄₀ of an inch. Button gauges are small, usually 3" sliding callipers, in either boxwood or brass, marked in this measure and metric. Many carry advertising for suppliers.

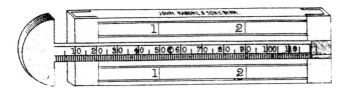

Button gauge marked with button lines [Rabone, 1892].

Paper rules

Thick laths with one edge bevelled, from 2ft. 6in. to 4ft. long. The edges are protected by brass strips. Some of these rules are marked with paper sizes and other information relating to paper. They were used for tearing or cutting up paper, and many examples bear knife scars.

Hatters' rules

For measuring the inside of hats. These are made of boxwood or occasionally ivory and are 5" long with an extending slide. Hat measures consisting of a flexible metal strip expanded by a scissors-like mechanism were also made. These would have given an accurate measurement without the skill needed when using the rule. Heads are not round so an interpolation of the length and width has to be made to establish the size.

Hatters making the harder types of hat – bowlers, top hats, etc. – also used the Conformature, a complicated instrument to record both the size and the shape of a customer's head.

Hatters' rule, available in wood, brass or ivory [Preston, 1909].

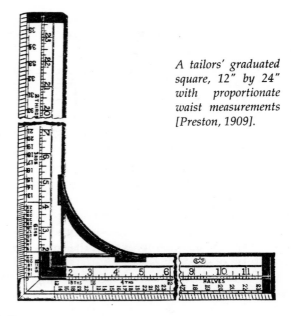

A tailors' graduated square, 12" by 24" with proportionate waist measurements [Preston, 1909].

Tailors

A 60" tape measure, draped around the neck, is *de rigeur* for any tailor and this has always been the principal, almost the only, measuring tool generally used, although the rule makers' catalogues are full of tailors' squares. Judging by the number that are to be found on market stalls today, they must have sold in large numbers but possibly, they were also bought by other trades, for they are a useful medium-sized square, whether fixed or folding.

Diagram square: A miniature (6" long) version of a folding square, used to draw out to scale any illustrations needed in the details of the order.

Trouser measures: Arch sided rules, 36" to 45" long. Quite rare.

Pattern cutters' rules: Pattern sizing is a true trade mystery known only to tailors' pattern cutters. Even after much study of **T.H. Holding's** graduate rule, we are little wiser.

Printers' rules

A rational or, at least, a somewhat rational measurement of type size was only adopted by the English printing trade in the last years of the 19th century when the point system (72 points to the inch) was adopted after wrangling that had lasted fifty years or more.

Originally, each size of type had a name – not too bad in the 17th and 18th centuries when there weren't too many sizes but by the 19th hopelessly inadequate. The usual form of printers' rule is a 9" 2-fold rule folding face to face and marked with 3 or 4 measures on each face, the sizes identified by name – pearl, nonpareil, brevier, long primer, small pica, etc. – in increasing size. Boxwood and ivory examples are found in equal numbers. The names are quaint – a reminder of times when if you didn't understand the trade-speak you were kept a real outsider.

Type sizes as illustrated in Chambers Universal Dictionary of Arts & Sciences of 1743. These faces are by the great English letter founder, William Caslon, of Chiswell Street, London. The heading on this page is a modern computer-generated face - Caslon Open Face.

Pearl.	Which when he knew, and felt our feeble hea
Nonpareil.	Emboſt with bale, and bitter biting grief
Brevier.	Which love had launced with his
Long Primer.	With wounding words, and
Small Pica.	He pluck'd from us all ho
Pica.	That erſt us held in lo
Engliſh.	The hopeleſs, heartl
Great Primer.	Perſuade to die t
Double Pica.	To me he le
Two lined Engliſh.	With whic

Jewellers

Rings are gauged with tapering, boxwood size sticks, fingers with sets of graduated metal rings.

Jewels are assessed with hole gauges of thin metal indicating the size in carats.

Finger gauge [Isaacs, c. 1900].

Price Guide

Shoemakers' measures, cont.

◆ An unusual early 19th C. European mahogany shoe stick, the callipers shaped as a shoe with ivory insets and brass scales. **£176** [DS 24/1288]

Opticians' rules

◆ A collection of 5 boxwood opticians' rules. G to **£94** [DS 24/352]

◆ An opticians' 6" ivory rule with ear hook (for measuring for length of sides). **£35** [DP]

◆ An opticians' boxwood near point focus tester with circular sight and sliding card holder. **£40** [DP]

Bristle rules

◆ 7" bristle rule with advertising for Shaw's Ltd., bristle merchants, London. G++ **£15** [DP]

Button gauges

◆ A 3" boxwood calliper button gauge. **£12** [DP]

◆ A 3" boxwood and brass button gauge by **Rabone. £16** [DP]

Paper rules

◆ A 48" brass edged thick section paper merchants' rule with table giving dimensions of 42 named paper sizes. **£55** [DP]

Hatters' rules

◆ A hatters' inside measure formed in flexible metal with scissors-type adjustment. **£36** [DP]

◆ An ivory hatters' rule with nickel silver mounts. **£66** [DP]

Tailors' rules

◆ A 6" boxwood and brass tailors' folding diagram square by **Rabone. £30** [DP]

◆ A tailors' square with brass arch at angle, proportional waist measure and other scales by **Rabone** No.1513 **£24** [DP]

◆ An tailors' boxwood and brass 19" rule with scales on the reverse with Natural Waist, Scye Depths and Shoulder Pitch by **T.H. Holding,** Maddox St. **£31** [DS 23/1191]

Printers' rules

◆ An early, nicely marked 12" brass printers' rule with shaped handle. **£45** [DP]

◆ A 9" 2-fold ivory printers' rule marked with eight print sizes including minion, pica and brevier. **£85** [DP]

Jewellers' gauges

◆ A boxwood ring stick and a set of fifty-two nickel silver finger gauges. **£40** [DP]

◆ Set of fifty-two brass "Wheatsheaf" brand finger gauges. **£14** [DP]

Ells

The ell was a traditional measure for cloth, only made obsolete in 1824. The British ell was 45 inches but most of the larger continental cities had their own sizes of ells, many of which continued in use till the end of the 19th century. Unlike the European inches and feet, which fell within a small range, the European ells were highly variable with some only half the length of the British one.

A typical ell rule, if there is such a thing, is a hardwood tapering lath around 3ft. long with a handle at one end – in many respects similar to a walking cane. The marking is simple with halves, quarters and eighths of an ell incised or inlaid. Some form of decoration is also often present.

The length of some European ells in imperial inches.	
Amsterdam	39.37
Austria	30.68
Bremen	22.77
Hamburg	22.54
Prague	23.20
Stuttgart	24.18

Drapers' yardsticks

Round yardsticks are perhaps a hangover from the old ell rules. Weights and measures legislation required that they be metal capped, verified and marked if used for retail purposes. Old-fashioned drapers were fast, almost balletic, when using their yardsticks to measure and display fabric.

Bowls

Taken seriously by the players, but not really a trade, the winning bowl is the one nearest the jack.

The measuring set for lawn bowls consists of a tape measure, adapted for taking inside measurements by means of a small horn on the case, and a pair of inside callipers.

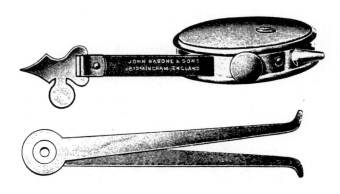

Bowls measure and callipers [Rabone, c. 1925].

Surveyors' laths

Not strictly a lath as almost all fold to some extent. Used when taking dimensions of buildings, the markings are simple and bold, it seldom being necessary to work to better than half an inch. They are usually 6ft. long. Earlier examples seem to have fewer folds than later versions which fold into six.

Drapers' round yardstick [Rabone, c. 1925]. This stick was available in both beechwood and lancewood, marked in inches divided into eighths and into parts of a yard.

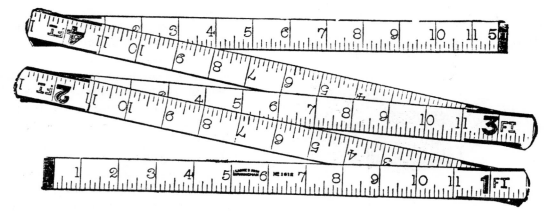

5 ft. 5-fold surveyors' lath [West, 1930].

Clinometers

A simple form of surveyors' instrument, principally intended for use by the military. The 6" boxwood stocks are hinged together. The hinge includes a protractor and one stock is fitted with a level.

There are many variations: both stocks with levels, a small compass fixed or swivelling, various angle/rise tables, military tables, folding sights and even small tubular sights.

The purpose was to take the angle, vertical or horizontal, between two objects, thus allowing calculation of the distance between the two. Made by **W. F. Stanley**, **Tree & Co.**, and the military suppliers **Steward**.

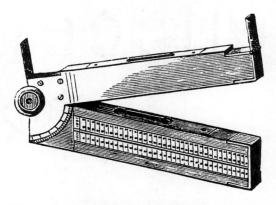

Clinometer with two levels and sights, inclination scale and scale of fathoms [Hicks, c. 1890].

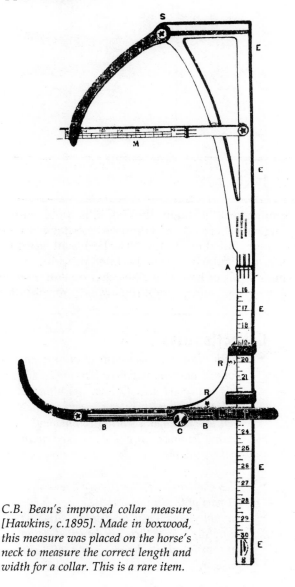

C.B. Bean's improved collar measure [Hawkins, c.1895]. Made in boxwood, this measure was placed on the horse's neck to measure the correct length and width for a collar. This is a rare item.

Price Guide

Ell rules

◆ A 27" Dutch ell rule with ebonised handle and ivory and mother of pearl fittings. **£115** [DS 24/42]

◆ A 20" Dutch ell rule in marquetry with turned handle & ivory finial. **£143** [DS 23/206]

Drapers' yardsticks

◆ A drapers' boxwood round yard measure with brass ferrules at each end marked in parts of a yard and inches in ⅛ths by **Preston**. **£10** [DP]

Bowls measures

◆ A nickel silver cased bowls measure and callipers in leather case by **Rabone**. **£22** [DP]

Surveyors' measures

◆ A surveyors' 6ft. 6-fold lath marked in feet and inches in ⅛ths by **Rabone**, No. 1613. **£10** [DP]

◆ A surveyors' 48" brass scale for plotting surveys with diagonal scale by **W.F. Stanley** in original mahogany case. **£150** [DP]

◆ A 12" boxwood alidade by **W.F. Stanley**, London in original leather case. F **£42** [DS 25/414]

Clinometers

◆ An oak drainage clinometer with German silver fittings by **W.H. Harling** also stamped Dr Bates Drain Inclinometer. G **£73** [DS 24/89]

◆ A 6½" boxwood and brass clinometer by **W.F. Stanley** with folding sights in original lined case. F **£240** [AP]

◆ A boxwood and brass clinometer by **J. Tree** with Indian Army mark. Marked with formulae for calculating the numbers of spherical and rifle shells in variously shaped piles. **£220** [DP]

◆ A Captain B.F. Jelley's military sketching protractor 5⅛" x 2⅜" with extension table of horizontal equivalents. **£24** [DP]

Miscellaneous tools

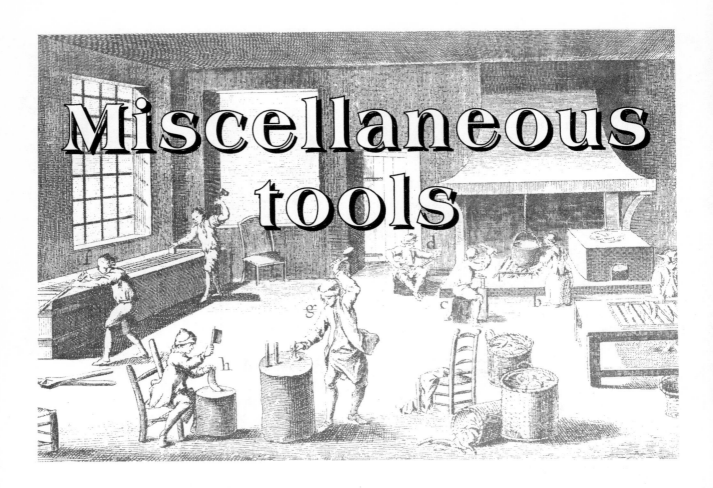

HAMMERS

I remember watching large chain links being welded up by two men using a two-handled hammer. The hammer must have weighed at least 28lb. and with each stroke it went over the top – one man working left-handed and the other right-handed. Incidentally, there is a popular story that left-handed boiler makers could get higher wages as there were places in which only they could work.

Collecting hammers

The hammer collector is presented with a bewildering range from which to choose. Many hammers are specific to one trade – but the differences are often small. This is an area where context can be all important and it is always worthwhile asking the vendor what he or she knows about the source.

Hammers combined with other tools are commonplace – wrenches, pliers and brewers gimlets come to mind but there are many others.

The king of tools

No collector of hammers should be without *The Hammer, the King of Tools* by Ron Baird & Dan

Commerford. Though the book is published in America, there is a common heritage in the hammer and there is much to be learnt from this book, particularly about the later proprietary and patent types of hammer. The other essential for the collector is some well-illustrated reprint tool

Collecting tips

➢ *With most hammers costing only a few pounds and the majority less than £15, this is a good area for the less affluent collector – but there are a few rarities that will command substantial money.*

➢ *Many hammers are not marked with makers' names, so do not let this put you off.*

➢ *The challenge of identifying the use is all part of the fun.*

➢ *Hammers are easy to display. Also, if hung on a wall they can be lifted down and weighed in hand – one of their pleasures.*

catalogues – the shape of a hammer is the clue to identifying its use and the differences are such as can only be defined in pictures.

Fixing the head

The apocryphal story of grandfather's hammer that had had three new handles and a new head but was still grandfather's hammer perhaps illustrates more than just the immortality of hammers. The parting of head and handle was always a problem. Measures to combat this include the adze eye, invented in America around 1840; strapped heads – expensive to make but usual on the best quality light pattern hammers used by trades such as upholsterers, saddlers and coach trimmers where the size of the eye is small; and, somewhat less effective, detached straps.

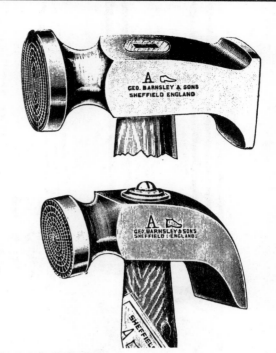

Right: Cobblers' hammers are varied and interesting. Top: Crispin shoe hammer. Bottom: Cordwainers' pattern shoe hammer [Barnsley, 1927]. Note the unusual screw head fixing.

Price Guide

♦ A joiners' framing hammer (double ended) by **Marples**. **£30** [DP]

♦ Two brass headed hammers, one large. **£16** [AP]

♦ An upholsterers' strapped hammer by **Howarth** with part of trade label. **£46** [DS 24/190]

♦ A coach trimmers' hammer and two strapped upholsterers' hammers. **£28** [AP]

♦ A comprehensive set of sixteen coppersmiths' hammers of varied shapes, some double ended, some modern. **£190** [AP]

♦ A fine very small watchmakers' hammer, the head blued on a rosewood handle. F **£24** [DP]

♦ Four watchmakers' hammers of varying size with beech handles with characteristic squared shoulder. **£28** [AP]

♦ A repoussé hammer of traditional form with one large flat face and a ball pein with a bulbous ended boxwood handle. **£30** [DP]

♦ A silversmiths' horn hammer with hickory handle. **£22** [DP]

♦ Sixteen panel beaters' and plumbers' hammers. G **£80** [DS 21/761]

♦ A Crispin pattern shoe hammer with knurled face and a wood heel hammer by **G. Barnsley**. **£24** [DP]

♦ Nine cobblers' hammers of various shapes, mostly in good condition. **£42** [AP]

♦ A London pattern veneering hammer (one end wide and flat for veneering). **£35** [DP]

♦ A glaziers' hammer with chequered face, the handle with turned grooves to give a better grip. **£26** [DP]

♦ A scissors makers' setting hammer. **£22** [DP]

♦ A large, 7¾ lbs., filemakers' hammer with well worn finger grip marks on handle. **£105** [DS 20/93]

♦ An early (probably 17th C.) small claw hammer head with some decoration. From a London archaeological source. Condition good considering age and source. **£40** [DP]

Steel claw hammer with adze eye extension to head [Millard, c.1910].

HAMMERS.

C. S. FARRIERS' SHOEING HAMMER.

28/ per dozen.

CANTERBURY HAMMER.

PLUMBERS' HAMMER.

IMPROVED HORTICULTURISTS HAMMER.

CRAMPING HAMMER.

SHOE HAMMER.

Shoe and Cramping do. 10/ 11/6 13/ 15/ 17/ 19/

1 2 3 4 5 6

KENT SOLID CHEEK.

IMPROVED SCOTCH HAMMER.

BRIGHT KENT HAMMER.

SCOTCH GARDENERS' HAMMER.

BRIGHT LATHING HAMMER.

	00	0	1	2	3	4	5	6	7	8	9	10	11	12	
Bright Kent Hammers......	6	8	9	11	13	15	17	19	22/	25	28/	31	34	38	doz.
Black Kent do.	7	8	10	11	13	16	18	21/	24	27/	30	33	37		"
Bright Solid Cheek do.	9	10	11	13	15	18	20/	23	26/	30	34	38	42		"
New Kent do.	12	13	15	17	19	21	23	26/	29	32	35	38	42		"
Imp. Scotch do., whole head	16	18	20/	22/	25/	28	31	37/	42	48	54				"
Canterbury do..........	13/	14/	16/	18/	21	24/	27	30/	33/	36/	40	44	48/		"
Bright Lathing do.	12/	14	16/	18	20/	21/									"
Black do. do.	11/	13	15	17/	19	23/									"

COACH TRIMMERS' HAMMER.

SADDLERS' HAMMER.

UPHOLSTERERS' HAMMER.

LONDON UPHOLSTERERS' HAMMER.

CABRIOLET HAMMER.

BENWELL'S UPHOLSTERERS' HAMMER.

LONDON GLAZIERS' HAMMER.

	No. 1	2	3	dz.		No. 1	2	3	dz.
Saddlers' Hammer ..	16	18	20/	dz.	Cabriolet Hammers ..	24	26	28	dz.
Upholsterers' do ..	13/	15/	17	"	Coach Trimmers' do ..	25/	27	29	"
Improved do. do ..	16/	18	20	"	Benwell's pattern, solid				
London do. do ..	17/	18	21	"	Upholsterers' do ..	24	26	28	"
Scotch Garden do ..	18/	20	22	"	Plumbers' do ..	15	17 6	20	"
Imp. Horticulturists' do ..			27	"	London Glaziers' do ..	20	22	24	"

Upholsterers' Hammers, with Hard Wool Handles, 6/ per dozen extra.

Hammers from the 1862 Sheffield List

The joiners' chest

Some time around 1770 it became normal for cabinetmakers to make elaborately fitted chests to contain their tools.

Whether such chests were also made at this time by joiners is less certain. However, joiners had stored their tools in chests for a century or more before this date but it would seem that these were simple unfitted chests. Today, it is impossible to tell if any surviving simple chests were used for the storage of tools.

During the 18th century the cabinetmakers' and joiners' tool kits expanded with many new kinds of tool. No doubt the need for better storage in the often rude workshops of the time triggered the improvement in chests.

Virtually all tool boxes of any trade or date were painted black on the outside for protection. No doubt the makers of well-fitted interiors relished the contrast with the outside.

Early cabinetmakers' chests of 1770-1820

Typically a large chest, 40" wide by 24" deep by 24" high, internally fitted with a single lift-out till containing small drawers. They may or may not have storage wells for the plough and sash fillister. Storage for saws can be in slots fixed to the lid. The moulding planes are stored on end in two rows in the bottom level. The internal fittings are mainly softwood but veneered with mahogany. Decorative features were inspired by the Sheraton and Hepplewhite design books.

Cabinetmakers' chests of 1820-1860

Still large chests, with front and back tills and a saw till under the front till. There may be a flat vertical slide at the front to accommodate the larger measuring and marking tools. The construction of chest and tills is similar to earlier chests but the veneering may incorporate the fashionable woods of the period – rosewood, ebony, etc. – and the design is less likely to be of 18th century inspiration. It is more probable that the top will have metal protection to the edges.

Cabinetmakers' and joiners' chests after 1860

After 1860 chests generally seem to have become a little smaller in all dimensions – less than 36" in width. The inside of the lid may be panelled rather than flat veneered. If the lid is veneered, the designs can be very original and appealing but often they are not based on pattern books and can appear untutored, even eccentric.

Price Guide

♦ A fine 18th C. cabinet makers' chest, 42" x 26" x 24½" high. There are three sliding tills with sixteen dovetailed oak lined operating drawers with matching dummy fronts to the back of the front till. The interior is veneered and cross banded in flame mahogany with ebony and boxwood stringing. The plane wells are covered with mahogany lids. The lid, which has a dust-proof roll joint, contains a slotted saw box to hold six saws. The exterior is painted black and has two rope handles. **£1800** [DP]

♦ A cabinetmakers' lockable pine chest, dovetailed throughout, 41" x 23" x 28" high, two tills with a total of sixteen drawers veneered in feathered mahogany to the tops and drawer fronts and ivory drawer pulls. Well for moulding planes. One crack in lid. **£990** [DS 24/896]

♦ A cabinetmakers' lockable pine chest, 34" x 20" x 20" high. Solid Cuban mahogany interior. One bank of three sliding trays over moulding plane well and brass hinged top. Interior lid divided by ebony inlaid mahogany into seven panels, two panels bird's eye maple, four figured oak and the central panel satinwood inlaid with JB. **£280** [DS 24/1330]

♦ A cabinetmakers' pine chest, 38" x 24" x 24" high, fitted with two banks of eight trays in Cuban mahogany, the tray lids with ebony inserts and brass hinges. Well for moulding planes and a sliding divider. Lacks one brass pull. **£240** [DS 24/1771]

♦ A cabinetmakers' lockable pine chest, 34" x 23" x 21" high with two banks of five trays, dovetailed mahogany throughout. Sliding divider. The lid with mahogany outer panel and raised moulded surrounding veneered panel. G++ **£710** [DS 22/1110]

♦ A cabinetmakers' lockable pine chest, 38" x 28" x 28" high, fitted with four sliding tills containing thirty drawers or dummy drawers, dovetailed throughout and with leather drawer pulls. Drawer fronts and tops in figured mahogany with ebony stringing and edging, over tills for saws, moulding planes etc. Veneered lid with mahogany panel, trimmed and inlaid with ebony and with mahogany surround. **£800** [DS 20/656]

♦ A cabinetmakers' dovetailed pine chest, 38" x 23" x 20" high, of unusual design with three sliding tills and inlaid mahogany lids over other compartments. The lid inlaid with over 3,000 pieces of marquetry in geometric designs in bird's eye maple, ebony, boxwood, rosewood, satinwood, tulipwood, walnut, etc. **£600** [DS 19/1002

Coachmakers' chests

These are likely to be painted internally employing techniques used in coach painting – lining, varnishing over and heraldic decorations.

Ships joiners' and railway carriage builders' chests – mid-19th century onwards

Chests made by men working in these two distinct branches of joinery seem to be very similar. The characteristics are size – typically, quite compact, around 33" or less wide. The insides are fitted with trays not tills. Apart from the top tray, which is covered by lifting flaps and is divided into sections, the trays are full length to accommodate the larger chisels and screwdrivers used by these trades. The chest is not tall enough to take moulding planes on end as those employed in these trades would have owned fewer moulding planes. Solid teak, much used in these trades, was a favoured material for the trays but mahogany was also used.

Shipwrights' chests

Shipwrighting is a hard trade and the chests reflect this. They are simple chests made of softwood with little in the way of fittings inside except perhaps one open tray. Most chests are of quite modest size, say 33" wide by 16" deep by 16" high.

Patternmakers' chests

Patternmaking as a distinct trade seems to have become commonplace only after about 1875. It flourished until the 1960s after which developments in engineering practices caused a decline.

Patternmakers' chests are distinctive – typically 30" wide by 16" deep by 16" high, made of and fitted inside with yellow pine. This was the patternmakers' wood and the finish inside is like patterns – clear varnished, glowing yellow. The owners often seem to take particular delight in making the box the absolute minimum size, giving a three dimensional puzzle to get everything in. Arched strengthening battens on the top to deter use as a site for work are typical. The chests are more useful than collectable.

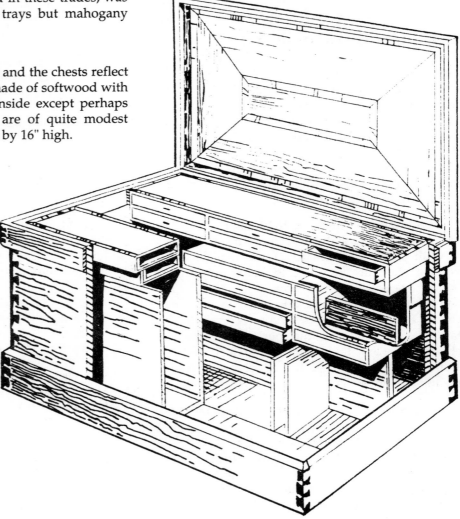

A typical "developed" tool chest of early 19th century form. The carcase is softwood painted black on the outside. The inside of the lid and the four tills are veneered onto softwood. Some later chests may have oak or mahogany drawer linings but softwood is more usual. Note the space at the front for storing large flat marking-out tools; in some chests these were fitted into a lifting slide.

Polished oak gent's wall cabinet of tools [Melhuish, 1912]. Advertised in two sizes with either twenty-five or forty-six tools. The top lifts to allow the longer tools to be taken out.

The gent's tool chest

The usual format is a chest about 18" wide by 10" deep by 5" high with a lifting lid giving access to a shallow section. Under this are one or two drawers. Unlike all the chests that we have previously described, the carcase is hardwood, most likely oak but occasionally mahogany. This type of chest, which was sold with a complement of tools, has a long history. An example sold by a London ironmonger in 1773 and still containing most of the original tools is in the Colonial Williamsburg collection. This chest, like most, has a vendor's label pasted in the lid. Chests of this sort were still being sold until the First World War.

The gent's wall cabinet

An alternative to the gent's chest, but intended for the same household market, was the gentleman's registered design tool cabinet. **Wm. Marples** sold these in four different sizes but most were around 20" high by 15" wide. A pair of doors opened to give access to the tools in racks. Like gent's chests, these wall cabinets were supplied with the tools.

Price Guide

• A cabinetmakers' lockable pine chest, 40" x 27" x 27" high with two banks of tills containing eighteen drawers and twelve dummy drawers and tills for saws etc. The drawer fronts and till tops in figured mahogany with boxwood and rosewood stringing and leather drawer pulls. Lid veneered in figured mahogany with oval rosewood insert surrounded by banding and stringing in various exotic hardwoods. Shaped brass plates to corners and iron strips to lid. Some damage to lid veneer. **£495** [DS 23/1350]

• A cabinetmakers' lockable pine chest, 39" x 26" x 26" high with fourteen drawers in four tills, with saw well and moulding plane well. The dovetailed drawers with oak fronts, rosewood surrounds and leather drawer pulls, the till tops in oak with satinwood and rosewood key pattern inlay. The lid veneered in oak with satinwood and rosewood inlaid small starburst and key pattern surround. Fitted with two brass carrying handles at each end. **£550** [DS 21/1278]

Price Guide

• A lockable tool chest, 35" x 18" x 16" high with five drawers in mahogany with brass pulls. The moulded mahogany lid and tray top decorated with floral and musical instrument transfers. Hinges and pulls replaced. Includes 30 **Sorby** tools. **£210** [DS 22/1161]

• A late 18th C. two drawer oak tool chest, 22" x 12" x 12½" high fitted with brass handles to drawers. The top lifts to reveal a fitted interior with the label of **James Buttall**, Ironmonger & Brazier, pasted onto the inside of the lid. **£400** [DP]

• A fine brass trimmed gentleman's fitted tool chest in figured mahogany, 22" x 12" x 10", undoubtedly by **Fenn & Co.**, circa 1870, with eighteen little used tools, including a beech plated brace, mostly stamped Fenn and original to chest. F **£600** [DS 17/1167]

• A coachbuilders' pine chest, 35" x 20" x 19" high, lined in mahogany with four sliding tills and divider, the till lid crossbanded in walnut. Lid veneered in mahogany with moulded walnut surround and oval centre piece. **£360** [DS 20/776]

• A kit of patternmakers' tools in lockable pine chest with seven trays. Includes brass and steel router, two removable sole planes in fitted tray, bow saw, panel gauge, brass mallet, nine turning tools, pad saw handle, thirty-eight chisels, gouges and carvers, dividers, bevels, two saws fitted in lid and an odontograph with owner's name, F. Hartley. F **£660** [DS 22/1238]

• A shipwrights' pine tool chest, 35" x 13½" x 10½" high, with JM painted on front and copper plate on lid stamped "J. Mulholland Shipwright" and tools including four mast and spar planes, solid teak jack plane, nine caulking irons and mallet, adzes, braces and bits, augers, etc. G– to G+. **£340** [DS 17/992]

• A painted softwood railway carriage makers' chest, 36" x 22" x 20" high fitted with three sliding trays, dated circa 1890. The lid veneered in a diamond pattern within an outer panel. The trays in plain mahogany. **£240** [DP]

MELHUISH'S PATENT LONDON. E.C.

Just one of the many differing designs of combined workbench and tool cabinets. This is the "Exonian". In 1912, in best wainscot oak it cost £11 9s 0d without tools and £20 11s 9d with tools [Melhuish, 1912].

The "compactum" tool cabinet

Not so much a tool chest but more a complete workshop, measuring 48" wide, 21" wide and 39" high. The construction is basically a workbench fitted with cupboard and drawers underneath and a lifting cover holding tools in clips. This formed a back in the open position with the tools conveniently to hand. These cabinets were supplied with or without tools and are prominently featured in many catalogues around 1910.

The bass bag.

Once used by woodworkers for carrying out tools to a job, nowadays they only seem to be used by plumbers. The modern ones are made from coarse canvas but originally they were made of bass, a fibre obtained from some types of palm tree.

Top: Joiners' tool bass, web bound and canvas lined. Bottom: Painters' nest hand bass, supplied in nest of six. [Ward & Payne, 1911].

Gent's polished oak tool chest [Marples, 1909]. This firm listed twelve different sizes of gent's tool chest, each with differing tools according to size. The smallest size (15¼" wide) had no drawer, the largest (33" wide) had two drawers.

Price Guide

- A gent's oak tool chest 21½" x 12" x 10" high with one drawer under deep top section. **£130** [DP]
- A dovetailed oak gentleman's tool chest, 16" x 9" with brass plate of **Wm. Marples**, Westland Terr., Sheffield, ivory escutcheon and brass handles with list of original tool contents. **£55** [DS 22/1449]
- A gent's mahogany tool box, 16½" x 8½" x 4" by **Timmins**, Birmingham, with label showing original tools. **£30** [DS 22/1392]
- A gentleman's dovetailed oak tool box with brass carrying handles by **Mathieson** with trade label and list of contents on inside of lid. **£70** [DS 19/1325]
- A gentleman's fitted oak tool cabinet, 25" x 14" x 6" No. 1 by **Army & Navy Stores**. **£85** [DS 20/1302]
- A gentleman's dovetailed tool chest with brass carrying handles by **Sorby** with label showing contents (22 tools) and two other **Sorby** labels. **£85** [DS 19/843]
- A **Marples** hanging tool cabinet, 19" x 7½" x 27" high with manufacturer's label and tool holders inside. **£60** [DS 22/1162]
- A combined work bench and tool cabinet by **Melhuish** in American walnut. Lifting lid to work surface, recessed central kneehole with six drawers and end cupboards. No tools and the vice is missing. **£600** [DP]

The carpenters' carrying case

The modern equivalent of the bass bag. This is a light case, usually made of plywood, of the minimum length to take a hand saw, 10" high and 6" deep. There is a carrying handle on the top but, if it is full, you would not want to carry one far! Part of the front opens and inside there is usually one small tray. Apart from one lovingly lettered "Joe Smith, Britain's best carpenter" – who worked on a job for me – none have seemed collectible, but they are useful.

Once upon a time, before everything came in cardboard cartons, the cellar was an important place for storing and readying goods for use. Often overlooked by collectors are a surprising number of tools used by the warehouseman and cellarman in opening and dispensing the contents of both wet and dry casks, sacks, wooden packing cases and baskets.

There and back

Suppliers charged for containers so it was normal practice to return them – or at least it was the case once the railways made transport cheaper.

To remove markings the warehouseman used the box scraper, actually a small plane, with the body pivoted in a fork. These were made by **Stanley**, **Record** and other makers and are unusual in that they are pulled towards the user.

Box scraper [Melhuish, 1912]

The speedy opening and closing of packing cases was effected with special hammers – combination tools that include lever edges and/or claws for opening, occasionally a hatchet and the hammer head for closing or knocking apart boxes that weren't to be returned.

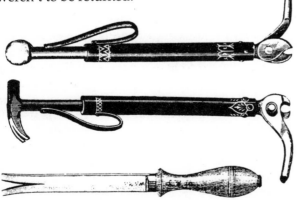

Two nail pullers [Melhuish, 1912] and a case opening chisel [Ward & Payne, 1911].

Sampling

When quality control and storage conditions were less certain than today, the sampling of goods was essential. Quite common are cheese tasters (testers), the half round blade about 6" long being

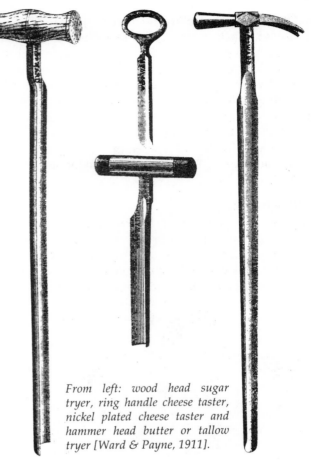

From left: wood head sugar tryer, ring handle cheese taster, nickel plated cheese taster and hammer head butter or tallow tryer [Ward & Payne, 1911].

used to withdraw a sample. **R. Timmins & Sons** listed no less than twenty-six different patterns in their c.1845 catalogue; indeed a specialist collection could be made of these alone. Points to note on the different patterns are the muslin rippers on the handles and the T-head pattern which includes ebony ends to the handle enabling the cheese to be tapped and the condition assessed by the sound made.

Similar to cheese testers but rarer are butter or tallow testers, again with a half round blade but longer – from 14" to 36". Flour samplers are distinguishable by transverse divisions within the blade. **Ward & Payne** 1911 listed sugar tryers – very similar in size and shape to a butter tryer but longer and the larger blade 1" across. Sugar tryers must be very rare objects as I have never seen one.

Cigar or confectioners' box opener [Melhuish, 1912]

Liquid refreshment

If small quantities are to be dispensed from a cask, it will need a hole for a tap in the head in addition to the bung hole in the side. Sometimes this was not made until it was required. In these circumstances the hole was bored using a cock (plug) bit (see p. 83) which automatically sealed the hole when the wood was penetrated. It was then quickly withdrawn and the cock fitted.

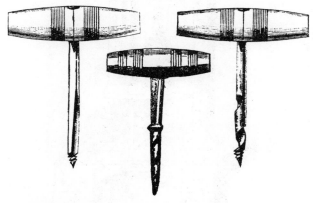

Left and right: Brewers' gimlets [Joseph Long, 1922]. Centre: Wine fret [Loftus, c.1870].

The most common cellarman's tool to be found today is the brewer's or spile gimlet though the true use of these tools is often unrecognised. This was to bore a hole through the bung to let in air as the contents were drawn off. The clue to identification is that the bit is shorter and fatter than ordinary gimlets and the handle is barrel shaped and sometimes made in brass. The handle had a secondary use as a hammer for knocking in the spile (a small round wooden plug) to close up the barrel. Similar is the wine fret, useful for making a small hole to take a sample and distinguishable by having a more acute spiral and a shoulder on the bit.

Surprisingly rare are floggers (bung starters) – a light pattern mallet head on a long handle – used to start (loosen) the bung. The handles, often of cane, are thin and therein must lie the reason for their rarity.

Bottling it up

Whether you were a chemist bottling one-off prescriptions or a brewer with thousands of bottles, a cork was the only stopper available until proprietary types of closure were developed in the later 19th century. Equipment for inserting corks extends from the simple cork squeezers to more productive lever presses. To improve keeping or to

Price Guide

◆ A No. 70 box scraper by **Stanley**. **£12** [DP]

◆ A No. 070 box scraper by **Record**. **£18** [DP]

◆ A Williams patent nail puller by **Wynn Timmins**. G **£18** [DS 24/928]

◆ A Cyclops nail puller by **Union**. **£20** [DP]

◆ A case openers' compendium tool in the form of a claw hammer with a tack lifter extension to the top of the handle, marked **Perry**, London. G **£30** [DS 20/46]

◆ Three retailers' packaging hammers, a strapped orange box hammer, an iron grocers' crate hammer and a fruiterers' hammer. G **£36** [DS 21/649]

◆ A steel cheese tester with brass T-handle with ebony tips in original leather case. **£26** [AP]

◆ A silver-plated cheese tester with deceptive fake silver marks and T-handle with rosewood tips. **£22** [DP]

◆ A steel cheese tester with ring handle incorporating muslin rip by **Timmins**. **£18** [DP]

◆ A 20" rosewood handled butter tester. **£35** [DS 22/1142]

◆ A fine plated cigar box opener in the form of a hammer with extended nail lifter by **Underwood**, London, with ebony handle and nickel silver ferrule. F **£50** [DS 25/1615]

◆ Two brewers' gimlets with barrel shaped brass handles, one stamped **Bass & Co**. **£22** [DP]

◆ A brewers' gimlet with barrel shaped boxwood handle. **£7** [DP]

◆ An all boxwood beer tap. F **£8** [DP]

◆ Three brass beer taps, one double ended. **£35** [AP]

◆ A cast iron, decorative pattern cork squeezer by **W. Bullock**. **£30** [DP]

◆ A 9½" high turned boxwood bottle cork inserter with handled plunger. **£24** [DP]

◆ A set of eight rosewood handled brass seals for wine bottles with sherry, etc. and the merchant's name on one. **£140** [AP]

◆ A fine boxwood handled bottle seal with six interchangeable heads marked with the names of wines including ginger wine and cowslip wine contained within the handle, for domestic use. **£150** [DP]

◆ A large bale hook, of New Zealand wool hook pattern. **£12** [DP]

◆ A six claw wheat bag hook. **£12** [DP]

◆ A rare folding bag hook with beech handle, two pronged. **£20** [DP]

identify and certify the contents, it was common practice to cover the cork with sealing wax and mark it with a seal.

Today both chemists' and wine merchants' seals used for this purpose are in demand from collectors, whilst much bottling, sampling and racking equipment from the cellars of the once-numerous wine merchants now decorates wine bars.

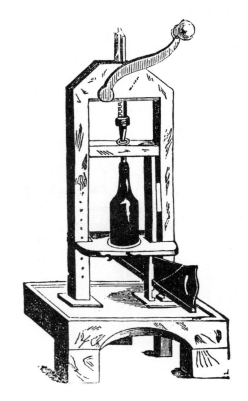

A corking machine [Loftus, c. 1870].

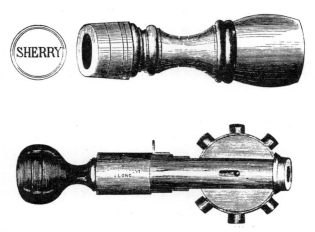

Bottle seals: top, with an interchangeable head and bottom, with a rotating head [Joseph Long, 1922].

Getting a grip

The bag hook, a favoured weapon in dockside brawls, really did have a proper use. Many employers tried to ban the use of hooks, which damaged goods and were less productively used to rip open containers in order that the goods could be pillaged.

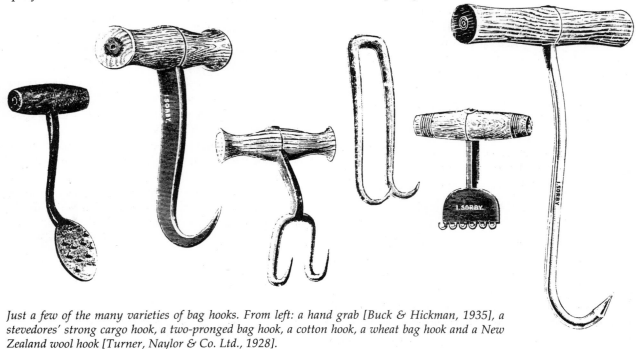

Just a few of the many varieties of bag hooks. From left: a hand grab [Buck & Hickman, 1935], a stevedores' strong cargo hook, a two-pronged bag hook, a cotton hook, a wheat bag hook and a New Zealand wool hook [Turner, Naylor & Co. Ltd., 1928].

When threatening nature started at the end of the village street, the gardener wanted nothing but strict formality to demonstrate man's dominance over the forces of nature. By the middle of the 18th century, nature was being seen as more friendly but not until the 19th century did the less formal approach of the park landscapers find its way into the domestic garden.

Early gardening tools

Smith's *Key* of 1816, which is the earliest illustrated Sheffield tool catalogue, devotes a whole plate to garden tools and there is also a handsome illustration of a pair of garden shears. Shown elsewhere are spades, hoes, sickles and billhooks, tools which would have been used both in the garden and in agriculture, but the existence of separate and distinct garden tools shows that gardening was, by 1816, seen as a specific trade and, perhaps, even a pastime for the amateur.

The English love of lawns can be deduced from the item "Grass plot edging knives" – a half-moon lawn edging knife. Metal rakes are there, a very modern-looking Dutch hoe and pruning shears but the garden trowels have more similarity to a bricklayer's pointing trowel than the hollow garden trowels of today.

19th or 20th century?

Collecting interest in garden tools is a very recent phenomenon fuelled not only by the general interest in everyday living in the past and garden history but also by "country living" decorative influences.

A problem for collectors is that in general gardening tools do not have a long life and, in the absence of much detailed literature, it is difficult to be certain which are the older tools. Perhaps this does not matter too much, as the designs evolved very slowly and the items illustrated in the early 20th century catalogues look much the same as those in Smith's *Key*. This is a field (or garden?) where interest and, most importantly, decorative potential are the factors that determine price.

Collecting tips

> As for many trades, smaller better finished tools for the gentleman were available. These are attractive and today command premium prices.

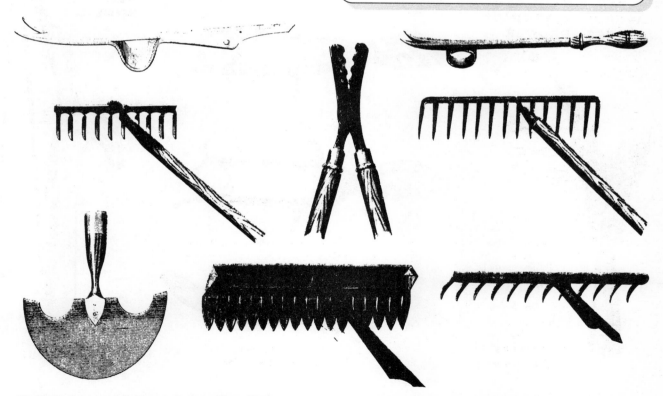

Clockwise from top left: Strong dock grubber, thistle tongs, improved daisy lifter, American pattern rake, gravel rake, daisy rake, grass plot knife and improved steel rake [John Harrison & Sons, 1902].

SECATEURS.

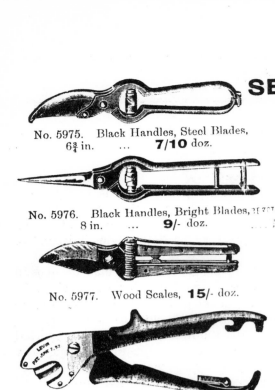

No. 5975. Black Handles, Steel Blades,
6¾ in. … **7/10** doz.

No. 5976. Black Handles, Bright Blades,
8 in. … **9/-** doz.

No. 5977. Wood Scales, **15/-** doz.

No. 5978. Levin Patent. American make, **28/8** doz.
„ 5979. Large size ditto … **40/-** „

THE "MARVEL."
No. 5986. Steel Blades, Wood Handle,
Length 7½ ins. **6/-** doz.

No. 5980.   Black
Handles,   Bright
Steel Blades.

7¼   8¼   8¾ in.
11/6 12/6 14/- doz.

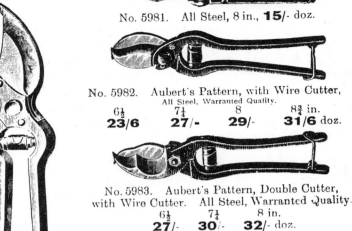

No. 5981.   All Steel, 8 in., **15/-** doz.

No. 5982.   Aubert's Pattern, with Wire Cutter,
All Steel, Warranted Quality.

6½	7¼	8	8¾ in.
23/6	**27/-**	**29/-**	**31/6** doz.

No. 5983.   Aubert's Pattern, Double Cutter,
with Wire Cutter.   All Steel, Warranted Quality.

6½	7¼	8 in.
27/-	**30/-**	**32/-** doz.

No. 5984.   All Steel, Fine Nickel Plated,
4¾ in. **25/-**   5¼ in **28/6**   6¾ in **31/6** doz.

No. 5985.   Chamois Bags for Secateurs and
Flower Gatherers.
5 in. **6/6**   6 in. **8/6**   7 in. **11/-** doz.

AVERUNACATOR OR
TREE PRUNER.

No. 5987.
Fine Quality Steel,
Red Japanned, with
Pole Socket and
Pulley Wheel,
43/- per doz.

No. 5988.
Cheaper article.
Saleable Line.
Mounted on Card.
24/- per doz.

No. 5989.   Japanned, with Brass Springs.   American
make, **20/-** doz.

No. 5990.   Japanned Handles, Brass Spiral Spring,
14/8 doz.

No. 5992.   Sabatier Pruning Knife, Horn
Handle. **15/-** doz.

TREE
PRUNER.

No. 5991.   New Improved Pattern, extra strong for branches
up to 1¼ in. Galvanized,
57/- doz.   **5/-** each.

*Just some of the secateurs and pruners from the **Millard Bros.** catalogue c. 1910.*

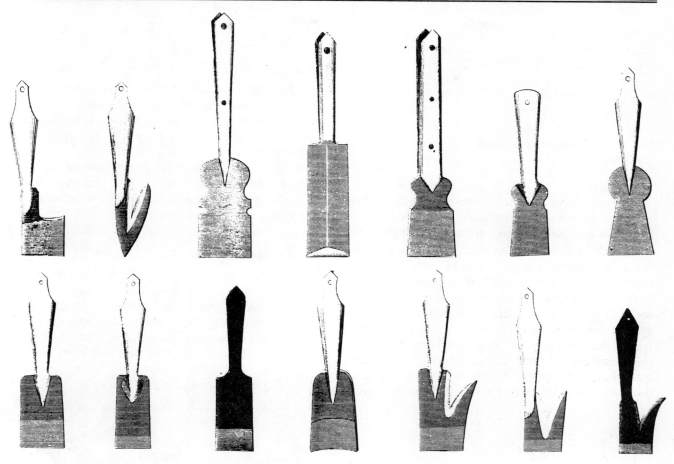

A variety of weed hooks and spuds [John Harrison & Sons, 1902]

Snip, snip

Amongst the tools that seem to have attracted a lot of attention from the innovators are long arm pruning and picking devices and pruning shears – secateurs in today's terms – although as late as 1911 the **Ward & Payne** catalogue uses the earlier name for the eleven types shown.

Mist and spray

That terror of the greenfly, the brass garden syringe, was available in a bewildering array of varieties, each distinguishable by some small difference in the valves or roses The best have a drip preventer to stop the spray running down the arm.

Price Guide

• An ebony handled saw edged asparagus knife by **Ward & Payne**, 18" o/a. **£45** [DP]

• A well shaped wrought iron garden line reel and peg. **£14** [DP]

• An early pair of large (21" o/a) wooden handled garden shears. **£20** [DP]

• A D-handled dock grubber of garden fork size with foot rests to both sides. **£35** [AP]

• A nicely made daisy grubber with round foot and turned ash handle. **£12** [DP]

• A 20" brass garden sprayer by **The Eclipse Spraying Co.** with various extensions and nozzles in original box. G++ **£66** [DS 24/174]

• A crutch handled turfing iron. **£22** [AP]

• A nicely shaped weed hook on 5ft. handle. **£14** [DP]

• A folding pruning saw with rosewood handle and brass fittings, 9¼" folded. **£36** [DP]

• A 4½" blade folding pocket pruning knife with stag horn handle by **Saynor**. **£12** [DP]

• A pair of steel thistle tongs with 3ft. wood handles. **£48** [DP]

• An iron bulb planter by **Skelton** with ash handle. **£75** [DS 21/1573]

• A gent's double edge billhook with nicely turned beech handle by **Sorby**. **£18** [DP]

• Five different pairs of secateurs of early design and a pair of flower gathering scissors. **£42** [AP]

From one perspective, the history of agriculture can be seen as a progressive replacement, in one process after another, of the hand tool as mechanisation became both practical and cost-effective. But there were many small marginal farms which could not afford expensive new machinery, and many specialist hand tools that had evolved over the centuries were still available until the 1940s.

Drainage

The construction and maintenance of open ditches required little more than spades. In wet conditions the wooden spade offered advantages in that the soil stuck less – as late as 1902, **John Harrison & Sons** of Dronfield near Sheffield included eight patterns of wooden spades in their catalogue.

Drainage using clay pipes also required special tools. Digging in clay soils was eased by the use of special "draining tool" spades with tapering, well-curved blades, up to 22" long. Needless to say, the handles were also specially reinforced with straps extending well up the handle, sometimes right to the eye. The bottom of the trench was prepared using a draining scoop before the pipe was lowered in with the pipe layer.

Hedging

In the days before barbed wire, the proper maintenance of hedges in a stock-proof condition was a regular winter employment with the hedges being cut and laid on a 10 or 12 year cycle. The tools needed were few – a slasher to remove higher branches, a hedging hook for trimming up the lighter stuff and a billhook for the majority of the laying work. A large wooden beetle to hammer in any stakes was the last essential.

Collecting tips

> *There are numerous patterns of billhooks, and, to a lesser extent, slashers, which are mostly named after the area or county where the design originated. The differences may be small but they mattered to the man working all day, just as today they matter to an increasing band of collectors interested in country crafts.*

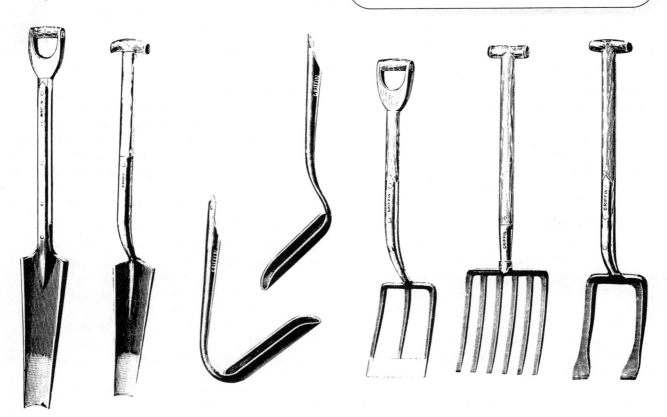

Draining tools, draining scoops and three agricultural forks: a clay fork, a malting fork and a cultivating fork [Swindell, 1903].

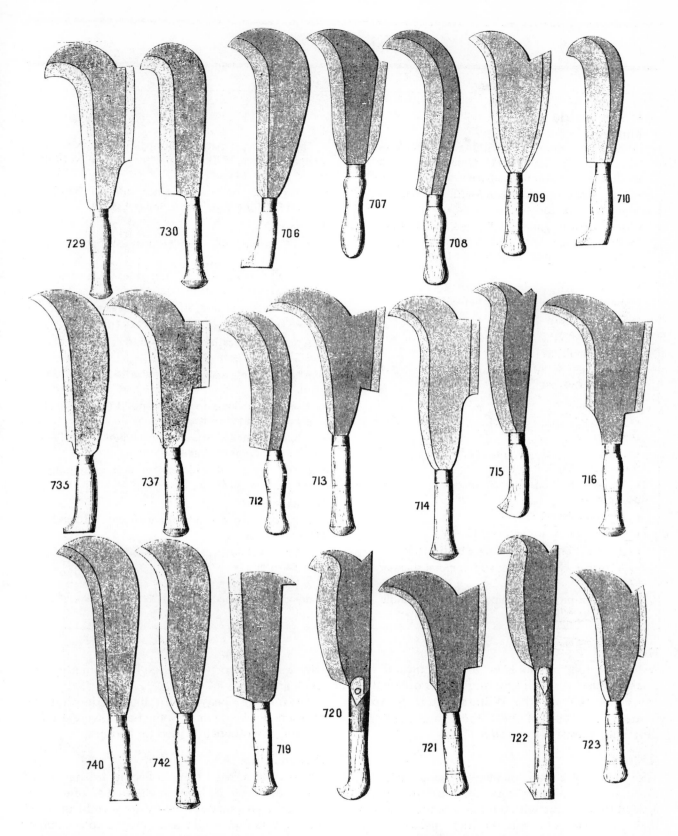

Billhooks from a catalogue of **James, Isaac & John Fussell Ltd.** of Mells and Frome, Somerset, dating from c.1890. Local manufacture of agricultural hand tools continued until the 20th century, by which time the large manufacturers in the Midlands had taken most of the market.

The patterns above are: Top row: Kent, Kent, Pontypool, Natal broom, Chichester, Liverpool pattern broom, Gravesend. Middle row: Llanelly, Stafford, Berkshire, Nottingham broom, bright Stratford broom, Norfolk, Lincoln. Bottom row: Knighton, Ledbury, block bill, strong Suffolk, bright Lincoln Fir, light Suffolk, bright broom.

Price Guide

Billhooks

Billhooks, medium size with original handles, late 19th/early 20th C. **£8–15**. Unusually large or small size or uncommon pattern **£14–20**.

- Two billhooks by **William Swift**, both with the large WS mark. **£38** [DP]
- An unusual pattern bill hook by **William Swift**, a straight back with hook at end and a straight edge (would suggest that this is a block bill). **£40** [AP]
- An unusually wide and short double edged billhook (probably Lincoln pattern). **£28** [DP]

Drainage tools

- An iron-shod wooden drainage (?) spade with straight handle, 54" o/a. **£32** [AP]
- Two scoop draining tools and a long bladed draining spade. **£40** [DP]

General tools

- A wrought iron breast plough with 83" oak shaft. **£45** [DS 21/1384]
- A long handled rabbiting spade with hook end by **Swindell & Co. £30** [DP]
- Three steel shod wooden spades with long handles. **£60** [DP]
- A potato shovel. **£24** [DP]
- A seven prong dung fork. **£14** [AP]
- A three prong manure drag. **£10** [DS 21/1574]
- A barbed wire tightener with integral spanner. **£24** [DS 21/1579]
- An eight-pronged turnip fork with D-handle. **£16** [DP]

- Three different pattern turnip knives, a Norfolk straight pattern, a Scotch pattern and a Norfolk round pattern. **£14** [DP]
- A six blade star turnip cutter with ash handle, marked IH (**John Harrison & Sons**). **£32** [DP]

The harvest

- A 42" scythe with a double bent ash handle. **£30** [DP]
- A small reaping hook with original trade label by **Castle & Turton**. G++ **£18** [DS 22/127]
- Three reaping sickles, all different styles. **£18** [DP]
- Two scythemen's stake anvils. G **£85** [DS 25/1476]
- A small reed scythe with 13" edge and beech handle. G **£26** [DS 24/582]
- A scythemen's stake anvil and a hammer. G **£44** [DS 24/806]
- A barley cocking fork, 4-prong, 15" wide with long handle and wire stop. **£28** [DP]
- A barley malt spade with crutch handle – excellent condition and patina. **£60** [DP]
- A threshing flail with leather and horn hinge, shafts replaced. G **£25** [DS 20/119]

Haymaking

- A 30-teeth wooden hayrake with divided handle. **£18** [DP]
- A three-section steel hay tester, 6ft. long in leather case. **£40** [DP]
- A huge Northern pattern hay spade, the blade 16" wide with a strapped T-handle 23" wide. **£45** [DP]

Hoes

A humble tool but look in any agricultural tool catalogue and you will find more types of hoe than any other tool. In the **William Hunt & Sons, Brades Co.'s** List of 1905 there are *ninety-five* different patterns of hoe listed.

Reaping

The picture of a team of men scything their way through the standing corn may be an icon of agricultural productivity but many sources suggest that the sickle hook was the more usual tool for hand reaping whilst the scythe was used for cutting hay.

Inspect any catalogue and you will find numerous patterns of sickles, reaping hooks and hooks. There is a difference – the sickle has a lighter thinner blade than a hook and is sharpened with a serrated or wavy edge by hammering on a small anvil.

Hooks are heavier in the blade and are sharpened with a stone. Some hooks were used for reaping and others intended for hedging.

Hay making

Always attractive to the collector are hay rakes. These are true country-made items, often with wooden teeth, made in many sizes right up to 8ft wide giants that would seem more in place behind a horse.

Hay knives and spades were used to remove small amounts of hay from the rick, and hay testers – eye-handled steel rods with a small hook on the end – were used to determine whether the rick was overheating.

Horses

The belief that bleeding was an effective treatment lasted until well into the 20th century. How backward is veterinary science, you might think – but then bleeding equipment for humans was still being sold in the 1930s!

The equipment used for animals was the fleam. The majority of fleams have three blades though there are one and two blade versions and a few also include a knife blade. The selected blade was driven in by a sharp blow on the back of the blade with the fleam (or blood) stick. Fleam sticks are a lot rarer than fleams; one supposes that often any suitable bit of stick was used.

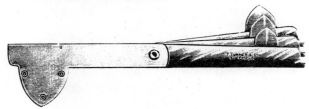

A brass handled fleam [Wingfield, Rowbotham, 1904].

The administration of medicines required the use of a gag – most commonly lyre shaped – and either a balling gun of some form to place a ball (pill) right to the back of the mouth, or a veterinary syringe for liquid medicine.

Commonly found, and still needed today, are tooth rasps used for removing sharp edges left where the animal has an imperfect bite.

Price Guide

Horses

- A three-blade brass cased fleam with ebony blood stick. **£50** [DP]
- A horn scaled three-blade fleam by **Turner**. F **£30** [DP]
- A larger than usual triple fleam with horn scales and integral knife by **Long**. **£50** [DP]
- A turned boxwood blood stick. G **£26** [DP]
- A pistol handled balling gun (probang) by **Arnold & Sons** with rubber mouthpiece. **£45** [DP]
- A horse tooth rasp with ash handle. **£14** [DP]
- A beech Rutter's horse twitch formed with two turned and grooved arms, linked with steel hinge and leather securing strap, marked with the maker's name. **£28** [DP]
- A beech handled horse tail docker with two searing irons. **£36** [DP]
- A rosewood scaled horse tail docker with brass fittings together with the searing iron all in original fitted leather case by **Arnold**, London. G++ **£85** [DS 22/1188]
- Four different shaped firing irons in rosewood (charred) handles. **£45** [DP]
- A gentleman's folding pocket combined hammer and hoof pick. **£24** [DP]
- A strap type horse sweat remover in brass with one beech and one leather handle. **£22** [DP]
- A turned lignum vitae stable log. **£10** [DP]

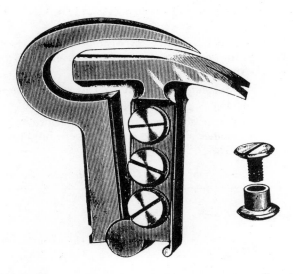

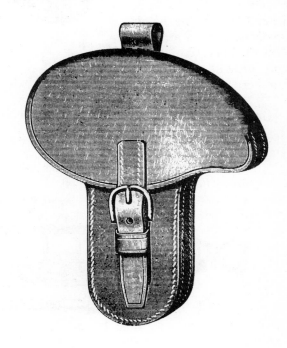

The "Multum-in-Parvo" Registered Hoof Pick, consisting of pick, hammer, nail claw, bradawl, screwdriver, and three strap repairing screws. One of the screws is shown centre and, right, is the leather case to strap to the saddle [Hawkins, c. 1890].

Patent balling gun (or probang) [Arnold, 1886].

Amongst the many less pleasant but now outlawed devices for horses are tail docking shears and searing irons for cauterising the wound and also firing irons used hot for treating leg troubles.

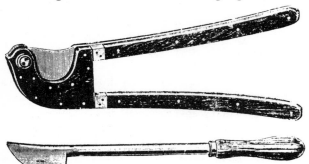

Docking machine and firing iron [Wingfield, Rowbotham,1904].

Amongst other horse accoutrements found are hoof picks, to remove stones from the underside of hooves, and stable logs, through which the halter rope is knotted in order to allow the horse to move its head without tripping on its tether.

Lignum vitae stable log [Hawkins, c.1890]

Collecting tips

➤ *Stable logs are often confused with plumbers' bobbins. The key is the size of the central hole and also signs of wear at the point where the rope is threaded through.*

Price Guide

Veterinary

♦ A large brass veterinary syringe to hold 3 pints, by **Day & Sons**, complete with a large wooden nozzle that stores inside the plunger and smaller brass nozzle. **£35** [DP]

♦ A large pewter clyster syringe to hold 1 quart, with wooden plunger and raised reinforcing bands. **£40** [DP]

♦ A lamb clam (castrating knife) by **Sorby** complete with brass rest. **£18** [DP]

♦ A pair of steel castrating clams, 10" o/a. **£16** [DP]

♦ A large (for cattle) rosewood handled trocar with nickel silver canula. **£20** [DP]

♦ A small ebony handled trocar and canular with screw-on ebony cover. **£35** [DP]

Cattle and sheep

Cattle holders (bull leaders): There are numerous designs of varying complexity **£8–14**.

♦ A good pair of sheep shears by **Sorby**, with good clear "Punch" and "Sorby" marks. **£18** [DP]

♦ A pair of bull nose-ringing punch pliers. **£12** [DP]

♦ A steel horn trainer by **Arnold** with rings to fit over the horns and wing nut adjustment G **£55** [DS 25/711]

♦ A well shaped wrought iron shepherd's crook on a 5 ft. ash pole. **£32** [AP]

♦ A pair of sheep ear pliers to cut centre of ear and another similar to cut edge, both bright finish. **£15** [DP]

*Remember - condition is **G+** unless otherwise shown.*

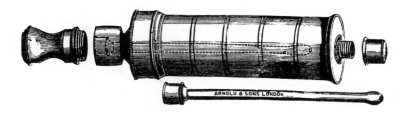

Pewter clyster syringe for cattle [Arnold, 1886].

Farriers' searcher [Wingfield Rowbotham, 1904].

Cattle

The best method of controlling a bull is to put a ring in its nose – in order to do so the nostrils must be punched through with a nose punch or a trocar. These simple tools often go unrecognised. More common are bull leaders, calliper form devices with ball-ends for clipping into the nostrils although, contrary to the name, the normal use was to restrain cows that needed attention.

Cattle, and sheep, are prone to bloat – excessive gas in the abdomen. The cure in severe cases was to release the gas by piercing the animal's side with a trocar (a sharp spike) inside a canula (a tube). A trocar intended for cattle is about ¼" diameter. Smaller versions for other purposes were also part of the vet's kit.

Castration

The instruments of past practice are numerous: knifes, irons, clamps and in more recent years, bloodless castrators.

The farrier

The man who shoes horses and, in the past, also attended to their minor ills was the farrier. He may have doubled as the village blacksmith but these are two distinct trades and were not always combined. For farriery, the tools were few but mainly distinctive: clench cutters, pincers and horn-handled searchers and drawing knives.

Clockwise from the top left: a lamb clam (castrating knife), a castrating clam, a horse mouth rasp, a curved trocar and canula for inserting a copper ring into a bull's nose, a tooth saw, a balling iron and a cattle holder [Wingfield, Rowbotham, 1904, and Arnold, 1886].

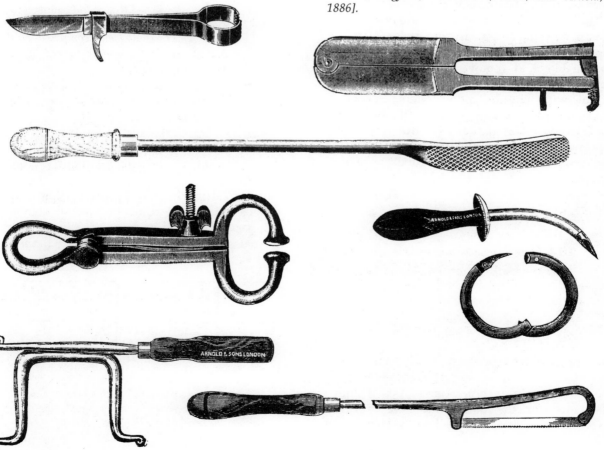

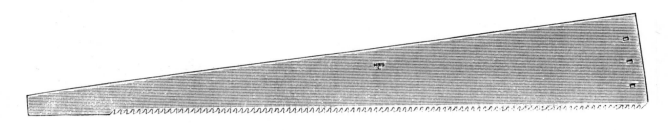

Pit saw, two forms of pit saw boxes (handles) and pit saw tiller [Harding, 1903]. The box, which is the handle at the lower end of the saw, is made to be quickly removable so that the saw can be withdrawn from and then replaced in the kerf when sawing past the supports.

Specialist tools of the timber trade

The basic tools of the timber trade were axes, saws and various dogs and hooks for handling timber. Cross-cut saws at present attract little interest, as only the pit saws and the large framed type of pit saws seem to appeal to collectors.

The main collecting interest in this field is in the testing, marking and measuring tools.

Timber marking hammer/axe: Some forms have an axe blade to remove a section of bark. The timber is then marked with the hammer end which is cut with one or more letters. Several fine examples of a gent's version of this tool by **Underwood**, Haymarket, London, have been reported. These are steel hafted with rosewood chequered scales and were originally in a leather holster.

Timber marking hammer [Harding, 1903].

Timber girthing tape: Used around the girth of a tree the tape is marked to give a direct reading of the diameter in inches. Some tapes include an allowance for the bark and there is also an improved type that is marked on the back to give a

Collecting tips

➤ *Farriers' knives and rubber tapping knifes both closely resemble some timber scribes and are sometimes passed off as such.*

➤ *Also mistaken for timber scribes are saddlers' racers – all-metal compasses with one point flattened and turned up to form a cutting edge.*

➤ *Timber testing gouges without their leather cases or covers are often overlooked.*

direct reading of the cubic content for one foot length.

Timber callipers: In Britain rarer than might be expected, these instruments have substantial iron callipers on a handled lath. Usually 3ft long they were used to gauge round and square timbers.

Improved type of timber girthing tape [Rabone, c. 1925]. Top: the front of the tape giving the diameter; bottom: the back, giving the cubic content.

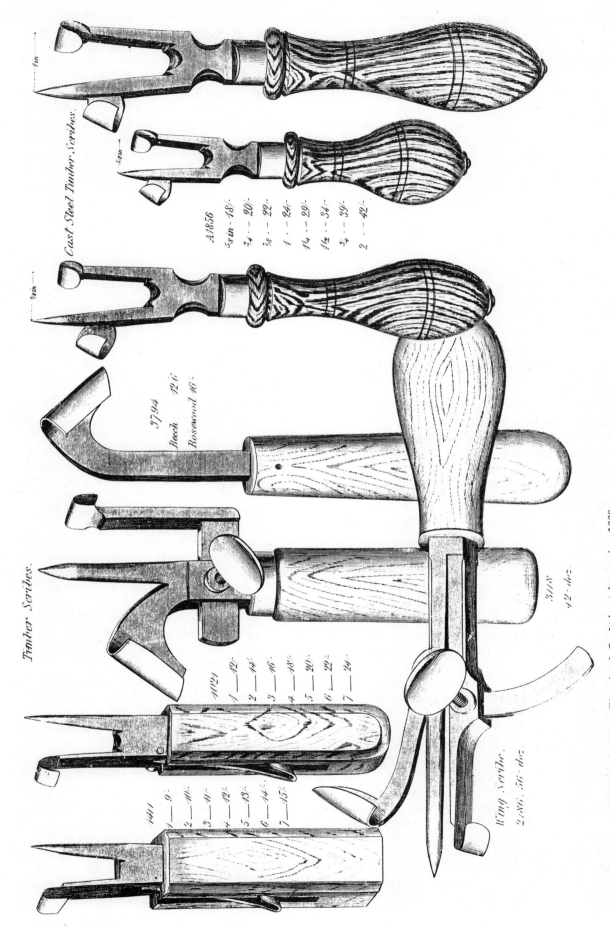

Cast Steel Timber Scribes.

4in

5-8in

7-8in

A1856
5-8 in – 18/-
3/4 – 20/-
7/8 – 22/-
1 – 24/-
1¼ – 29/-
1½ – 34/-
3/4 – 39/-
2 – 42/-

3794
Beech   12 6
Rosewood 16 -

Timber Scribes.

1021
1 – 12/-
2 – 14/-
3 – 16/-
4 – 18/-
5 – 20/-
6 – 22/-
7 – 24/-

1401
1 – 9/-
2 – 10/-
3 – 11/-
4 – 12/-
5 – 13/-
6 – 14/-
7 – 15/-

Wing Scribe.
2/86, 56/- doz.

3/18
42/- doz.

Some of the timber scribes from the Wynn Timmins & Co. Ltd. catalogue, circa 1880.

205

Timber scribe: Not only a tool of the timber yard, the timber scribe was used by many trades for marking barrels, cases etc. In 1843, **Wynn Timmins** illustrated about thirty different types in their catalogue – needless to say, you would have to look pretty carefully to see the difference between many of them. Variants to look out for include folding arms, triple and quadruple scribing edges and sliding arms.

Timber testing gouge: Of short shallow curvature, this was a lightly swept (curved in its length) form of gouge intended for the timber merchant to carry in the pocket. It was used for cutting into timber to assess quality. There are two kinds: the simpler has a leather sheath to cover the blade; the rarer form has a boxwood cover which screws to the handle.

Barking spud: Used for peeling bark (particularly oak bark which was of considerable value in tanning), most spuds seem to have been locally made. They vary in size from 9" long up to shovel size – indeed the working edge of some patterns, 2" to 4" across, resembles a very small shovel.

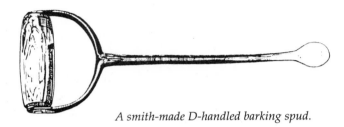

A smith-made D-handled barking spud.

Price Guide

Pit saws

* A 7ft. pit saw blade (no filler or box), usable condition (most aren't!). **£75** [DP]

* A 7ft. 6in. pit saw with iron filler – pitted and not usable – no box. **£80** [DP]

* A pit saw box of turned form in ash. Well worn and patinated. **£50** [DP]

Barking spuds

* Two early wooden handled barking spuds. G– **£24** [DS 24/1423]

* An early barking spud with wooden T-handle. G **£40** [DS 21/948]

* Two D-handled smith made barking spuds, one 21" long, the other 26" long. **£20** [DP]

Measuring

* A timber girthing tape by **Rabone**, Patent No. 119189-86 with Contents Feet and Inches/parts. **£35** [DS 20/329]

* A timber girthing tape by **E. Preston**, in leather case calibrated in ⅓ ft. and allowing 1" for the bark. **£60** [DS 25/1020]

* A girthing "sword", the round steel sword with a hooked end and beech handle. **£18** [DP]

* A pair of timber callipers, measuring to 30". Iron stock with wooden handle. **£24** [DP]

* The "Vade Mecum" slide rule designed for the timber trade by **George Bousfield**, Grimsby, Patent No. 12806. Boxwood and brass, 13" long. **£125** [DP]

* A timber pricing slide rule in boxwood and brass, 13" long. Designed, registered and made only by **Aston & Mander**, London. Includes scales for timber cubing and pricing and for conversion to the St. Petersburg Standard. **£160** [DP]

Marking

* A gentleman's timber marking combined axe and hammer by **Underwood**, Haymarket, London, with rosewood scales, riveted to the steel haft, contained in leather holster to hang on belt. **£180** [DP]

* A gentleman's timber marking compendium tool by **Underwood**, Haymarket, London, as above but also with a saw that folds out from the haft. **£280** [DP]

* A timber marking hammer to mark "PTT", probably French. **£38** [DP]

* An 18th C. timber scribe (race knife) marked **T. Symonds** on a rectangular wooden handle. Has point, fixed cutter and folding drag knife, 4½" o/a. Similar to two examples dated 1737 and 1758. **£220** [DS 25/661]

* A double blade timber scribe with turned rosewood handle by **Timmins**. F **£44** [DS 25/671]

* A folding timber scribe of penknife form with antler scales by **Marples**. **£18** [DP]

* A timber scribe with early rectangular pattern beech handle with one fixed blade and one folding blade. **£28** [AP]

* A fine and rare boxwood handled timber scribe with two folding drag knives and a screw-on boxwood top. G++ **£209** [DS 24/218]

* An early timber testing gouge by **Green** with ebony handle and ebony screw-on cover. **£95** [DS 19/87]

* A timber testing gouge with beech egg-shaped handle in a leather case. **£26** [DP]

* A timber testing gouge by **Howarth** with turned boxwood knob handle and boxwood screw-on cover with brass ferrule. **£65** [AP]

* A timber testing gouge by **Ward** with bone handle and bone screw-on cover. **£110** [DS 24/1524]

The importance of tools to mankind is demonstrated in the many images adorning everything from letter headings to trade union banners – indeed, the elevation of the tool to the status of an icon is not unknown.

Ceremonies

But the image was not all; the importance of the tool was recognised in ceremonies – fancy mauls or trowels for the laying of foundation stones, elaborate scissors for cutting the tape at opening ceremonies and cased sets of hammers and chisels for the laying down of a ship's keel.

The symbolism of tools is nowhere stronger than in the regalia of some societies. The Ancient Order of Foresters is perhaps the most obvious – the ceremonial axe is at the centre of their proceedings – whilst the Masons have a rich heritage of tool related imagery.

In the case of the friendly societies, the tools are often carved from wood and painted to resemble the original – we have seen axes, spades, forks and others in this genre.

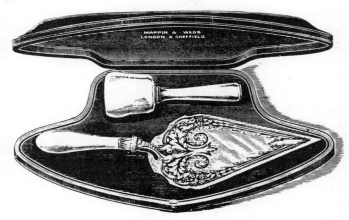

Foundation stone laying set in ivory and silver by Mappin & Webb, 1900, who also listed presentation spades and wheelbarrows.

Presentation

For the collector, presentation items can also be an area of interest. These tools are usually standard manufactured items – of the best quality, naturally – but enlivened with a presentation inscription including an all-important date. However, for many ceremonial tools, the utility has been lost in a tide of decoration.

Foundation stone laying trowel, ivory and silver by Thomas Turner & Co., c.1900, and opening ceremony presentation scissors.

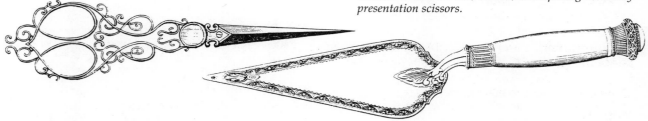

Price Guide

- A keel laying set consisting of a very elaborately turned and carved boxwood mallet and a boxwood handled chisel. **£350** [DP]

- A presentation spill plane, in rosewood, stamped twice **Currie**, Glasgow and inscribed on a silver plaque "Presented to Mr Richard McCan from Daniel Currie 8 May 1872". F **£1100** [DS 16/1252]

- A presentation carved rosewood plumb square with a silver plaque engraved "Presented to Aldn. W.H. Bailey on his Laying of the Corner Stone of New Baths. Lower Broughton July 22 1890" with a silver plated plumb bob. **£290** [DS 18/50]

- An unusual Masonic brass triangular plumb square with ivory bob. **£235** [DS 20/317]

- A 12" parallel level in ebony and brass with a well engraved leaving inscription dated 1910. **£75** [DP]

- A solid silver decorated Masonic plumb square, hallmarked Birmingham 1864, approximately 5" high in a shaped leather covered case. G++ **£110** [DS 25/1377]

- An ivory handled presentation Masonic silver trowel engraved "Presented to M.W. John Stewart, Grand Master, Upon the occasion of the laying of the Corner Stone of the Masonic Temple at Cohores N.Y. August 21st 1895 by the brethren of Cohores Lodge BO. 116". F **£200** [DS 18/388]

- An all wood felling axe carved from softwood and painted to resemble steel, probably Ancient Order of Foresters regalia. **£85** [DP]

- A presentation silver hammer with rosewood handle profusely engraved "Presented to Mrs H.D. La Touche and Bellary Kistna State Railway, Madras 1889 etc." F **£195** [DS 19/1106]

Drawing instruments are delicate and for this reason have mostly been supplied in cases. John Robertson writing in *A Treatise of such Mathematical Instruments as are usually put into a portable case* (1775) makes it clear that the pocket case was adequate for most purposes and that the larger magazine cases contained items that were hardly ever used.

In the 18th century, clearly, most people bought pocket cases – tapering, oval, fish-skin cases with the instruments dropped in from the top; cases of this type are comparatively common but boxed sets of 18th century date are rare, commanding high prices.

Makers

Until **W.F. Stanley** commenced in business in 1853, drawing instruments – usually called mathematical instruments – had been made by the general scientific instrument makers and there seems to have been no firm exclusively making drawing instruments.

Famous Makers

Elliott[1]	London	c.1800–1940s
C.W. Dixey[2]	London	1838–62
W.H. Harling	London	1851–1961
W.F. Stanley	London	1853–1960s
A.G. Thornton	Manchester	1880–present
J. Halden & Co.	Manchester	1880–1950s

1. 1800: William I then II; 1850: & Sons; Later: & Bros.
2. Mathematical Instrument Maker to Her Majesty, Bond St. London.

Materials

From around 1835, nickel silver (sometimes called electrum) started to replace brass for the best quality items. **Stanley**, **Harling** and the other principal English makers of the late 19th century always used nickel silver. More importantly, they established a practical style of instrument intended for the professional to use on a daily basis.

By the 1920s simpler cheaper ranges in brass and, particularly, chromium plated brass were becoming common. None of these are of much interest to collectors.

Instruments – an outline

Drawing pens: 18th century pens are customarily brass-handled with wing screw adjustment and pricker in handle. 19th century pens are ivory

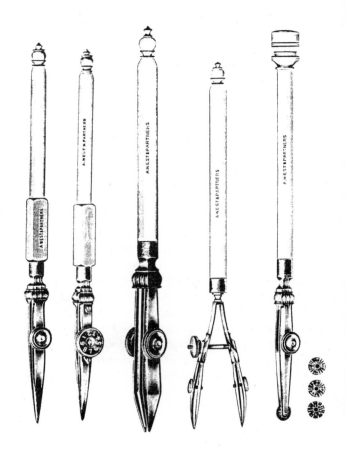

Drawing pens [West, 1930]. From left: drawing pen, drawing pen with index screw, bordering pen, road pen and dotting pen.

handled with a nickel silver neck and knurled adjusting screw. There are many variations – plain, hinged nib, square-on handle, indexed screw adjustment, thick line (three leaf), double head, dotting (extra wheels are stored in top of handle), and curve pen (the handle rotates).

Proportional dividers: Included in larger sets but more often found individually cased. Those with hooked points, a fine adjustment link or a fine

Collecting tips

➤ *The products of all the later 19th century makers look very similar – until you try to replace a missing instrument in a box, when the variations in knurling, chamfers etc. will be quite apparent.*

➤ *Items do go missing from sets and get replaced so check very carefully – completeness is important. So is originality.*

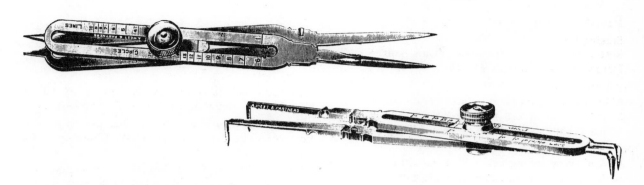

*Proportional dividers, straight and hooked point types
[West, 1930].*

adjustment rack and pinion are rare, commanding significantly higher prices.

Compasses: Compasses mostly formed part of sets – the number of different types and sizes of compasses/dividers being a good indicator of the quality of the set. Sets of three spring bows and Napier compasses (a folding/rotating end type) are frequently individually cased.

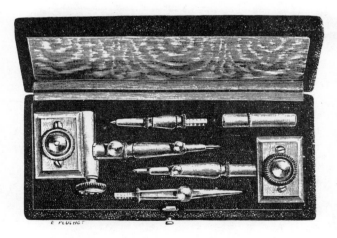

Cased set of nickel silver beam compasses [Melhuish, 1904].

Trammel heads: Also known as beam compasses, these were supplied cased on their own, usually without the beam, complete with interchangeable needle points, pencil points and pen points. The larger sizes of cased sets of drawing instruments included trammel heads.

Erasing knives: Rosewood, ebony or ivory handled knives with a curvaceous blade used to

Price Guide

Drawing pens
Ivory handled steel drawing pens, prickers and tracers, mid 19th C. and later, with makers' name **£3–6**. Broad line, road and dotting pens etc. **£10–15**.

◆ A set of five different ivory handled drawing pens and a steel tracer in a japanned metal case by **Harling. £35** [DP]

Proportional dividers

◆ An early and unusual pair of steel tipped brass proportional dividers with fine adjustment and bar by **W. & S. Jones**, Holborn, London. G **£253** [DS 23/1216]

◆ A pair of 7" nickel silver proportional dividers by **A.G. Thornton** in shaped case. **£34** [DP]

◆ A pair of 10" nickel silver proportional dividers in original leatherette shaped case by **Elliott Bros. £65** [DP]

◆ A pair of 7" nickel silver proportional dividers with rack adjustment by **West. £45** [DP]

◆ A 7" pair of nickel silver proportional dividers with hooked points by **W.F. Stanley** in original case. **£90** [AP]

Compasses

◆ A set of three steel spring bow compasses in a fitted cloth covered case by **Watts.** F **£22** [DP]

◆ A nickel silver pocket compass of pillar (folding) type with interchangeable pen, pencil and points in leather covered case. **£55** [DP]

◆ A three legged compass in nickel silver by **W.F. Stanley**, uncased. **£154** [DS 25/390]

◆ A pair of nickel silver trammel heads by **A.G. Thornton** in leatherette case with interchangeable pencil and ink ends. **£38** [AP]

*Remember - condition is **G+** unless otherwise shown.*

Price Guide

Erasing knives

Rosewood or ebony handled erasing knives £8–12. Ditto with ivory handles. **£12–15**

Parallel rules

6" ebony bar parallel rules with brass fittings £5–8. Ditto but 9" or 10" **£10–14.** Ditto but 12" to 15" **£16–20.** 6" ivory bar parallel rules with nickel silver fittings **£18–25.**

♦ A 6" ivory rolling rule with brass fittings, marked with degrees and drawing scales. **£75** [DP]

♦ A 12" brass rolling rule in a mahogany case with brass dividers by **Halden. £60** [DP]

♦ A rare ebony navigators' bar parallel rule by **G. Wilson** London with ivory insert marked "U of C Local Devn", the rule revolving on a square brass compass base. Small chip. G++ **£252** [DS 24/102]

♦ A set of three ebony and brass rolling rules by **Cassartelli**, 9", 12" and 18" long. **£95** [DP]

♦ An unusual 15" ebony rolling rule with ivory scales to edges and two other scales mounted on face. **£192** [DS 24/1286]

♦ An unusually large 24" brass rolling rule by **Adie & Wedderburn** in mahogany box.. F **£77** [DS 23/583]

♦ A 24" Capt. Field's Improved bar parallel rule in boxwood with brass fittings by **W.F. Stanley.** F **£50** [DP]

♦ A large 30" ebony bar parallel rule, the edges with brass reinforcement and three links. **£48** [DP]

♦ A 10" ivory triple bar parallel rule with pieced nickel silver fittings. **£85** [DP]

Curves

♦ A set of twenty pearwood yacht curves stamped 1Y to 20Y in mahogany box, unused. F **£247** [DS 25/216]

♦ A set of fifty pearwood railway curves in original mahogany box by **W.F. Stanley. £95** [DP]

Erasing knives [West, 1930].

scrape ink from both parchment and tracing paper. Some types include a blade for pencil sharpening.

Parallel rules: There are two styles: bar parallel rules, consisting of two (occasionally three) parts connected by brass or nickel silver strips, and rolling parallel rules, a single stock with a brass or nickel silver roller set in. These were supplied in sets of drawing instruments and were also frequently sold as individual cased items. Both types are found in boxwood, brass ebony, ivory, nickel silver and ebony with ivory facings.

Curves: Cased sets of curves, intended for architectural (known as French curves), ship, yacht or railway work can be found. Usually made in pearwood but also occasionally boxwood.

Solid brass rolling rule, sold in a mahogany case [West, 1930]. By the 20th century rolling rules were superseded by the T-square in drawing work but they continued in use for navigation. Some of those intended for drawing work include ivory scaled strips on the edges and ivory scaled wheels.

Cased sets

The value of drawing instrument cases is, in general, proportional to the size of the set, and wooden boxed sets are preferred to morocco or leatherette cased sets.

In the late 18th and early 19th century the usual size of a fish-skin pocket case was 6" though smaller 4½" cases are to be found. The latter often have a more fancy covering, shagreen or even tortoise shell.

By the 1850s most sets were being supplied in wooden boxes. Simplest were single layer arrangements, though these are few in number. The most commonly found arrangement is a box around 7" wide with a single tray. The flatter items – parallel rule, protractor, scale and, sometimes in earlier sets, a sector – are housed under the tray. Between this 7" size and the largest boxes, which are 13" long to contain 12" scales, there is a wide range of layouts of increasing size and complexity. The largest boxes, sometimes referred to as "magazine cases", can have up to three lift-out trays and a drawer.

In the 20th century it was more usual for boxes to be made in oak rather than mahogany, although the largest boxes were sometimes veneered in burr walnut or other fancy woods.

Instruments intended for use in tropical conditions were supplied in black japanned copper boxes. Such boxes sometimes include a Marquois set – a boxwood triangle and two thick square-edged scales – used by field officers for simple drawings. These are more usually found by themselves in flat mahogany boxes.

Price Guide

Cased sets

• A magazine case by **Halden**, the walnut veneered case 14½" x 6" x 4½" deep containing two trays. The top tray holds twenty-eight nickel silver and ivory drawing instruments, the lower tray fitted for paints and paint brushes. The bottom of the box holds six boxwood scales and offsets, a 12" ebony rolling rule with ivory edges and an ivory protractor and ivory 6" bar parallel rule. **£660** [AP]

• A Marquois military set by **Cary** consisting of a boxwood triangle and two boxwood scales in a mahogany sliding top box. **£35** [DP]

• A military set in a mahogany box, 14¼" x 5¼" x 2" deep with one tray, fitted with twelve nickel silver drawing instruments, an ivory protractor, an ivory bar parallel rule with three piece boxwood Marquois set under. **£95** [DP]

• A set by **Watson & Sons** in an oak case, 8¼" x 7¼" x 2" deep, with nickel silver reinforcements with one tray containing fourteen nickel silver instruments with four boxwood scales and a protractor under. **£80** [DP]

• A large pocket case, 8½" x 5¼", by **B.J. Hall & Co. Ltd.** covered in green leatherette containing twelve nickel silver drawing instruments. **£55** [DP]

• An early 19th C. pocket set contained in an oval fish-skin covered case containing brass instruments and an ivory sector, scale and bar parallel rule by **Cary**. **£275** [AP]

Further reading: *Drawing Instruments, 1580–1980* by Maya Hambly, London, 1988.

Two cased sets of drawing instruments [West, 1930]. The right-hand set is in a japanned copper case.

Floating by

Although files can be float cut, which is usually a single coarse cut, the term float normally refers to a *very* coarse-cut serrated tool.

Commercially-made slotted tooth floats, hollow and round like moulding planes, some also compassed, are to be found but the use of these is uncertain – gunstocking has been suggested.

Floats, which are often craftsman-made from old files, are extremely useful tools, I have several small ones in the workshop for opening out round holes when fitting square chisel tangs into handles.

A rarely found form is a grisail – a wide thin float used in horn working.

A typical cabinetmakers' float.

Price Guide

• Four tapering floats of different size, mostly made from old files in beech handles. **£32** [DP]

• Three planemakers' floats originally from the **C. Nurse** workshop. **£99** [DS 24/795]

• Three pairs of hollow and round straight floats, one pair compassed. **£66** [AP]

• Four small tooth cranked floats (possibly used by tinsmith). **£42** [DP]

Collecting tip

➤ *Floats are often sold as **planemakers' floats**. Without provenance or a known planemakers' mark, it is unlikely that they were used for planemaking – many other trades used floats.*

Floored

Portland cement is so called because it resembles Portland stone – there must have been wily marketing men even in the 1830s to have sold us this one. It is a hydraulic cement, that is, a cement that will set under water and once set is unaffected by water.

This wonder material made it possible, for the first time in history, to construct seamless floors which were water resistant. Evidence of this once specialist trade, eventually called granolithic flooring, is to be found in the special finishing and

Price Guide

• A 12" wide brass dot cement roller by **Gilchrist**, Glasgow with brass tip to handle. **£36** [DP]

• A 2" wide brass concrete edging roller with flange and ribs. **£30** [DP]

edging rollers – heavy brass patterned rollers about 3½" diameter – which appear from time to time in the market. Incidentally these tools are almost always misidentified and have attracted some of the most inventive stories I have ever heard.

A driveway cement roller and a dot cement roller [Tyzack, c. 1910]. Narrower rollers 2" wide for edges, were also made.

In tune

Making and tuning organs is a highly skilled trade that nowadays combines woodworking and metalworking crafts with electrical and even electronic skills. Few of the tools are however specific to the trade. Salaman describes metal cutting planes with near vertical irons and a "scoring" plane but there is only one item that seems to appear on the market with any regularity.

These are tuning cones, the most common form being double ended, an external cone at one end and an internal cone at the other. Made either of boxwood or brass, they are used to slightly open or close the ends of the metal organ pipes, thus fractionally altering the pitch.

The boxwood examples look particularly fine on the shelf, always attracting questions, but don't be tempted to use them. I am informed that many good organs have been damaged by their injudicious use.

Price Guide

* A graduated set of three brass double-ended organ tuning cones and three brass single-ended cones. G £75 [DS 24/991]

* An unusual organ builders' tuning cone in boxwood with ivory ends. G++ £55 [DS 21/290]

* A fine set of six double-ended brass organ tuning cones by **Goddard**, sized ½", ¾", ⅞", 1", 1¼" and 1½", being the full set available in the double-ended sizes. £160 [DP]

* A graduated set of four boxwood double-ended tuning cones. £105 [AP]

Price Guide

Cast iron glue pots of small or medium size in good usable condition by well-known makers £8–14. Ditto but larger (over 3pt. capacity) £22–30 according to size. (Note: glue pots are sized by the capacity of the inner pot.)

* A small copper glue pot in very little used condition. F £35 [DP]

* An early copper glue pot, 1½ pt. capacity with castellated seams. £55 [AP]

* A large electric copper glue pot by **Premier**. Pat. Safety Device No. 240910. G+ £33 [DS 23/416]

* Two craftsman-made all copper glue spoons of narrow form, of different size, with tubular handles. £18 [DP]

* An 8 pt. (the largest size made) cast iron glue pot. £44 [AP]

Collecting tips

➤ *If buying a glue pot for use, inspect the outer pot very carefully as holes are often covered by flakes of rust. Better still, buy from an established dealer and get him to promise to an exchange if it leaks.*

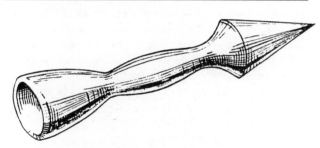

Organ tuning cone.

Glue pot and ladle

The cast iron double pot with boiling water in the outer part was a 19th century development – just one of many products that were made in cast iron as the problems of quality and price were resolved. The earliest glue pots were lead (Moxon, 1680s) and evidence suggests that pots of this material were used into the 19th century.

The only alternative to cast iron is double skinned copper pots. Usable examples are today keenly sought by woodworkers as the glue from them cannot become contaminated with iron.

Not common and usually causing difficulties of identification is the glue ladle or spoon, which was used to pour glue into joints. Many are craftsman-made in copper and are of a very narrow spoon-like form.

A glue ladle and iron and copper glue pots [Harding, 1903].

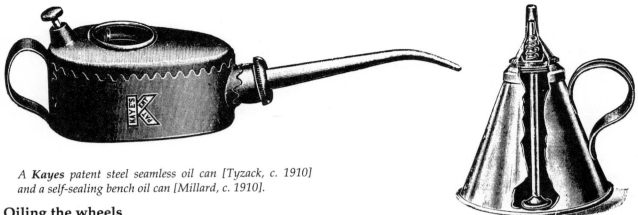

*A **Kayes** patent steel seamless oil can [Tyzack, c. 1910] and a self-sealing bench oil can [Millard, c. 1910].*

Oiling the wheels

The need for the oil can is not what it once was. Time was when heavy machinery was lubricated with copious amounts of "the real Mc Coy" (in the 19th century, the best available oil for locomotives) but even small machines like sewing machines all came with a suitable oil can. Not that a lot of attention was paid to keeping the lubrication where it would do most good – in the bearings – but the makers of oil cans were ingenious in their search for a "leakproof" design. But such is the penetrative quality of oil that some always ends up where it's not wanted.

A leading maker was **Kayes** of Leeds, which marked their cans with a brass stamped logo.

Price Guide

Most oil cans sell for less than **£10** and, more usually, less than **£5** but, as in many fields, there are collectors who will pay substantial prices, in some cases surprisingly large sums, in order to obtain the rarest and most interesting examples.

* A **Kayes** miniature oil can, 4" long, with brass fittings, stamped "K" on the side and also "Meccano" on the bottom. **£150** [AP]

* A very rare ⅛ pt. **Kayes** Patent Force Feed oil can with brass fittings. **£70** [DS 25/696]

* A ⅓ pt. **Kayes** Reg. Design oil can with copper rim round lid and Kayes, Leeds, device on side. G+ **£44** [DS 25/697]

* Four **Braines** Patent Oilers Nos. 03, 09, 12 and B2, all with double slide grit excluders. **£55** [DS 25/1307]

* A conical tin plate self-sealing oil can with spiral ribbing and brass bottom with **Preston** label on side. **£24** [DP]

* A miniature brass oiler in the shape of a watering can. **£71** [DS 23/663]

Collecting tips

> *The more information on oil cans the better: manufacturer's name, patent date, owner, e.g. railway company, or the machine intended for – all are plus points.*

Make it legal

Perhaps the most misidentified of all tools is the scriveners' wheel. I know of no catalogue that illustrates one and they mimic dressmakers' pricking wheels. The most usual explanation is that they are for pricking out stitching but a moment's thought would show that this was highly unlikely: the points mark ¼" or more apart – well, if you only want three or four stitches to the inch!

Whereas...until well into the 20th century, in the backwaters of the legal profession, there were still scribes at work with a pen, writing onto parchment and then, as if once wasn't enough, making a counterpart. The scriveners' wheel was run down each side edge of the document and the pricks joined up with feint pencil lines to guide the scrivener.

Price Guide

* A scriveners' wheel, with boxwood handle and brass wheel with steel points. **£18** [DP]

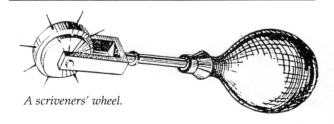

A scriveners' wheel.

Red tape

Still used today to secure the pages of legal documents together, red tape was inserted into parchment with the lawyers' bodkin, really a packing (eyed) needle in a finely turned boxwood handle.

Price Guide

Boxwood handled lawyers' bodkins **£12–15**.

A boxwood handled lawyers' bodkin [Preston, 1909].

Graining

Particularly popular in the late Victorian and early Edwardian periods, this style of paint finish was used to disguise cheap woods as something better. It is now enjoying a revival.

The grain effect is achieved by applying a tinted glaze over a base colour. This is then patterned by combing with a metal comb or by graining with a tool. These tools varied from simple feathers to later complex and patent tools. Leather rollers, cut to a pattern resembling wood grain, were an early form, later replaced by tools made of rubber or rollers with cut metal discs.

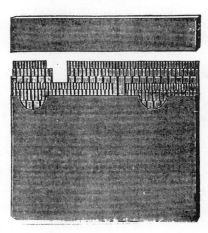

Price Guide

◆ An early leather covered graining roller with an ash handle. The profiled leather covers a wooden roller. **£16** [DP]

◆ An interesting collection of seventy graining tools including seventeen bone combs, twenty leather combs, brushes, etc. G **£60** [DS 25/1604]

◆ A set of unused graining rollers in original fitted box by **Hamilton**. London. G++ **£60** [DS 24/1402]

◆ Two sets of graining combs and twenty-six brushes. **£42** [DS 22/501]

◆ A set of seven **O'Brian's** Patent Graining Rollers in original softwood box with label in lid. The small diameter steel rollers are held in brass yokes. Complete with original instruction leaflet **£70** [DP]

◆ A set of three unused graining rollers, Patent No. 16440. F **£22** [DS 21/907]

◆ A set of twelve graining combs in a tin case. **£14** [DP]

◆ A bench standing, hand cranked cast iron paint mill in good usable condition. **£45** [DP]

*Remember - condition is **G+** unless otherwise shown.*

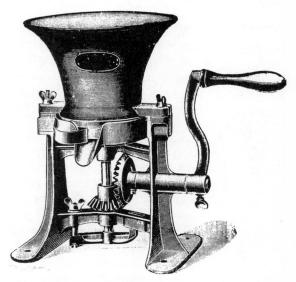

Above: Improved paint mill [Farmiloe, c. 1905]. As late as the 1920s, some decorators were still making their own paint from white lead, pigments and oils, combined in a paint mill.
*Left: A set of 12 steel graining combs [Millard, c.1910]. **Love's** improved double line oak combing roller and feed brush and **Hamilton & Co.'s** patent oak grain finisher [Farmiloe, c. 1905].*

Recently a copy of the two volumes making up the first edition of Moxon's *Mechanick Exercises* sold at auction for £22,000. Published between 1677 and 1683 in part-work form, Moxon's book is perhaps the first English "do-it-yourself" manual. The first volume is probably of more interest to the tool collector for it illustrates contemporary tools and trades and gives an insight into the use of tools. It was reprinted several times before 1703, and the cheapest of these editions is available for around £1,000.

It is the second volume, in which Moxon describes the process of printing in minute detail, that is rare as only about 60 copies are known so it must have represented most of the £22,000 price.

This rather extreme example shows that books on tools sometimes have value as well as interest. As far as I know, only one tool, a Bergeron lathe, has sold for more than these two volumes of *Mechanick Exercises*.

Original sources

Whilst copies of Moxon are not to be found every day, other books on woodwork and building, dating from the early 19th century onwards, are more plentiful and affordable.

Peter Nicholson: Author of quite a few books on carpentry and building, starting in 1805 with *The New Carpenter's Guide*. This continued in various editions until the 1860s. By 1823 he had produced the *New Practical Builder* but this, like most of his books, was a reworking of his earlier works with additions. Nicholson is detailed in his descriptions of tools, giving a good insight into carpentry,

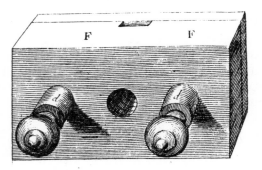
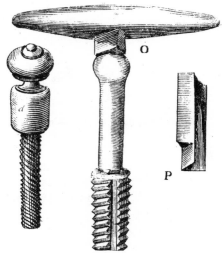

A wood screwbox and tap from Plumier's L'Art de Tourneur, *2nd ed. 1749.*

joinery and building practices of the early 19th century.

Robert Riddell: *Carpenter, Joiner and Handrailer* (1870s). A good seller over a long period so copies are fairly easy to procure. But watch out, the cardboard models of the handrails are often defective.

Lister Sutcliffe: Author of the eight volume set *Carpentry, Joinery and Cabinetmaking* (1902) – a detailed description of the trades at a period when trade skills were the finest.

Holtzapffel: *Turning and Mechanical Manipulation*, five volumes written over thirty years in the mid 19th century. Contains much information applicable to work other than turning.

The French connection

France has been a source of many good books with excellent tool illustrations. The first book on turning was Plumier's *L'Art de Tourneur en Perfection* (1701). This two volume book contains plates showing many tools.

Diderot and D'Alembert's *Encyclopédie ou dictionnaire raisonné des sciences, des arts, et des*

Collecting tips

➤ *Like tool collectors, book collectors are attracted by fine condition and original packaging – which, in the case of books, means original binding. Books in fine condition will fetch high prices. If you, as a historian of technology, want the book to use, a copy in well-used condition will satisfy you, but make sure the price is appropriate if you are buying books in this condition.*

➤ *Repairs to books are expensive. Re-binding a quarter leather bound (i.e. leather spine only) volume will cost about £55 and a cloth bound volume about £20.*

métiers, published between 1751 and 1772, covers some 200 trades and includes around 3,000 illustrations. Today, every historian of 18th century technology finds himself inevitably drawn to this great work. Complete sets of this many-volumed publication command multi-thousand pound prices, but over the years many sets have been broken into individual sections, which may be as little as two pages but are mostly of around six. These can be bought for between £5 and £10 per page but be sure that you are buying a complete section.

The third 18th century French book that contains numerous illustrations of tools is Roubo's work on joinery and cabinetmaking, *L'Art du Menusier* (1769–75).

Modern sources

The first modern book to treat tools as historical objects was Henry Mercer's *Ancient Carpenters' Tools*, published in America in 1929 and last reprinted in 1957. In Britain, the earliest books to tackle this subject were W.L. Goodman's *Woodwork, from the Stone Age to Do-it-Yourself* (1962) and *The History of Woodworking Tools* (1964). The latter book is still the only general history of woodworking tools. It went through several

THE NEW PLUMMET.

MESSRS. C. NURSE AND Co., are just putting before the public a newly-invented line-winding plummet, which is the invention of Mr. Green, of Camberwell, and for which provisional protection has been obtained. The plummet contains a cavity, in which is placed a rotary bobbin, which can be worked from the outside for the purpose of winding in and unwinding out, and also storing up its own line, in contradistinction to the ordinary solid ones which can only be used by a

line kept separately, and which is consequently very liable to get entangled with the other tools when put into the basket. The illustrations show —(1) a sectional view, showing position and working of bobbing; (2) exterior view of plummet. The advantages of this improvement will at once be apparent to all our readers, and as the prices (for which see our advertisement columns) are but a small advance upon the old-fashioned ones, Messrs. Nurse and Co. should find a considerable demand for so ingenious and useful an article.

Review of new plumb bob in The Illustrated Carpenter and Builder, *20 Jan. 1888.*

editions and, although now out of print, is reasonably easy to find in the second-hand market.

Periodicals

The following periodicals contain interesting contemporary information. Loose copies have little value, even in quantity, as bound series of most can be found for less than the cost of binding.

The Illustrated Carpenter and Builder: From 1878. A trade magazine. The contents are much as might be expected from the title. Interesting because of its early date of commencement.

Work: From 1889. Intended for amateurs, it contains instructions for making a whole range of items – a wooden camera, for instance.

Building World: From 1887. A trade magazine, very similar to the *Illustrated Carpenter and Builder*.

The Woodworker: From 1901. A magazine for amateurs, it is still published today. Contains some of the earliest articles on the history of tools. Apart from the early years, bound copies can be purchased for less than £5.

Price Guide

Books

♦ Diderot and D'Alembert. *Encyclopédie*. 1969 reprint of the complete 3,000 plates (plates ¼ size). hb. Quarter leather bound. **£100** [DP]

♦ Goodman, W.L. *The History of Woodworking Tools*. 1964. hb. **£35** [DP]

♦ Holtzapffel, C. and J.J. *Turning and Mechanical Manipulation*. Vol. I, 1843 (CH). Vol. II, 1846 (CH). *Vol. III, 1850* (CH), *Vol. IV, 1879* (JJH), *Vol. V. 1884* (JJH). hb. All 5 vol. **£450**. *Vol. III* was revised and enlarged by JJH in 1894. A set containing this volume. **£550** [DP]

♦ Mercer, Henry. *Ancient Carpenters' Tools*. 1957. hb. **£60** [DP]

♦ Nicholson, Peter. *The New Carpenter's Guide*. 1825 (original 1823 ed. enlarged by John Bowen. hb. **£110** [DP]

♦ Proudfoot, C. & Walker, P. *Woodworking Tools*. 1984. hb. **£20** [DP]

♦ Riddell, Robert. *The Carpenter, Joiner & Handrailer*. c.1875. hb. **£120** [DP]

♦ Spon, E. *Workshop Receipts*. 1875. hb. **£35** [DP]

♦ Sutcliffe, Lister. *Carpentry, Joinery & Cabinetmaking*, 1902, (8 volumes). **£95** [DP]

♦ Young, F.C. *Every Man His Own Mechanic: A Complete Guide for Amateurs*. 1891. hb. with embossed cover. **£40**. This book went through dozens of editions. 1929 ed. **£10** [DP]

Receipts

The old-fashioned term for recipes. Books giving details of concoctions for lubricating, polishing, cleaning, etc. have a long history. The earliest likely to be found at an affordable cost is *The Handmaid to the Arts*, published circa 1760, attributed to Robert Dossie. Later, and less expensive, is Spon's *Workshop Receipts*, available in several different editions dating from 1880 onwards.

NEW PLUMMET.

Improved Line Winding Plummet, Best Quality, London Made, is the only Plummet which will store up its own line and thus avoid the trouble always attaching to loose lines.

PRICES IN BRASS.

6ozs., 2s. ; 8ozs., 2s. 6d. ;
10ozs., 3s. ; 12ozs., 3s. 6d. ;
14ozs., 4s. ; 16ozs., 4s. 6d.
each.

Post free, 3d. each extra.
From the Sole Makers—

C. NURSE & CO.,
182, WALWORTH ROAD, LONDON, S.E.

Nurse's advertisement for the new plummet from The Illustrated Carpenter and Builder, *20 Jan. 1888.*

Catalogues

The conditions that led to the development of the early tool catalogues are explained in *Marketing*, pp. 42–43. Suffice to say that by the 1880s, developments in printing and paper making had made the production of catalogues vastly cheaper and thereafter most leading firms, whether manufacturers, wholesalers or retailers, produced catalogues. Those dating from before the 1880s are much fewer and consequently command considerably higher prices.

The majority of blocks used in the preparation of wholesalers' and retailers' catalogues were provided by the tool manufacturers and therefore

Collecting tips

➤ *The size of books emanates from the size of the paper on which they were printed, traditionally about 22½" by 17½". If the paper is folded once (giving four sides) the book is folio (11¼" by 17½"). Folded again (giving eight sides) quarto; folded yet again (giving sixteen sides), octavo. Because paper size and trimming were variable these are not definitive sizes.*

the same illustrations are to be found in many different catalogues – often with some attempt to conceal their origin.

Fortunately for collectors, as I have pointed out, most manufacturers of any substance did produce catalogues and you may well come across a *Marples* or *Rabone* catalogue from the 1880s. A *Howarth* 1884 catalogue has turned up and there are several dating from the turn of the century from *Preston, Mathieson, Ward & Payne, Sorby,* and *Spear & Jackson* as well as a vast range of *Stanley* catalogues published both here and in the USA, the latter starting in the 1850s.

These mostly cover a wide range of tools. There are also many more specialised catalogues such as *Barnsley* (leather working tools) and *Harrison, Brades* and *Gilpin* (heavy edge tools) dating from the early years of this century.

The style of trade catalogues did not change appreciably until the 1940s and there are numerous catalogues to be found dating from the 1920s and 30s covering everything from a wide range of tools, *Buck & Hickman,* to surgical instruments, *Allen & Hanbury.*

Price Guide

Catalogues

• Buck & Hickman, *catalogues.* hb. 1923: **£70**; 1935: **£35**; 1953, **£20**; 1964: **£15**; 1970 (reprint of 1964): **£15** [DP]

• John Harrison & Sons, *1902 catalogue of edge tools including axes, scythes, bill hooks and garden hand tools.* hb. **£75** [DP]

• Jas. Howarth & Sons, *1884 catalogue.* hb. **£250** [DP]

• William Marples, *1928 (Centenary) catalogue with 1930 price list.* hb. **£55** [DP]

• William Marples, *1938 catalogue with price reductions list.* hb. **£35** [DP]

• Alex. Mathieson, *1899 catalogue.* hb. **£225** [DP]

• Ed. Preston & Sons. *1909 catalogue, No. 18.* hb. **£275** [DP]

• Record Tools. *1950 catalogue, No. 16.* (1957 price list). pb. **£30** [DP]

• Stanley Rule & Level Co. *1915 catalogue No. 34 (Tradesman's).* pb. **£35** [DP]

Periodicals

• *The Illustrated Carpenter & Builder. Vol. I, 1878, Vol. II, 1879–90.* hb. The first 12 years of this important publication. 4,912 pp. in all including advertisements. All 12 Vols. **£220**. Individual volumes. **£15** each. [DP]

• *Work.* 1887, bound volume of one year's issues. **£25** [DP]

Trades

PLUMBERS' TOOLS

The plumbers' trade has, in this century, become divided into lead workers, who install sheet lead to roofs, and plumbers, who install plastic, copper, cast iron, clay and steel pipework – but today, on pain of a large fine, never a foot of lead water pipe.

In the 18th century, lead pipes were made by rolling and soldering cast sheet, and at this time, the plumbers' work included the construction of the often-decorated lead cisterns that were installed in the basements of town houses.

Leadworkers' tools
The principal tools of interest to the collector are the dressers and various forms of "sticks" – all used to shape sheet lead – and the bossing and shaping mallets which were used for areas of greater complexity. The traditional material for all sticks was boxwood though other hardwoods including hornbeam and lignum vitae were sometimes used.

Well-patinated (but not battered) boxwood examples are sculptural, tactile and appeal to many collectors. There is also a lively demand from lead workers for the wooden tools which, I am told, are better than the plastic versions made today.

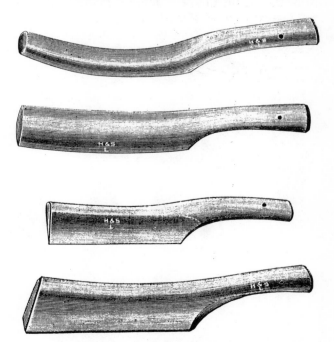

Plumbers' sticks and dressers [Harding, 1903]. From the top: bending stick, bossing stick, setting up stick and dresser.

219

Plumbers' tools

The advent of manufactured lead pipe, generally available by the 1840s, simplified the plumbers' work in one respect, but the availability of mains water at higher pressure meant that water could be piped to a tank in the roof and from there distributed throughout the house – so now the plumber had to deal with higher pressures.

For lead pipe there are no fittings. T-joints are formed by opening the pipe: in smaller sizes, with a pipe opener, and in larger sizes, after marking the opening with compasses and a scribing gauge, with snips. The parts are carefully pared, fitted and then soldered together with a neat wiped joint. All the more skilful when it is realised that, until the end of the 19th century when the first blow lamps became available, the heat came from a brazier and was transferred to the work with a plumbers' soldering iron.

To make a good running joint it was necessary to work the ends of the pipes to get a degree of mating one within the other. This could be effected by paring away the lead or by expanding the female end – or a mixture of the two. Plumbers' knives, hooks and shaving devices were used for the paring and cleaning and boxwood or lignum vitae turnpins for shaping and expanding.

Turnpins seem to have survived in large numbers, as also have plumbers' bobbins, though the larger sizes of 3" and over, are less common.

Plumbers' iron and pipe trimmer [Harding, 1903] and plumbers' turnpins and bobbins [Millard, c.1910]. Pipe trimmers are usually misidentified as woodworking tools.

Bobbins were driven through a pipe after bending to restore the bore. A brass follower of smaller diameter than the bobbins was threaded on the cord behind them and then jerked forward to drive through the string of bobbins. Also used for removing dents were plumbers' dummies – a lead tip on a cane handle. These are frequently misrepresented as coshes or "priests" for dispatching fish.

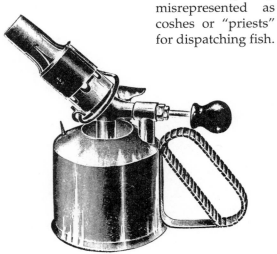

The "Ever Ready" brass benzoline blow lamp [Harding, 1903].

Price Guide

Lignum vitae plumbers' bobbins and turnpins, 1" and under **£1**; 1½" **£2**; 2" **£4**; 3" **£7**; 4" **£10**.
Boxwood dressers and sticks **£6–8;** ditto if fine and well patinated **£10–12**

♦ A 6" lignum vitae plumbers' coring ball. **£40** [DP]

♦ A plumbers' boxwood reducing mandrel for 3" and 4" pipe. G++ **£15** [DP]

♦ A 6" lignum vitae plumbers' turnpin. **£30** [DP]

♦ A set of forty boxwood, lignum vitae and ash plumbers' tools, including bobbins and turnpins, mandrels, bossing mallets, dressing sticks, chase wedges, etc. **£253** [DS 23/981]

♦ A collection of approx. twenty-eight boxwood and lignum vitae plumbers' bobbins, mallets, mandrels, dressing sticks etc. G **£60** [AP]

♦ A **Bladon** brass blow lamp with trade label and in original box. **£22** [DS 25/1304]

♦ A collection of six different brass blow lamps by **Sievert, Optimus** and **Monitor**. G++ **£80** [DS 22/115]

♦ A plumbers' iron (for soldering). **£12** [DP]

♦ A steel pipe trimmer with an ash handle. **£6** [DP]

Mention coopering and most people think of beer or wine barrels, in trade terms, "tight work". But coopering was far more than this: coopers made tubs and pails, also watertight, and dry coopers made casks used to transport the widest range of goods – from nails to dried fish.

Unlike the packing case, the cask could be rolled – a very important characteristic in an age with no mechanical handling equipment. American tobacco came into the London docks in huge dry-coopered casks, each weighing 800 lb. or more, which were little more than a wrapping around the compressed tobacco and were taken apart to release the contents.

A specialised branch of coopering was vat or back making – the construction of the huge containers used in distilleries, soap factories and other 19th century processing plants.

A hard life

The coopers' trade differs from most woodworking trades in that the tools used were owned by the employer and not by the tradesman. Whether this was always so is not certain but it was the norm from the late 19th century onwards.

This is bad news for the collector – tradesmen invariably looked after *their* tools but coopering is a hard trade and if the tools belonged to the boss... Another consequence is that coopers' tools often

Price Guide

◆ A comprehensive coopers' kit from the Bass brewery at Burton-on-Trent including chiv and croze, sun plane, side axe, brace, adzes, shaves, scorp, drawing knives, hammer, hoop driver, etc. all original to the kit. G **£396** [DS 25/1492]

◆ A collection of seventeen coopers' tools including side axe, reamer, shaves, hand adze, sun plane, croze, flagging iron etc. G++ **£275** [DS 24/1773]

Axes and adzes

◆ A coopers' l/h side axe by **Greaves** with ash handle and 10" edge. **£110** [DS 25/1058]

◆ A coopers' r/h side axe by **Walker** with ash handle. **£40** [DP]

◆ A coopers' r/h side axe by **Greaves** with offset ash handle. **£46** [AP]

◆ A coopers' No. 2 hand adze by **Walker**, Burton-on-Trent. G **£26** [DS 24/929]

◆ Three different coopers' adzes in varying sizes. **£32** [DP]

Sun planes

◆ A fine and probably unused coopers' sun plane by **Langley**. **£85** [DP]

◆ A coopers' sun plane by **J. Dolman**, Burton-on-Trent. **£82** [DS 23/784]

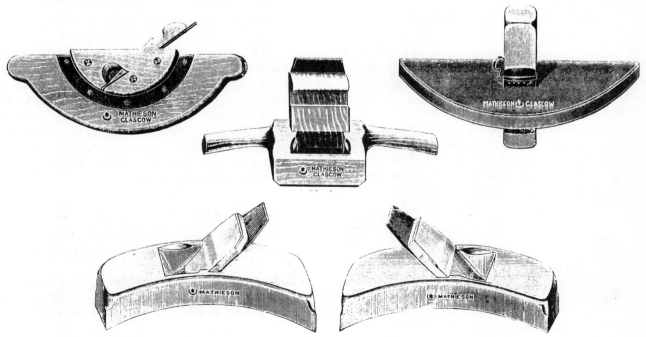

Coopers' tools [Mathieson, c.1900]. Top: a cheve (chiv), a head plucker (swift) and a croze. Bottom: right- and left-hand sun planes.

"Sending out Warehouse" at Beaufoy's Vinegar Works, Vauxhall, London from the series "A Day at a Factory" in The Penny Magazine, *October 1842.*

come to the market in large quantities – what does one do with forty crozes, forty drawknives, forty... all dirty from many years' storage in a damp shed.

But coopering is an ancient trade, with its own mysteries, special tools and appeal to collectors.

The tools

The first process in making a cask is getting out the staves. For this purpose, a special wide form of hand axe was used, followed by a hollowing draw knife to shape the inside face and the coopers' jointer to shape and bevel the edges. The coopers'

Coopers' (head) vice and a croze iron [Mathieson, c.1900]. The vice is used to hold the head whilst it is being fitted. Croze irons – the cutting part of the croze – are often found without the wooden part.

jointer is unusual in that it stood upside down and the staves were pushed over the plane – just as well as the planes are between 4 and 6 feet long – and we have seen even larger examples.

Even after "raising up", the barrel was far from finished; the ends were chimed (cutting the small slope at the top of the stave) with the coopers' adze and the chiv and croze grooves were cut; the heads were assembled from boards doweled together and then cut and edged (bevelled) before the hoops were fitted and driven home.

Collecting tips

➤ *Coopers' axes are broad in the blade – and quite thin – the result is that many are cracked or have welded repairs.*

➤ *Sun planes were made right and left hand. We have been looking for a pair in good condition for longer than we care to remember!*

➤ *Coopers' tools in good or fine condition carry a considerable premium.*

This is but the briefest description of the process. There are many variations depending on the type and size of cask, and the names of the tools vary between regions.

Collectors of coopers' tools are, in general, knowledgeable about the craft and are usually interested in building up complete kits of coopers' tools to include all the less common items. Most commonly found are cheves and crozes which were used to make the two cuts needed to receive the head. Whilst most coopers' tools – crozes, cheves, swifts and the drawing knives – can be bought for small sums, particularly if they are in well-used condition, the rarer items – sun planes, jointers, axes, braces, and the more interesting bung borers will command better prices – particularly if they are in above average condition.

A sharp adze and a coopers' axe [Mathieson, c.1900].

Price Guide

Drawknives and shaves

◆ A coopers' plucker marked Peter Hackwick. G **£33** [DS 24/1111]

◆ Four different types of coopers' drawknives including a jigger and heading knife. **£45** [DP]

◆ A group of three beech swifts and downwrights, all with brass reinforcement to top of mouth. **£44** [DP]

Cheves and crozes

◆ An unused cheve and croze, both by **Langley**, with original trade labels. **£60** [DP]

◆ A rare coopers' flincher (Scottish cheve) with solid brass stock and wood fence. **£99** [DS 25/912]

Other tools

◆ A pair of French coopers' compasses with a one piece wood arch with steel points and ferrules with wooden screw adjustment. **£220** [AP]

◆ A large coopers' bung reamer by **Walker**, Burton-on-Trent, with turned beech handles. G **£18** [DS 24/968]

◆ Two large, unused coopers' bung boring augers by **Kaye**, Hull, with original paint. G++ **£55** [DS 25/1277]

◆ Two coopers' flagging irons. **£20** [DP]

Braces

◆ A coopers' beech brace by **Dolman**, Burton-on-Trent, with brass ferrule. G++ **£154** [DS 25/1639]

◆ A large coopers' beech brace with exceptionally thick wooden webs and large diameter wooden head with turned decoration. F **£125** [AP]

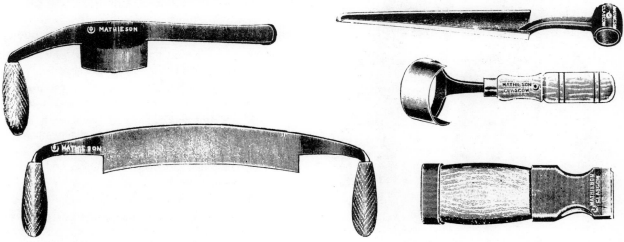

Top: a crum knife or jigger, a scillop or bung hole borer. Below: a circle-edged heading knife, a one-handled inshave and a hose (hoop) driver [Mathieson, c.1900].

Although from the 1850s onwards iron, and then steel, started to replace wood, a century later some wooden ship hulls were still being built. Even today, in the less developed parts of the world, construction of wooden vessels continues.

The collector should not think that all the tools used by the shipwright are special to the trade, as most are not. Generally, the tools used were those of the heavy woodworking trades and most of the kits of shipwrights' tools that I have seen contain many tools – axes, adzes and the like – which, according to the catalogues, were intended for wheelers or plate layers.

Basic tools for building the larger type of hull are adzes, augers and axes. Augers seem to have little attraction for the collector and the adzes sold for shipbuilding differ but little from many other trades. But the axes include several rather different forms – blocking, mast makers', and shipwrights' axes – all characterised by long, almost ungainly blades. Used on large timbers, they are specific to the trade, and they are difficult to find.

Another tool specific to the shipwright is the slick or slice – a large socketed chisel with a 3" or larger edge – used to pare away the face of ribs so that the planking will lie tight to them.

Caulking tools

Caulking may have been a separate trade or just something that the shipwrights got round to when finishing the job, but the job required a special set of tools.

Best ships' axe and best ship carpenters' adze [Sheffield List, 1888].

Price Guide

• A caulking mallet and a set of 13 caulking irons, a roll of 30 shipwrights' bits and a **Brades** No. 2 ship adze. **£148** [DS 25/716]

• A shipwrights' caulking mallet with unused spare head, six caulking irons, two deck dowelling bits, plug chisel, punches, augers etc. G **£132** [DS 25/1487]

Axes and adzes

• A Scottish pattern shipwrights' axe with 4½" edge by **Spear & Jackson**. G **£55** [DS 25/1052]

• A Newcastle pattern shipwrights' axe, marked K&R DUNSTON. **£75** [DP]

• A shipwrights' masting axe with 6" edge and ash handle. G **£104** [DS 25/1053]

• A No. 1 ship carpenters' adze by **Mathieson**. **£25** [DP]

• A large ship carpenters' adze. **£18** [DP]

Slicks

• A 1¾" shipwrights' slick with hooped ash handle by **Mathieson**. **£44** [DS 23/326]

• A 3" shipwrights' slick by **Gilpin** with ash handle. **£36** [DP]

Caulking tools

• Sixteen caulking irons of various shapes and sizes in a heavy duty canvas bag. **£56** [AP]

• A shipwrights' lignum vitae caulking mallet, well patinated but little used. **£25** [DP]

• A shipwrights' caulking mallet with eight irons and three joint scrapers, in a painted wooden box, with carrying strap, which can double as a seat. **£85** [AP]

• A shipwrights' caulking mallet with four different sized heads and four irons. G– **£93** [DS 23/954]

Other tools

• A 12" rosewood boatbuilders' bevel with two brass blades and brass fittings. **£18** [DP]

• A 12" boxwood boatbuilders' bevel marked with inches on the sides with two blades. **£28** [DP]

• A 7½" rosewood stocked bevel with single brass blade with owners' name nicely punched onto blade. **£20** [AP]

• A combined deck dowelling bit and tap by **Mathieson**. **£38** [AP]

• A deck dowelling bit. **£7** [DP]

• A 7½" long beech mast (or spar) plane by **Mathieson**. The upper part of the stock is boat shaped, the lower part straight-sided. **£20** [DP]

• A 14" mast makers' drawknife. **£24** [DP]

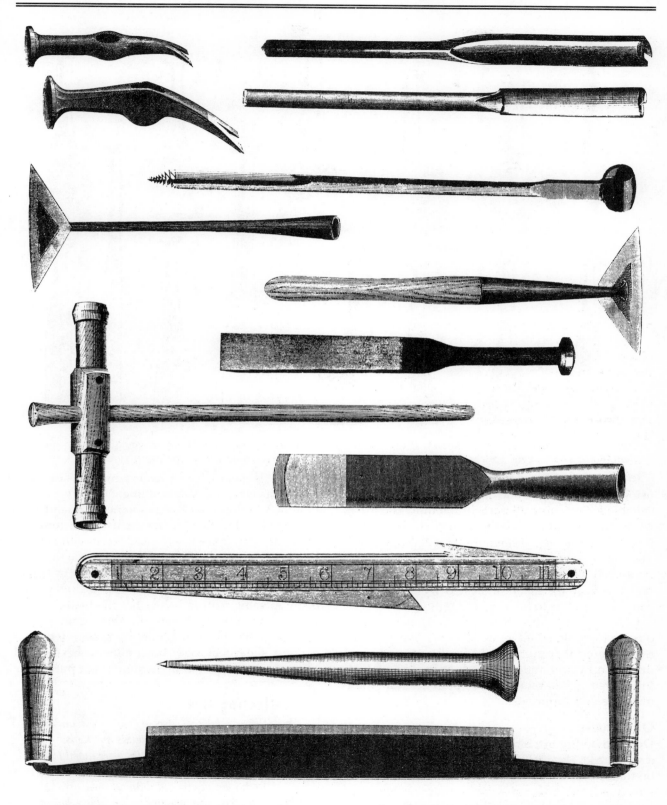

Shipwrights' tools [Sheffield List, 1888]. London coppering hammer, ship carpenters' shell auger, beat coppering hammer, Dod's pattern shell auger, eyed ship gimlet, cast steel socket ship scraper, cast steel handled ship scraper, iron head boat chisel, London pattern caulking mallet, cast steel ship slice, boxwood rule marked boatbuilders' bevel, coppering punch, mast shave.

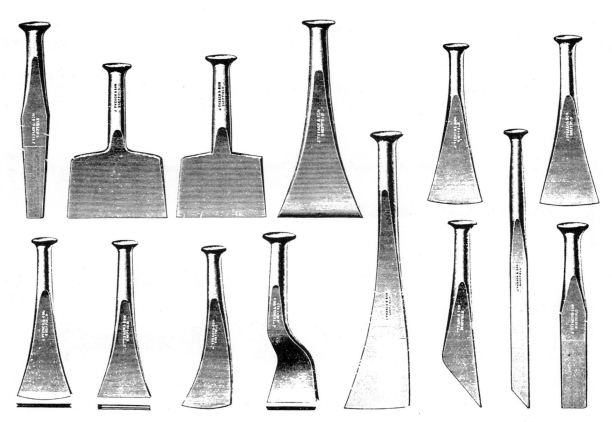

Cast steel bright caulking irons [J. Tyzack, 1913]. Top row from left: spike, deck, square reaming, fantail reaming, sharp, blunt. Bottom row: single crease, double crease, bent blunt, double bent crease, long bent, jerry, long jerry, trenail.

The oakum, strands of old rope, was punched in using irons which came in a variety of shapes suitable for reaching all parts of the hull. The irons were driven using a caulking mallet. These are long-headed mallets little more than 2¼" diameter. The head is cut through with two slots and reinforced with rivets and heavy bindings. The note given by the mallet indicated when the seam had been caulked to the required tightness.

Caulkers often kept their tools in a small square box that could double as a seat. When it is remembered that much of the caulkers' work was driving upwards under the hull, it is obvious that the degree of skill required was perhaps more than might be first supposed.

Other tools

Deck doweling bit: Similar to a centre bit but with a solid nose. Used to cut out a recess around a hole on the deck of a ship to receive the head of a fixing which is then covered with a plug or bitumen caulking. A rarely encountered variant has a parallel threaded nose for screwing into the metal deck beam for ships of composite construction.

Planes: The spar or mast plane is perhaps the only plane specific to shipbuilding. This is a top mouthed hollow plane of smoother size and thus very similar to other planes such as the forkstaff plane used on land. It was used to smooth smaller spars and other cylindrical items such as oars.

Other tools: Mast makers' drawknives – in length similar to other drawknives, 10" to 16" edge, the distinguishing feature being that the blade is both wider and thicker; block makers' chisels and gouges (both of heavy pattern socket type for making sheave blocks); boatbuilders' bevels and coppering tools, including hammers and punches.

Collecting tips

➤ *Slicks (slices) seem to be underappreciated. They are handsome tools.*

➤ *Bevels with brass tongues were mostly intended for ship or boat building and are found both with and without measures.*

➤ *The real shipwrights' spar plane has a boat top and a square lower section to the stock – although many planes used seem to have just boat shape.*

The piano tuner/adjuster at work at the Broadwood factory. From The Penny Magazine, *April, 1842.*

Price Guide

◆ Five different piano tuning hammers and levers with rosewood handles. **£42** [DP]

◆ A set of three cane tuning wedges (look like sprung tweezers in bamboo), each in a box. **£24** [DP]

◆ A piano cleaning bellows (looks like a wooden syringe), 3" in diameter. **£20** [DP]

◆ A toning needle with an ebony handle. **£7** [DP]

◆ A kit of piano tuners'/adjusters' tools in a cheap cardboard attaché case consisting of twenty-three items including mute wedges, music wire cutters and gauges and tuning hammers, damper cranks, benders, etc. **£120** [DP]

◆ Large piano tuning hammer with rosewood handle. **£8** [DP]

◆ A group of six different action regulators, three by **Buck**, in oval boxwood handles. **£36** [DP]

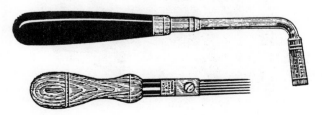

Time was when every well-to-do home contained a piano. The tuning and maintenance of these instruments, each with eighty-seven sets of delicately poised levers and hammers forming the action, was a minor trade in itself.

Piano adjusters' tools seem to cause more bewilderment than those of any other trade. The only manufacturers' names that seem to occur with any regularity are **Buck** and **Goddard**. Both firms were situated in Tottenham Court Road, centre of piano selling in London. Whilst **Buck** was a tool seller and planemaker, **Goddard** sold pianos and, as can be seen from their catalogues, were merchants for all piano parts and tools.

Tuning hammers and levers, although of many types, are readily identified, but the many varieties of action regulators (Goddard lists 18 different ones) cause confusion. Rarities are the casting irons, likely to be mistaken for bullet moulds, and the felt iron, that resembles a wooden handled spanner. Commonly misidentified as medical items are the toning needles whilst tuning wedges and cleaning bellows seem to challenge the inventive imagination of the antique dealer.

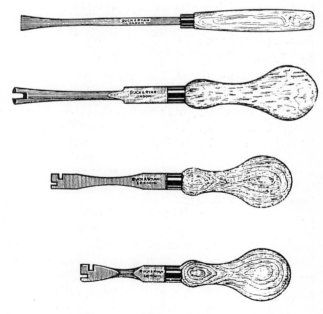

From top: Hopper tuner or set off tool, key spacer, long check bender, short check bender. Left: Cranked tuning hammer, toning needle [Buck & Ryan, 1930].

BOOKBINDERS' TOOLS

Much of the bookbinder's work is accomplished with few and simple tools – knives, folders and needles. Only in the finishing stage, when lettering and other decorations are applied, is there a requirement for many and varied tools. The demand for second-hand bookbinding tools comes from users who want good finishing tools.

Sewing frame

This holds the cords to which the sections of the book are sewn. It consists of a base board into which two wooden screwed columns are set. Wooden nuts support a bar. The cords are held between this and the base.

Presses

A book needs to be pressed at various stages of the binding process. Firstly to tighten the sections before sewing, for which a flat press is needed. Today many bookbinders use the cast iron presses originally sold for copying for this purpose. Their drawback is that they will not take large books.

Secondly the book is held in the cutting press while the edges are trimmed with a bookbinders' plough, and lastly in the finishing press which holds the volume after it has been bound whilst the decoration and lettering are applied.

In amateur workshops the same press will often double for cutting and finishing. Both types consist of two chunky beechwood bars which can be brought together with (usually) wooden screws. In the combined type, one side has battens for guiding the plough, the other is chamfered away to give better access when tooling the spine.

Bookbinders' plough

Used on the cutting press, it has two stocks, one with two arms that slide through the other, the two parts brought together with a central screw. The plough rests on the jaws of the press and as the cut

From "A day at a Bookbinder's" The Penny Magazine, 24 September, 1842, showing the use of a sewing frame.

proceeds the screw is tightened, bringing the two stocks together thus working the blade into the cut.

Lettering tools

Letters of 14 point and under are usually in the form of individual pieces of brass type which are used in a type holder. The sets should have a minimum of four of the common letters, two of the others and punctuation.

Larger sizes of letters are mounted individually into wooden handles. Sold in sets these should include the full alphabet, numbers, an ampersand and, hopefully, a diphthong. Roman is the most common face, followed by sans serif.

These letters, like all finishing tools, are made of brass and are used hot to fuse the gold foil to the leather of the binding.

> ## Collecting tips
> ➤ *Finishing tools are used hot. Don't worry about the handles which inevitably char away – they are easily replaced.*
> ➤ *Even one missing letter will reduce the value of a set by two-thirds. It is possible to get replacements made but these are expensive.*
> ➤ *Some centre tools are handed, so for these there should be a pair.*

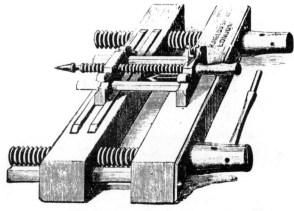

A bookbinders' cutting press with plough positioned on top [Melhuish, 1912].

Price Guide

Finishing tools: Centre tools: **£4–15** according to size and complexity. Pallets: plain line **£5–8**; patterned **£10–20**, according to size and decoration. Rolls: plain line **£8–12**; patterned **£25–75**, according to size and complexity. Sets of letters: Roman face **£30–60**; decorative and less common faces **£45–80**.

◆ A set of twenty-eight bookbinders' Roman face letters with ampersand, ¾". **£33** [DS 25/1299]

◆ Six bookbinders' brass rolls of single, double and triple lines. G **£52** [DS 25/1586]

◆ Two pairs of matching decorative corner tools, ½" and ¾" o/a. **£45** [DP]

◆ A set of thirty-six unusually decorative bookbinders' letters and numbers, ¾" (approx. 72 point) high, including ampersand and diphthong, in original box. G++ **£137** [DS 23/624]

◆ A large collection of seventy brass bookbinders' tools including nine agate burnishers, ten liners and fillets, a twenty-six Gothic letter set, twenty-six small letters and a bookbinders' plough. **£200** [DS 21/863]

◆ A bookbinders' sewing frame in mahogany and beech with elegant turned finials. **£45** [DP]

◆ A large and early bookbinders' plough with elegant finial and wing nut. **£60** [DP]

◆ An amateur pattern combined finishing and ploughing press complete with a small size bookbinders' plough by **Dryad**. **£100** [DP]

◆ A cast iron copy press, 14" within frame, with original black paint and gold decoration and gold painted knobs. **£40** [DP]

◆ Eleven agate burnishing tools. G++ **£154** [AP]

◆ A bookbinders' beating hammer (sold as a gold beating hammer!). **£48** [AP]

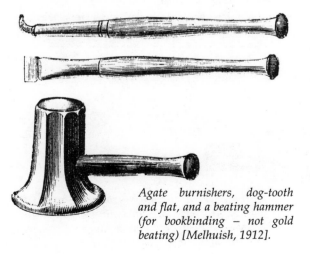

Agate burnishers, dog-tooth and flat, and a beating hammer (for bookbinding – not gold beating) [Melhuish, 1912].

Finishing tools

Brass tools used hot to decorate bindings.

Gouges: For curved lines – surprisingly versatile. Used in combination they will make many patterns.

Pallets: These can be plain lines or decorated lines. The tools are no longer than 3" – their purpose was to decorate the spine.

Line rolls: Wheels held in a fork. Varieties include double, triple and unequal thickness lines.

Pattern rolls: These can be anything from a row of dots up to a 1" wide highly complex pattern which is not only cut but also surface engraved.

Centre tools: Highly variable; at the simplest a few dots, at the most complicated a 1" square highly decorative pattern or coat of arms.

Corner tools: Similar to centre tools. If asymmetrical there must be a pair.

Burnishers

Long handled steel or agate burnishers were used for various purposes on binding and paper fore edge.

From "A day at a Bookbinder's" The Penny Magazine, 24 September, 1842, showing a roll in use.

At the beginning of the 19th century, with few exceptions, most engineering was still to woodworking type tolerances. Where this would not do, accuracy was achieved by hand skills in fitting one part to another but this was not standardised production.

Engineering to fine and repeatable tolerances requires accurate measuring instruments and these were not available at prices that could be afforded by machinists and fitters until the 1880s. Today the skill of the old-style machinist has largely been replaced by the accuracy of the machine but, as in many fields, as the tools of the trade become obsolete they become of interest to the collector.

In the late 19th and early 20th century the market leaders for engineers' measuring tools were American – **Brown & Sharpe** and **The L.S. Starrett Company** were the largest producers. By the 1920s, both these firms were making a huge range.

Micrometers

The term "micrometer" was originally applied to devices for measuring in optical instruments. The micrometer made by James Watt in 1819, now in the Science Museum, London, is almost certainly the earliest micrometer made for engineering measurement. But there is much misinformation about: according to **Brown & Sharpe** (1926 catalogue) the micrometer was invented in France as late as 1848 but only became a practical proposition after *they* had improved it in 1867!

Once available in quantity the micrometer rapidly became *the* accurate measuring tool for engineering and has become an icon of precision.

Price Guide

All precision measuring instruments, particularly larger ones, are very expensive to purchase new so most amateur and many professional mechanics will buy second-hand. The general prices quoted are for items by the best makers – **Starrett**, **Brown & Sharpe** or **Moore & Wright**.

Micrometers: ½" or smaller **£24–30**; 1" size **£12–18**; sizes over 1" **£10–15**. **Vernier callipers**: Up to 12", user condition **£14–20**; collector condition, **£25–35**. **Combination squares**, with centre finding head: 4" and 6" size **£20–24**; 9" and above **£15–20**. **Combination sets** (including protractor head): **£40–60**. **Firm joint callipers**, all shapes: **£3–6**. **Spring dividers and callipers**, all shapes: **£5–10**.

Micrometers

♦ A fine *"Mesure Metrique"* early digital micrometer dated 1903 by **Ciceri Smith,** London, in original lined and fitted mahogany box. **£902** [DS 25/827]

♦ Two outside micrometers by **Brown & Sharpe**, 1" to 2" and 2" to 3", both in original boxes and with instructions. **£22** [DS 23/20]

♦ A **Moore & Wright** inside micrometer set No. 902, 2" to 8" capacity in original box. **£35** [DP]

♦ A 1 cm. watchmakers' micrometer with nickel silver frame by **S. Leunig. £28** [DP]

♦ A ½" micrometer with pad jaws (for measuring paper). **£24** [DP]

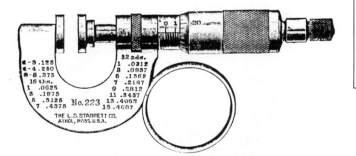

Two special purpose micrometers from **The L.S. Starrett Co.** *50th anniversary catalogue No. 25, 1930. Top: a paper gauge micrometer. Bottom: a hub micrometer.*

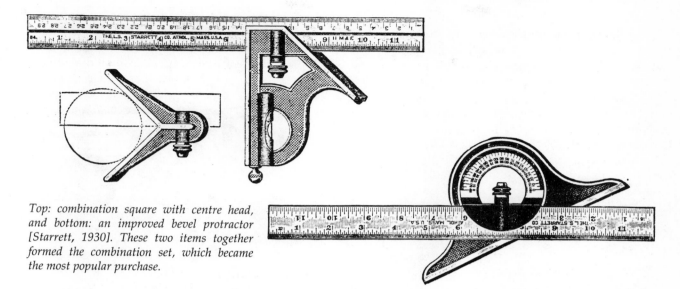

Top: combination square with centre head, and bottom: an improved bevel protractor [Starrett, 1930]. These two items together formed the combination set, which became the most popular purchase.

In 1926 **Brown & Sharpe** were making 400 different models of micrometer, including 24" monsters, types with different anvils for measuring everything from screw threads to soft materials and a range for internal measuring.

The collector will find micrometers fitted with ratchet stops, an improvement added by many manufacturers, intended to ensure that the same amount of pressure was always applied. However, these may be disabled – I have often been told by old machinists that, when apprentices, they were forced to do this as they were told no *worthwhile* engineer would use anything other than feel. In so doing, they have rendered their micrometers worthless – at least as far as collectors are concerned, for in this field condition and completeness are highly valued.

Vernier callipers

Pierre Vernier published his invention of a second scale alongside the main scale in the 1630s. It was intended to give greater accuracy when reading the quadrants on scientific instruments and it remained in the realms of science until well into the 19th century when it was adapted for use on engineers' measuring tools.

Most vernier callipers found are of light all-steel construction and are of 20th century date. English makers were producing verniers in the 19th century – heavy brass framed devices with deep jaws in sizes up to 36", handsome tools which today command handsome prices.

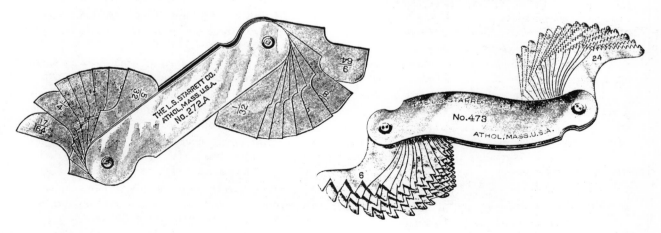

Above: a fillet or radius gauge; right: a positive stop, screw pitch gauge [Starrett, 1930].

Collecting tips

➤ *A tool that was originally sold in a case or box should have a case – its lack will affect the price.*

➤ *Matt chromium plate was used as an improved finish on many types of measuring tools from the 1940s onwards.*

➤ *Engineers' measuring tools are not particularly old and are quite common, so buy only the best.*

Faster! Slower!

Puzzling to many are speed indicators. Almost every manufacturer of engineering tools seems to have made these devices which are, in fact, revolution counters. When lathes were driven from overhead shafting which could run at widely differing speeds it was important to know how fast your workpiece was going. The speed indicator was driven by a small shaft with a triangular point end which was applied to the centre of the workpiece and was rotated by it. By means of a worm gear, the number of revolutions in the chosen time was indicated on a small dial.

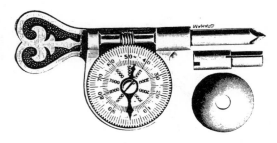

Revolution counter [Melhuish, 1904].

Famous names

Brown & Sharpe Mfg. Co. Providence, Rhode Island, USA. The origins of this firm go back to 1833 but it was only in the 1870s that it really started to expand. The extent of the product line, before business pressures forced some

Price Guide

Callipers

• A 24" heavy brass inside and outside vernier calliper with steel jaws by **Elliott Bros.**, London, in original softwood box. **£220** [AP]

• A No. 270 24" vernier calliper by **Brown & Sharpe** in original box. F **£60** [AP]

• An 8" vernier calliper by **Darling, Brown & Sharpe** and another inside and outside metric calliper by **Kugler**. **£37** [DS 23/428]

• A 12" outside calliper with turned ivory handle by **Elliott Bros.**, London, with calibrated German silver insert in a large brass dial. The pointer revolves freely, suggesting some repair may be required to the internal mechanism. G **£308** [DS 25/948]

• A set of three **Moore & Wright** toolmakers' round leg dividers and callipers, 3", in leather case. **£30** [DP]

• A craftsman-made pair of steel inside callipers formed as shapely legs, 6¼" long. **£35** [DP]

• A fine quality large pair of inside/outside callipers in steel with brass disc joint, 13" long. Unnamed, but professionally made of Holtzapffel quality. **£65** [DP]

Speed indicators

• An Improved Speed Indicator by **Starrett** in original box with instructions. G++ **£20** [DS 22/480]

• A speed indicator, nickel plated on fancy cast stock. Woodman's Patent, Sept. 12, 1876. **£18** [DP]

Other tools

• A No. 493 6" bevel protractor by **Brown & Sharpe** in original case. **£30** [DP]

• A craftsman-made bronze and steel surface gauge incorporating a fine adjustment by means of a lever mechanism with finely knurled nuts. **£55** [AP]

• Two engineers' steel plumb bobs by **Moore & Wright** and **L.S. Starrett**. G++ **£22** [AP]

• A **Starrett** No. 59F trammel set complete with four pairs of alternative legs and four sets of centres in a craftsman-made softwood box. **£46** [AP]

• A **Starrett** No. 85C improved extension dividers complete with three sets of interchangeable legs. **£36** [AP]

• Three engineers' squares, the smallest 2", a small steel T-square and a steel rule, all by **Preston**. **£32** [AP]

• A **Moore & Wright** set of tools in a manufacturers' folding leather case, including four pairs of callipers, a square and other tools. F **£75** [DP]

• A group of engineering tools, dating from the 1920s, including five surface gauges, callipers, dividers, squares etc., at least 30 items. **£181** [AP]

• A No. 323 feeler gauge with six blades and two engineers' squares, 9" and 2½", by **Preston** and two small engineers' squares by **Buck**. **£20** [DS 22/1338]

• A 14" vernier height gauge, No. 369 by **Chesterman** in manufacturers' original wooden box. **£60** [DP]

rationalisation, can be seen in their 1926 Catalogue (No. 30) which has no less than 448 pages. **Darling, Brown & Sharpe** were a satellite of the main company between 1866 and 1892 with some products bearing this name.

The L.S. Starrett Co. Athol, Mass. USA. Founded in 1880, this firm seems to have grown steadily right from the start. Before he started in business, Mr Starrett had already designed the engineers' combination set, a combined square and mitre square head, a centre finding head and a protractor head, all locking onto a grooved blade. Starrett tools, which are both inventive in design and made with the finest quality of finish, deserve more attention from collectors.

Moore & Wright Sheffield UK. Founded in 1909, the company made a variety of products including

oil cans and jewellers' blow lamps but by the 1930s precision measuring equipment had become the principal product. Rearmament in the late 1930s and the subsequent war were the making of this business – the demand for precision engineering tools was enormous and imports were difficult.

Apprentice pieces

As with most tools, the occasional craftsman-made engineering tool of both quality and flair is to be found. Callipers, scribing gauges etc. were made by apprentices as part of their training. Named and dated pieces are desirable. In general, however, this is not an area where many craftsman-made tools of quality are to be found.

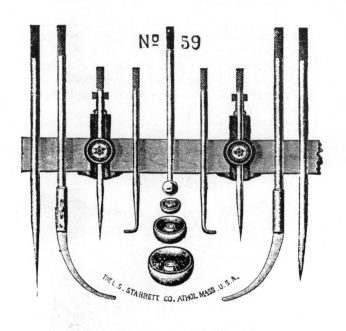

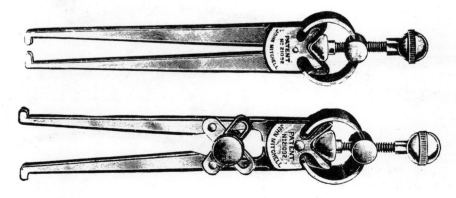

Clockwise from top left: Trammel set, including trammel heads, balls and holder, small and large calliper legs and divider points [Starrett, 1930]. Ciceri Smith's direct reading micrometer [Melhuish, 1904]. Two inspecting gauges for boiler and other plates (designed to be inserted through a hole) [Starrett, 1930]. Outside and inside screw adjustable callipers, John Mitchell's patent [Millard, c.1910].

A builders' rubbish basket [Harding, 1903].

Willow (withy) baskets, which have been made in Britain since pre-historic times, were, until the 1950s, an important form of trade container. The Kentish apple baskets, always prominent in period pictures of the old Covent Garden market, and the rectangular laundry basket are but two examples.

The tools used in basket making are few, with some, particularly the knives and bodkins, being difficult to identify unless found with other basket-making tools. From the collector's point of view, most desirable would be a kit but these very seldom seem to appear.

Appealing items are the cleaver (a three- or four-edged splitter with integral handle, invariably made in boxwood) and the two forms of basket-makers' shaves, used to make regular-sized strips for bindings and light baskets.

Price Guide

♦ A nicely patinated boxwood three-edged basketmakers' cleaver. **£25** [DP]

♦ A basketmakers' shave and a boxwood cleaver with a brass tip. G++ **£55** [DS 23/87]

♦ A wrought iron withy stripper made from a single strip of iron bent to provide sprung jaws through which the withy is drawn to strip the bark. G **£20** [DP]

♦ A fine pair of basketmakers' shaves (a shave and an upright shave) in beech and brass by **Marples**, probably unused. **£56** [AP]

♦ An all brass basketmakers' upright shave by **J. Buck**. G++ **£50** [DS 22/688]

♦ A basketmakers' picking knife with a boxwood cleaver, a brass and boxwood cleaver and a solid bodkin. **£45** [DP]

♦ A good set of basketmakers' tools comprising two shaves by **James Oxley**, Sheffield, two boxwood cleavers, two pairs of cutting shears by **Bamforth & Moss**, two wrought iron kinking tongs, a round commander, a hollow bodkin and six other bodkins, two picking knives and a hand knife. G **£150** [DS 16/1375]

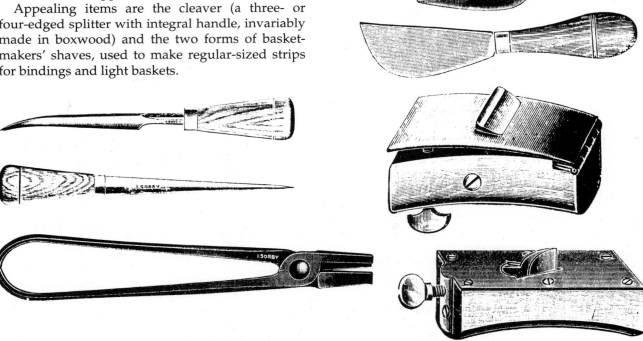

Left from top: Hollow bodkin, solid bodkin, cane squeezers or kinking tongs [Turner, Naylor, 1928]. Right from top: Cleaver, picking knife, shave and upright shave [Melhuish, 1912].

English clogs have wooden soles and leather uppers. For centuries they were the footware of the poor and were sold until as late the 1950s. Some are made today but these are for dancers and other heritage activities. For the soles the most commonly used wood was alder but any light close-grained wood was suitable with sycamore and birch also being frequent choices.

The clogmaker used three specialist knives to "carve" the soles. These are easily recognised, being about 3ft. long with a large hook at one end and a sweeping iron body terminating with a wooden cross handle. The cutting edges are straight (the stock knife), curved (the hollower) and V-shaped (the gripper, which is used to cut out the rebate to receive the leather upper). In use, the hook is engaged in a large ring on the cloggers' stool, a low bench, with the long handle giving considerable leverage.

The other tools used by the clogmaker are leatherworkers' tools: hammers, knives, punches and awls. It is not normally possible to ascribe these tools to clogmaking unless they come from a known source.

Top: Open tab clogs. Bottom: Capped pattens [Harding, 1906]. It is surprising to find pattens, a form of shoe raised on a metal ring to keep the wearer out of the mud, were still being sold in 1906. There were many other forms of clog. The "common" type were open-toed, being secured to the foot with a single leather tie across the upper, as in the pattens above, and laced versions of the boot type were also sold.

Price Guide

♦ A set of three cloggers' knives by **Henry Carter**, High Burton also marked James Matley, Agent, Oldham. **£230** [DP]

♦ A country-made cloggers' bench well used but appealing for this reason. **£110** [DP]

♦ A cloggers' stock knife. **£70** [DP]

♦ A group of cloggers' tools including a gripper knife and twelve other items, punches, awls and knives. **£80** [DP]

Collecting tips

➤ *Coopers and some of the other wood trades used a straight bench knife similar to the clogmakers' stock knife but shorter.*

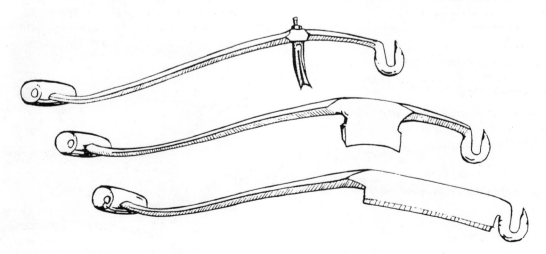

A cloggers' gripper, hollower and stock knife.

The raw material of the straw hat manufacturer was "plait", that is, straw plaited up into a long braid that was subsequently sewn up to form the hat. Straw plaiting was a cottage industry which enabled women to earn a little money. As in many home-working industries, travelling merchants supplied the wheat straw and collected the finished plait. The trade was particularly common in Bedfordshire, supplying the Luton hat makers.

Splitting
The straw, which had to have a length of at least 9" between joints and was always worked wet, was split lengthwise into the required size – which might be into anything from three to nine divisions. This was effected with a straw splitting "machine", a small hand-held tool with a point that went down the middle of the straw with fins behind doing the splitting. These devices were made in bone, iron and brass but are something of a trifle. Many must have been thrown away, for they aren't as common as one might expect. An alternative form has the splitter fixed in a small wooden stand.

Finishing
When the plait had been made, and there were many different patterns and sizes, it was rolled through a plait mill. This was a simple hand-cranked machine, usually made of beech with boxwood rollers. There are two types, the *splint mill*, with smooth rollers, used both to flatten split straw and to roll the finished plait, and the *plait mill*, with grooved rollers, for rolling plaits with fancy edging without flattening the edges. The mills were fixed to the frame of a door, allowing their use in the doorway, and are the principal surviving artefact of this once-extensive trade.

A splint mill and a plait mill.

Price Guide

- A bronze splitting machine. **£35** [DS 19/904]
- Five brass and one steel straw splitting machines. G **£125** [DS 19/905]
- Two yew wood-frame splitters, one to split the straws into five, the other into three. G **£150** [DS 19/906]
- A beech splint mill with plain beech rollers, some worm in frame. **£135** [DP]
- A beech plait mill with grooved boxwood rollers. **£185** [DP]
- A mahogany wood-frame splitter to split the straws into seven. **£75** [DP]
- A bone splitting machine to split into eight, in a crude ash handle tapering to a point. **£28** [DP]

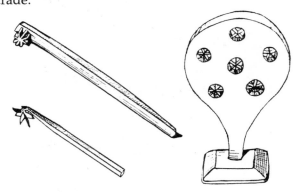

Straw splitting "machines" and a wood-frame splitter.

With few exceptions, the illustrations in this book have been taken from manufacturers' and distributors' catalogues. In the captions to the illustrations, only the name of the firm, in brief, and the date of the catalogue are given. The following list gives the full names of the firms as given in the catalogue, the dates of the catalogues used and the briefest of notes about the firms. Obviously over the years the title of firms did change as sons joined fathers and partners came and went.

George Adams, 1914. London. Situated in High Holborn, the firm were suppliers of tools to the light engineering trades, including clock and watchmakers. They catered for both professionals and amateurs.

Allen & Hanburys Ltd., 1930. London. Probably best known to the general public for their cough pastilles, the company were not only druggists but also the largest suppliers of surgical instruments and hospital equipment of the time. This catalogue extends to more than 2,000 pages.

Arnold & Sons, 1886. West Smithfield, London. Catalogue of veterinary instruments. The firm also sold medical instruments and equipment.

George Barnsley & Sons, 1927. Sheffield. Specialist manufacturers of tools for shoemakers and other leatherworkers.

Buck & Hickman, 1923, 1935. Whitechapel, London. See *Famous Names*, p. 41.

Buck & Ryan, 1930. Tottenham Court Road, London. See *Famous Names*, p. 41.

James Chesterman & Co., 1880. Sheffield. See *Famous Names*, p. 36.

Charles Churchill & Co. Ltd., 1935, 1939. London and other branches. Established 1889. Wholesale and trade suppliers, the firm were pioneer importers of American machine tools and also sold a wide range of tools for wood and metal workers.

Evans & Wormull, 1889. Stamford Street, London. Manufacturers of surgical instruments, apparatus and appliances. Prize medal winners at the Great Exhibition 1851.

George Farmiloe & Sons Ltd., c. 1905. St John Street, West Smithfield, London. Manufacturers of lead and paint and merchants of glass, sanitary ware and an extensive range of tools for the painter, plumber and glazier.

James, Isaac & John Fussell Ltd., c. 1890. Frome, Somerset. Fussell's Ironworks started in 1744 and continued until 1894 when the firm was taken over by Spear & Jackson of Sheffield who continued to use the name and issue Fussell's catalogues. They made agricultural edge tools.

G. Harding & Sons., 1903, 1906. Long Lane, Borough, London. Established in 1835, Hardings were wholesalers to the hardware, ironmongery and tool retail trades. The wooden plane illustrations in their catalogue, in spite of being marked with the Harding initials, were in fact from David Kimberley & Sons Ltd. of Birmingham.

Philip Harris & Co., 1896. Birmingham. Manufacturers of chemical and physical apparatus. Included are saccharometers, hydrometers, balances and weights.

John Harrison & Sons, 1902. Dronfield, nr. Sheffield. Manufacturers of light and heavy edge tools, the firm was founded in 1795.

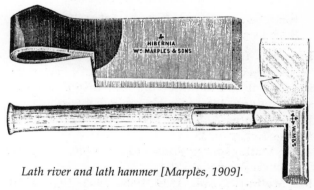

Lath river and lath hammer [Marples, 1909].

J.H. Hawkins & Co. Ltd., c. 1895. Walsall. *"The Equine Album"*. Wholesale manufacturers of saddlery, harness and all the other requisites for horses. The firm incorporated Hampson & Scott which had been established in 1794.

James J. Hicks, c. 1890. Manufacturers of meteorological instruments and chemical and philosophical apparatus which included hydrometers, saccharometers, surveying and drawing instruments and gauging rules.

Holtzapffel & Co. 1847. London. See *Famous Names*, p. 36.

William Hunt & Sons, The Brades Co., 1905. Birmingham. A leading Midlands maker of light and heavy edge tools.

M. Hunter & Sons, 1916. Sheffield. "Established 1760", a leading Sheffield cutlery manufacturer. Best known for their "Bugle" brand table, pen and pocket knives.

Henry Isaacs, c. 1900. Leeds. Established in 1880. Suppliers of tools and materials for watchmakers and jewellers, jewellers' sundries and optical goods.

Lawson & Heaton, c. 1930. Birmingham. A less well-known maker of cold chisels, screwdrivers etc.

W.R. Loftus, c. 1870. Oxford Street, London Suppliers of everything needed by the wine and spirit trade – not only instruments and rules but everything needed to run a pub.

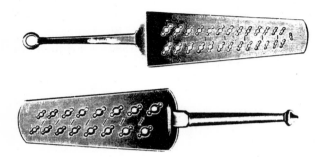

Watch screwplate and clock screwplate [Wingfield, Rowbotham, 1904].

Joseph Long, 1922. Eastcheap, London. "Established 1818". Makers of scientific instruments for the brewing and distilling trade, sugar refiners, tea and rubber merchants etc.

Wm. Marples & Sons, 1897, 1909. Sheffield. See *Famous Names*, p. 35.

Alex. Mathieson & Sons Ltd., c. 1900. Glasgow. See *Famous Names*, p. 39.

Rd. Melhuish, Ltd., 1899, 1904, 1905, 1912, 1925. Fetter Lane, London. Established in 1828, this firm were leading retailers of ironmongery and tools for many trades until bombing destroyed the premises during the war. The late 19th/early 20th century catalogues were clearly intended for amateurs as well as professionals.

Millard Brothers Ltd., c. 1910. Houndsditch, London. Wholesalers of a wide range of tools and household goods. Many of the lines were cheap imports.

Edward Preston & Sons Ltd., 1891, 1909, 1914. Birmingham. See *Famous Names*, p. 38.

John Rabone & Sons, 1892, c. 1925. Birmingham. See *Rule Making*, p. 154.

The Sheffield Standard List Illustrated, 1862. A manifestation of the price fixing system operated by the Sheffield manufacturers, it contained prices for a wide range of goods including machinery, files, rasps, joiners' tools, light and heavy edge tools, saws, Lancashire tools, machinists' and plumbers' tools, spades, shears, anvils, bellows, and wire.

The L.S. Starrett Co., 1930. See *Machinists' Tools*, p. 233. Although an American company, their tools were widely available in the UK. The 1930 catalogue was published for their 50th anniversary.

Swindell & Co. Ltd., 1903. Dudley, Worcestershire. Manufacturers of heavy edge-tools, "Established over 100 years". This catalogue includes agricultural tools, plumbers', smiths' and farriers' tools, including horse shoes, a variety of items made from chains and hinges, hooks, and traps for everything from rabbits to wolves and lions.

Turner, Naylor & Co. Ltd., 1928. Sheffield. Although by this date owned by Wm. Marples, this firm still produced a catalogue under its own name.

Joseph Tyzack & Sons, 1913. Sheffield. *List of Shipwrights' Tools.*

S. Tyzack & Son Ltd., c. 1910. London. A principal London tool seller.

Ward & Payne, 1911. Sheffield. See *Famous Names*, p. 38.

A. West & Partners, 1930. London. Established in 1888, the firm sold mathematical, surveying and drawing instruments and produced dye line materials and other copying and photocopying apparatus. It is unclear to what extent they manufactured the goods they sold but they certainly marked them.

Wingfield, Rowbotham & Co., 1904. Sheffield. Steel converters and refiners and manufacturers of saws, files etc. The interlock between Sheffield firms, not always clear, is illustrated in this catalogue where it is stated that the principals of the firm are the same as those of Thomas Turner & Co. and that some of the blocks illustrating the catalogue are the same as those used in the Turner catalogue.

Wynn, Timmins & Co. Ltd., c. 1880. Birmingham. Successor to the late 18th century firm of Richard Timmins & Sons. Many of the plates in this catalogue were the same as those used in the 1845 catalogue and probably dated from even earlier.

Cloth knife [Wingfield, Rowbotham, 1904].

MAKERS' INDEX

TOOL INDEX

adzes, 58, 60, 222, 224
Armstrong scales, 177
augers, 78, 224
axes, 57–9, 204, 222, 224

barking spuds, 206
beam compasses, 98, 209
beetles, 198
bench planes, 110–13
bevels, 50, 93, 95
billhooks, 195, 198
bits, 34, 82–3, 226
bobbins (plumbers'), 220
bodkins, 215, 234
bolting irons, 71
books, 217–18
braces (see also Ultimatum braces), 34, 78–81
burnishers, 229

callipers, 165, 174, 175, 204, 231
carving tools, 31, 38, 74–5
catalogues, 33, 41, 42–3, 218
caulking tools, 224–6
centre tools (bookbinders'), 229
cheves, 223
chisels, 34, 38, 66–71, 188, 224
cleavers (basketmakers'), 234
clinometers, 183
clockmakers' tools, 33, 42
corking devices, 193–4
corner tools (bookbinders'), 229
crozes, 223
cutting gauges, 100, 102

drawing instruments, 168, 208–11
dressers (plumbers'), 219
drills, 88
dummies (plumbers'), 220

files, 41, 212
fleams, 201
floats, 212
food testers, 192
forks, 53

gauges, 100–102, 174–5, 180, 181
gimlets, 78, 193
glue pots, 213
gouges, 38, 66, 72–3, 74, 206, 229

hammers, 53, 184–6, 192, 204
hatchets, 58
hayrakes, 200
hedging hooks, 198
hoes, 195, 200
hooks, 194, 200

jointers, 222

knives, 21, 55, 56, 74, 234
    drawknives, 223, 226
    for clogmaking, 235

lathe tools, 76
lathes, 36, 51, 76
lettering tools (bookbinders'), 228
levels, 51, 89–92, 93, 95, 97

mallets, 74, 219, 226
mapping scales, 177
marking gauges, 100–101
measures (see also rules: tape measures), 36–7, 149–151
metal braces, 80–81
metal planes, 21, 29, 52, 144–8
micrometers, 230–1
mortice gauges, 100–101
moulding planes, 28–9, 39, 106, 107, 114–23

offsets, 177
oil cans, 214
Orthops rules, 179

pallets (bookbinders'), 229
periodicals, 217
plait mills, 236
planes,
    bench, 110–3
    combination, 141–3
    grooving, 124–6, 134
    metal, 21, 29, 52, 144–8
    moulding, 28–9, 39, 106, 107, 114–23
    plough, 134–6
    rebating, 128–31
    reeding, 123
    router, 127
    spar, 226
    thumb, 111, 132–3
    wooden, 21, 33, 39, 48–9, 103–9, 138–40
ploughs, 228
plumb bobs, 96–7
plumb frames, 97
pruning shears, 197
punches, 74

reaping hooks, 200
revolution counters, 232
rifflers, 74
rollers, 212, 215,
rolls (bookbinders'), 229
rounders, 137

routers, 74
rules, 30, 49, 54–5, 152–5, 178–82
    barrel, 164–7
    Coggeshall (carpenters'), 106–62, 183, 170
    engineers', 163
    excise, 151, 172–3
    folding, 156–7, 160–62, 163, 167, 168
    scale, 176–7
    slide, 151, 163, 164, 169–73

saw sharpening/setting tools, 64–5
saws, 31, 38, 41, 61–5
scale rules, 176–7
scrapers, 76, 192
screwboxes, 74, 77
screwdrivers, 86–7
scribes, 206
scriveners' wheels, 214
seals, 194
secateurs, 197
sectors, 168
sewing frames, 228
sickles, 195, 100
slashers, 198,
slicks, 224
slide rules, 151, 163, 164, 169–73
spades, 53, 195, 198
spokeshaves, 84–5
squares, 50, 51, 93–5
"sticks" (plumbers'), 220
straw splitting "machines", 236

tape measures, 37, 158–9, 178, 182, 204
taps, 77
thread chasers, 76
thumb planes, 111, 132–3
tool chests, 114, 187–91
trammels, 98–9, 209
trowels, 195
tuning tools, 212–13, 227
turning tools, 76
turnscrews, 71, 86–7
twybills, 58

Ultimatum braces, 7, 54, 56, 78–9

veterinary instruments, 201–3
vices, 74

watchmakers' tools, 33, 42
wooden braces, 34, 78
wooden planes, 21, 33, 39, 48–9, 103–9, 138–40